SANTA ANA PUBLIC LIBRARY

D0387450

# Space Oddities

# Space
# Oddities

## Our Strange Attempts to Explain the Universe

## S. D. Tucker

AMBERLEY

First published 2017

Amberley Publishing
The Hill, Stroud
Gloucestershire, GL5 4EP

www.amberley-books.com

British Library Cataloguing in Publication Data.
A catalogue record for this book is available from the British Library.

ISBN 978 1 4456 6262 6 (paperback)
ISBN 978 1 4456 6263 3 (ebook)

Typeset in 10pt on 13pt Sabon.
Typesetting and Origination by Amberley Publishing.
Printed in the UK.

# Contents

Outer space is within us inasmuch as the laws of space are within us; outer and inner-space are the same.

Joseph Campbell (1904–87)

The human soul is so alienated from us in our present culture that we treat it as an extraterrestrial.

Terence McKenna (1946–2000)

Let my lamp at midnight hour
Be seen in some high lonely tower,
Where I may oft outwatch the Bear
With thrice-great Hermes, or unsphere
The spirit of Plato to unfold
What worlds or what vast regions hold
The immortal mind that hath forsook
Her mansion in this fleshly nook.

John Milton (1608–74), *Il Penseroso*

# Introduction: Black Mirror, White Ice

The eternal silence of these infinite spaces frightens me!
Blaise Pascal (1623–62), *Pensées*

Space, as you may have noticed, is dark black in colour – except, in actual fact, it is not. Actually, space is beige. To see it as being black is a trick of human perception. Space was traditionally thought of as being infinitely old, infinitely large, and full of stars. Given this, we might expect the night sky to be a bright place, due to the cumulative effect of the light from all those stars. Everywhere within the night sky we glance, in whichever direction, we should be looking directly at a star; and yet we see only inky blackness, interspersed with thousands of tiny twinkling lights. Why? It is a conundrum known as 'Olbers' Paradox' after the German astronomer Heinrich Wilhelm Olbers (1758–1840), who described the puzzle in 1823. One possible answer to the mystery was provided by an Essex vicar and Cambridge graduate named P. H. Francis, in a pair of self-published booklets dating from 1968 and 1970 entitled *The Mathematics of Infinity* and *The Temperate Sun*. Francis proposed that the sun was not actually hot at all (a 'childish' and 'primitive' belief), merely a bright object, possessing some form of electric charge. Our sun, it transpired, was also the only such self-luminous object in all existence. Therefore, Olbers' Paradox can be solved simply by saying that the night sky isn't full of white starlight because there are actually no such things as stars at all. Instead, those twinkling things we *think of* as being stars are really reflections of our electric sun, being beamed back from the edges of the infinite universe itself (quite how infinity can have edges I cannot pretend to understand). Furthermore, it turns out that infinity is not equal in all directions; in some places, it may be mere inches away, and in others, millions of light years distant. Hence, the reflective boundaries of infinity become like a set of never-ending fairground mirrors, with the sun's original reflection being beamed back off the edge of infinity in one place, and then back again to another opposing surface of infinity elsewhere, and so *ad infinitum*. The stars we see in the sky, therefore, have no real existence to them, 'any more than the image of a candle in a mirror has a real existence'. Because infinity is further away in one direction than another, it follows that the reflections of the sun from such areas will be less bright than others, or even not visible from Earth at all, thus solving Olbers' Paradox.[1]

Thankfully, there are now some slightly more plausible alternatives available to Reverend Francis' ideas. Today, unlike Olbers, we know that the universe is not infinitely old, but around 14 billion years old. Light travels at a finite speed, and some stars in the universe are so far away from Earth that their light has not yet had time to reach us. Combine this with the fact that stars have life-cycles and ultimately die, and you are left with one reason why space appears to be black; there is a kind of visual 'horizon' out there in the vacuum of space, over which we cannot see. Another reason space looks black, however, is simply because of the way our eyes work. When our universe was born in the Big Bang, a massive amount of heat, energy and light was emitted, which at one point would have filled all of space with bright illumination. However, as the universe expanded over the next 14 billion years, this light and energy became 'red-shifted', being reduced down to a much lower temperature and stretched out massively until it ended up being dragged to the wide end of the electromagnetic spectrum, becoming converted into microwaves. Such microwaves are invisible to the human eye, but are still there, present in every direction in space, being detectable via radio telescopes; cumulatively, these microwaves are known as Cosmic Background Radiation, and provide proof of the reality of the Big Bang, being effectively its echo. However, if the human eye was structured differently, and capable of seeing microwaves, then we would observe that the night sky is not in fact truly black, but full of a kind of light.

In 2002, a team of astronomers from Johns Hopkins University in Baltimore analysed the spectral range of such light radiation being emitted from the stars within over 200,000 galaxies, then performed calculations to determine what the average shade of light would be if human beings were actually able to see all of this cosmic energy. Their answer was that outer space, viewed from a vantage point on Earth, was actually a light shade of beige, or ever so slightly off-white. Think of space as being a bit like a lawn, which appears to be green when viewed overall, even if you can find areas where the grass is brown or yellowish; it is not as if every part of space is beige, but the average effect, when viewed cumulatively, would make it seem so. News of the discovery made the mainstream media, and a competition was announced to find a name for the colour. Unsuccessful entries included 'Skyvory', 'Cosmic Khaki' and 'Primordial Clam Chowder', but the winning suggestion, 'Cosmic Latte', was provided by a Starbucks customer named Peter Drum who happened to be reading about the contest whilst drinking a coffee, and noticed the shade of his drink, and of Creation itself, were rather similar. Had Drum ordered a black coffee, however, mimicking the false blackness we perceive in the universe around us, perhaps the colour of the cosmos would have ended up being christened something rather different.[2]

## Black *vs* White

It is not only the physical structure of our eyes which lends us a false picture of the endless realm beyond our Earth, however; so do our minds, cultures, belief systems and, often, our own personal foibles and eccentricities. But does it really matter what colour the universe is? For some people it does. Consider the case of certain American black nationalists who have devised a whole mythology around the deluded idea that, just because space happens to look black, this means that black persons are inherently superior children of the cosmos, whereas the white race in particular are vastly inferior beings. America's black nationalist Nation of Islam (NoI) sect, for instance, teach a bizarre Creation myth for humanity, in which, some 76 trillion years ago, a living atom named Allah spun out from the womb of total blackness that was space, thereby giving birth to the concept of time. This jet-black 'atom of life' then slowly incubated itself, accruing flesh, blood, bones and brains to its holy nature. Once Allah the Atom had achieved the appropriate stage in its growth, this fleshy super-particle later took upon itself the form of a man – a divine black man, perfect in limb and mind, who then proceeded to create other, lesser, gods in his own beautiful black image, named 'imams', who helped Him construct the endless parade of planets and moons in the universe.[3] This is how one contemporary researcher has summarised the NoI's Creation myth, in imitation of the Book of Genesis:

> **The Creation of the Heavens and the Earth**
> In the Beginning there was blackness, a triple blackness of space, water, and divinity. The One Supreme God [Allah] came into existence at the origin of the universe, 76 trillion years ago. He willed himself into being in the form of a black man, cell by cell, in a process that took him six trillion years. The Supreme God created Himself out of the blackness of space and the blackness of water, which composes three quarters of the body. Before the creation of the world, the black man was in existence ... we took our colour from the darkness out of which we emerged ... All that exists in the universe is created by the Black Intellect, or Divine Wisdom ... 'Man is God and God is Man'. The black man has no birth record: he has always existed, making the black race a nation of gods, originating in The Originator, descending from the base of His wisdom.[4]

Eventually, Earth was peopled with black men modelled after Allah; thirteen black tribes, united under one Black Nation, that of Islam. This original black race, taught the NoI, were collectively 'the maker, the owner [and] the cream of the planet', virtually gods in human form. For example, the brilliant black scientists of this long-gone era were responsible for shaping the globe as we know it today, using special burrowing-bombs packed with dynamite to drill down into the depths of the Earth's core to raise up mountain ranges like the Himalayas. On other planets like Mars, black god-men of this original type still rule in peace and justice, say the NoI, free from the oppressive post-colonial yoke of the white man which prevails down here on Earth.[5] Over time, however, the original black gods created by Allah slowly died – some from sheer boredom, whilst others suffered the consequences of 'poor diets' – leaving their Earthly kin vulnerable when a renegade black scientist with an abnormally large head named Mr Yakub created white men on his sinister Aegean island of evil. These unnatural pale-faced people – the 'blue-eyed devils' – then spread out across the world, turning Allah's natural order upside-down. The idea of evolution was later invented by white Victorian racists like Charles Darwin (1809–82), to cover up Caucasian shame at having been created in a laboratory by the colonial slave races they had wrongly deemed inferior. The white man was designed by Mr Yakub to be 'the skunk of the planet Earth'; deprived of that cosmic space seed which made black men be black and thus participate within the nature of the black space God Allah, the white man had little capacity for love or justice, being intrinsically evil like his twisted big-headed creator.[6] Seeing as Mr Yakub created white people via a process of selective breeding, however, slowly isolating the 'germ' of the white man from out of the black man's genes, it is said that he actually did blacks a favour by removing all possible capacity for evil out of their bodies, in a kind of surgical separation of yin from yang.[7] The traditional Western concept of white equating to goodness and black equating to evil was thus inverted by the NoI; phrases like 'the black sheep of the family' should actually be considered compliments.

## Black Lives Dark Matter

Absurd though it sounds, thousands of people since the NoI's founding in 1930 have believed this yarn, with major elements of the myth having been spread through preaching, pamphlets and song, garnering many fans and many subsequent additions to the basic core narrative. For example, several of the teachings of the NoI's current leader Louis Farrakhan (b. 1933) centre around the weird idea that a giant flying saucer piloted exclusively by undead black Muslims (and which, bizarrely, is also full of cows)[8] is hovering over America, just out of sight, waiting to drop more drill-bombs onto the supposedly racist nation and destroy it, if the white folk in Washington don't mend their ways. The formidable weapons of this so-called 'Mother Wheel', however, need not scare any Afro-Americans, as they will be given asylum on the spacecraft and whisked away to the safety of a new black homeland *en masse* before the likes of President Trump are forced to eat dynamite.[9]

But what precisely is this special cosmic substance which makes pure Nubians be black as the holy night sky? Several persons have concluded that it is nothing other than the chemical pigment melanin, which black people possess more of than white people do, a fact which does indeed help account for the darkness of black skin. However, a whole field of black American pseudoscience has now sprung up aiming to link melanin back to the blackness of the cosmos, thereby allowing the claim to be made that black people are inherently closer to God than whites are. Consider the black psychology professor T. Owens Moore, whose 1995 book *The Science of Melanin* and its 2002 sequel *Dark Matters – Dark Secrets*, revealed the startling news that black people's melanin-infused skin could act as a kind of free cosmic radio service, allowing blacks to pick up intelligent vibrations from the universe itself. Seeing as Moore bills himself as being a member of the Neuroscience Institute at Atlanta's Morehouse School of Medicine, and holds an actual PhD, he must presumably know his stuff.[10] As such, it is quite startling to see blurbs for his books saying things like 'Astrophysicists use [the term] "dark matter" to describe [an] intriguing material in the external universe, and … [Moore] uses the term "dark matter" to describe the neuromelanin in the brain of the internal universe (the human body [of a black man]).' Rather than science as such, this seems to be yet another restatement of the old NoI idea of (as another blurb for Moore's books puts it) 'the material spiritual connection between [the black] man and the universe', with melanin becoming a physical, neuro-chemical 'conduit to keep people in tune with the cosmic elements' rather than something that just turns your skin brown.[11]

Moore's reasoning for this is that the famous but mysterious dark matter which astrophysicists talk of as permeating the universe is so called not simply because scientists cannot properly detect it and are unsure of what precisely it is, but because it is actually melanin! In Moore's own words, melanin has 'special bioelectric properties' which foster and encourage the constant transmission of mystical energies to and from black people's bodies and the limitless, and equally black, depths of outer space. Apparently, black people's 'skin and nervous system link our internal melanin to the cosmic melanin to allow us to tune in with cosmic nature'. It is because the current establishment of majority-white scientists do not have enough melanin within them that they are unable to understand the true nature of dark matter, it seems; black people like Moore, however, can divine the truth about the substance more instinctively. After all, 'melanin in the skin and nervous system helps to provide metaphysical experiences', he says.[12]

## Bad Hair Days

And, if you're not quite dusky enough to talk to Creation like this, then why not just stop combing your hair? According to Moore:

> African [i.e. black] people naturally have kinky or wiry hair. Non-African people [i.e. whites] have matted or animal-like hair. In other words, kinky or wire-like hair is an evolutionary advance, since very few animals (sheep, buffalo, yak, etc) have hair similar to African people. Kinky or wire-like hair is constructed like an antenna to absorb more readily those naturally occurring electromagnetic waves in nature. The 'conscious waves' [of the living melanin-based universe] ... are also more attractive to kinky or wire-like hair.[13]

This news would have come as something of a surprise to another group of racial theorists who enjoyed spending their days making contact with the cosmos via their hair, however. The *Vrilerinnen* or *Vril-Damen* ('Vril-Ladies') were one of the most obscure groups of the Third Reich, being a band of pure-bred Aryan women of supermodel-like appearance whose unusual task it was to keep their hair combed straight in order to be able to tune in to telepathic radio waves being beamed out to them by a race of super-advanced alien racists from the distant star system of Aldebaran. A cabal of five highly psychic maidens with skin as pale and chaste as the driven snow – Maria, Traute, Sigrun, Gudrun and Heike – they bravely ignored the fashions of their day, refusing to wear short-cut bobs and instead letting their tresses hang long like horse tails, considering their luscious locks to be a form of 'cosmic antennae' to rival even the powers of T. Owens Moore with a big frizzy afro. Well, that's the story, anyway. In truth, it is debatable as to whether or not the *Vrilerinnen* even existed. You can easily find photographs purporting to be of them online, but (appropriately enough, I suppose) they look suspiciously like glamour shots of random models taken from glossy magazine adverts for shampoo brands than snapshots of members of a secret society.[14]

The tale of the *Vrilerinnen* appears to have its origins in a 1992 book, *Das Vril-Projekt*, by the Austrian writers Norbert Jürgen-Ratthofer and Ralf Ettl, who between them helped promote the notion that there was a secret occult organisation called the 'Vril Society' active in both Nazi and pre-Nazi society, which aimed to harness the power of a wonderful, universe-pervading source of cosmic energy named 'Vril'. Very much the Nazi version of T. Owens Moore's melanin-infused dark matter, only pure Aryan types could gain full access to it, thus making them the natural rulers of the world, and the supreme breed of humans on Earth. Needless to say, in reality there is no more such a thing as Vril than there is such a thing as 'The Force' from the *Star Wars* films (which it superficially resembles). The idea has its origins in an action-packed 1871 novel by the Victorian

English aristocrat and occultist Edward Bulwer-Lytton (1803–73) called *The Coming Race*, which concerns a race of superior beings named the Vril-Ya, who live inside the Hollow Earth and have full mastery over Vril, allowing them to build a super-advanced technological society. This trashy book, in its own peculiar way, was one of the most influential of all time, seeing as it attracted several readers who thought that it was not really a normal novel at all, but a true account, lightly fictionalised by its author. The full story of how Bulwer-Lytton's tale of wonder began to be transformed into the idea that the Nazis were in cahoots with a secret society of Aryan supremacists in order to harness the power of Vril, thus making the Third Reich invincible, is too complex to explain here, but suffice to say if you look up the term 'Vril Society' on the Internet today, you will find that the original notion has now grown into a complex, and almost wholly untrue, mythos involving everything from Nazi flying saucers to Adolf Hitler (1889–1945) surviving the Second World War and living on in a secret base somewhere beneath Antarctica.[15]

## Space Racists

By the time of Jürgen-Ratthofer and Ettl's 1992 book, the so-called 'Vril Society' – which in reality was a very small group of isolated cranks who met together in Berlin between the two World Wars hoping to explore possible means of tapping into free energy sources – had become magnified into a cabal of well-connected psychic Aryan occultists with close links to the likes of Deputy Führer Rudolf Hess (1894–1987). This band of obscure fantasists were not even actually called the Vril Society, and the most sinister activity in which they indulged was to meditate upon an image of the globe as a giant bisected 'World-Apple', whose North and South Poles represented a positive electrical Vril-charge and negative one respectively. This unappealing image of a sad gang of nuts sat around staring at pictures of apples all day was transformed into a much more sellable account of beautiful long-haired *Vrilerinnen*, led by the half-Croatian, half-German medium Maria Orsic (1895–?), meeting up in castles with top Nazis and channelling down messages from a race of 'light godmen', or pure Aryans, who lived on the planet Sumi-Er in the star system of Aldebaran, some 68 light years away from Earth in the constellation of Taurus.

The Vril Society had supposedly begun communication with these so-called 'Alpha-Aldebarans' as far back as 1919, which had allowed them to construct a picture of a perfect alien civilisation which resembled 'a kind of National Socialism on a theocratic basis' – with the Aryan aliens being the gods to be worshipped by other inferior alien races. Seeing as these white Alpha god-men were 'the Germans in the Sign of Taurus', it made sense that they should become allied with the proto-Nazis who, in this version of the story, made up the Vril Society. Wishing to conquer Jewish forces throughout the entire galaxy, the Aldebarans kindly allowed Orsic to receive channelled instructions for constructing both a time machine and an entire fleet of what were later to become Nazi flying saucers dubbed the 'Haunebu', all fuelled by the power of Vril. Unfortunately, not quite enough of these craft could be manufactured prior to Allied victory in the Second World War, so the Nazis are supposed to have retreated to secret bases in Antarctica (or even back through time!), whilst simultaneously sending a diplomatic mission off to Aldebaran in a special, dimension-shifting flying saucer, to seek military aid. Sometime soon, said Jürgen-Ratthofer and Ettl, the space Nazis would be back and, after linking up with their fascist brethren hiding out beneath the frozen South Pole (or in ancient Babylon, if you prefer) be ready to conquer and cleanse the world in a saucer-powered Third World War. According to Jürgen-Ratthofer, to ensure their combat-readiness, Nazi saucer pilots had even helped out that modern-day Babylonian Saddam Hussein (1937–2006) during the Gulf War of 1990/91, but if so then they don't appear to have performed very well.[16]

## I'm Dreaming of a White Cosmos

The parallel between the white supremacist myth of Nazi aliens waiting to rescue Aryan mankind using saucers powered by the inherently Aryan force of Vril and the black supremacist myth of radical Black Muslims waiting to rescue black humanity using their own 'Mother Wheel' and the power of cosmic melanin should be clear enough, as should the weird similarity between the notion of being able to contact either God or aliens via the unique racial properties of contrasting hairstyles. However, for obvious reasons it is most unlikely that either group copied and adapted ideas directly from the other. Perhaps it is simply the case that racial fantasists the world over have an inherent tendency to think alike?

In truth, Adolf Hitler, in spite of many popular legends to the contrary, had little abiding interest in the occult or flaky subjects such as Vril. His underling Heinrich Himmler (1900–45) may well have been different, but accounts of the Third Reich which attempt to portray the whole country as being in thrall to mad mystical fantasies about German rune magic, conjuring up demons or contacting alien beings are somewhat exaggerated, to say the least, as the scholar Nicholas Goodrick-Clarke (1953–2012) has comprehensively shown.[17] One element of all these Nazi-related space myths which does come close to hitting its target, however, is the idea that, following their final defeat, elements of the Nazi High Command might have sought refuge in the frozen Polar regions. This is not because such icy wastelands actually harbour any hidden Nazis, but because it is where many of them, Himmler in particular, thought the Aryan race originated. What the modern-day conspiracy theorists got wrong was the Pole in question – as an inherently Nordic people, the Nazis of course proposed that the Germanic race was born near the *North* Pole.

One of the most unreliable (though entertaining) books about the Nazis and the occult – and one which helped spread the whole legend of the Vril Society – was Louis Pauwels' (1920–97) and Jacques Bergier's (1912–78) 1960 counter-culture classic, *The Morning of the Magicians*.[18] Here, the authors take the Nazi fondness for ideas of the Germanic race once having had to grow up in icy conditions to an absurd extreme, claiming that Hitler disastrously pushed on with his invasion of the USSR during the bitter Russian winter of 1941 purely in order to demonstrate to the world that he was the undisputed 'Master of Ice' and that, as the authors put it, 'Winter would retreat before his flame-bearing legions.'[19] Curiously, the idea that white people are inherently wintry in nature has also been picked up on by some contemporary black supremacists, with the black New York professor Leonard Jeffries Jr (b. 1937) becoming infamous for his claim that Caucasians are cold-hearted 'Ice People' and blacks warm-hearted 'Sun People',[20] but there is no doubt that some of the stranger Nazi leaders and their fellow travellers really did look up into the night sky and see a universe not of deepest black, but of frostiest white, looking back down at them.

## Mr Frosty

The next time you cast your eyes up towards the Milky Way some clear and cloudless night, take a moment to stop and ask yourself what precisely it is you are seeing. The standard answer is that you are observing a twirling, milky band of light which stretches out across the heavens in a series of spiral arms, caused by the illumination given out by the innumerable distant suns of our galaxy. In short, you are looking at the stars. The renegade Austrian astronomer Hanns Hörbiger (1860–1931), however, didn't believe in stars, and in a hugely influential 1913 book made the rather startling assertion that, far from being the result of starlight, the Milky Way was in fact made entirely out of ice. According to Hörbiger, a series of massive, planet-sized ice blocks was floating around up there in space, encircling our entire solar system in an impenetrable white ring. Light from a few actual suns lurking beyond the ice ring then shone through this frozen barrier, reflecting off its massed ice crystals and giving observers on Earth the mere illusion of billions of stars twinkling down at us from the inky blackness. Various other astronomers might well object to this proposal, admitted the Austrian, and even attempt to show off photographs of the Milky Way's alleged 'stars' to prove their case, but he had an easy answer to hand for these arguments – all such images were simply fakes. As to any tedious mathematical objections which sceptical astronomers might have made to his proposal, Hörbiger had an even more emphatic response in store. 'Mathematics,' he once pronounced, 'is nothing but lies!'

Hörbiger could justify this assertion by pointing back to his successful career as an engineer, during which one of his most appropriate achievements was to have helped develop new cold compressors for use in manufacturing artificial ice. In 1894, he had also invented a special kind of low friction, automatically opening and closing steel disk valve for use in blast furnaces, blowers, pumps and engines, a genuinely helpful invention without which various industrial processes and methods of gas exchange would simply not have been possible. However, Hörbiger's invention of this valve was not something he had worked out laboriously at a desk in his workshop, through calculations and technical drawings; instead, it had simply 'come to him' whilst on the job. As such, for a qualified engineer, he had little time for mathematics. 'Instead of trusting me you trust equations!' he would harangue those who tried to point out to him the various reasons why his ice ring theory could not be true. 'How long will you need to learn that mathematics is valueless and deceptive?'

Hörbiger's full, entirely unmathematical, theory was termed the *Welteislehre*, or 'World Ice Theory' ('WEL' for short). Basically, it held that at some distant point in our galaxy's past there had been a gigantic super-sun, millions of times the size of our own, next to which had orbited a massive planet, many times larger than Jupiter, covered by layers of ice hundreds of miles thick. Eventually, this ice planet fell into the super-sun, melted, and transformed into jets of super-charged steam which blew the sun apart, spewing out lumps of rock and fire which ultimately settled down to become our own current solar system. Vast clouds of oxygen were also released from the explosion, and reacted with thin layers of hydrogen gases already swirling through space, creating masses of space water which, space being cold, soon froze into the gigantic ring of interstellar icebergs which now encircled us all. Sometimes, said Hörbiger, one of these ice blocks breaks away and floats into the pull of our sun's gravitational field, falling into it and creating sunspots, which are really colossal melting ice cubes. Occasionally the Earth happens to be orbiting in the path of one of these falling space-bergs, causing severe hailstorms before it finally drops into the sun. Our moon is less lucky; being higher up and thus exposed to more ice, it is continually accumulating more and more frozen layers of water on its surface. Eventually, it will get so heavy that it simply falls down to Earth and kills us, claimed Hörbiger. Apparently, such a catastrophe had already happened several times in the past; the Earth used to have other smaller moons, which became so heavy with cosmic ice that they crashed down onto our planet thousands of years ago, destroying Atlantis and making Noah feel glad he had built that Ark. If you thought that the giant iceberg crashing into the *Titanic* had been a disaster, implied Hörbiger, then just wait until the giant moon-berg finally collided with SS Planet Earth.

## Ice in His Veins

That's quite a bold theory, and in order to support it Hörbiger had to have amassed a huge amount of evidence, didn't he? Well, much of Hörbiger's 'proof' for his premise amounted to the fact that he had had a few strange dreams or visions which had revealed the 'truth' about our frozen universe to him. Just as he had created his Hörbiger-Valve entirely through intuition, so he had created his infamous World Ice Theory. As a small child, Hörbiger had owned a telescope. Through this, he liked to look at the moon. He thought its surface looked cold; and, all of a sudden, realised that this simply *must* be because it was covered with ice. That was Hörbiger's first revelation. His second came when he had a strange dream in which the Earth became transformed into a giant pendulum swaying on a luminous string. This apparently revealed to him the secrets of gravitation, showing how icebergs in space could be attracted towards the sun. Thirdly, whilst working as an engineer one day in 1894, he witnessed some molten iron falling onto a pile of snow, causing bits of soil beneath to explode under the pressure of the jets of steam which had been released by the snow suddenly melting. This caused Hörbiger to immediately understand that an ice planet had once dropped into a super-sun, thus giving birth to our solar system. Coincidentally (or not), the basic principles of World Ice Theory coincided perfectly with the physical laws relating to water, gas, freezing and pressure which Hörbiger had studied and made use of throughout his entire professional life. At last, the WEL was all falling into place; all that now remained was for Hörbiger to write his book – all 790 pages of it – telling the world about his discovery.

Even though it had been partly co-authored by a leading German mapper of the moon named Philipp Fauth (1867–1941), *Glacial Cosmogony*, as the book was called, didn't trouble the bestseller lists initially. Astronomers either rubbished or ignored it, so, following the end of the First World War in 1918, Hörbiger decided to go over their heads and appeal direct to the general public instead. Noticing how the German people had been 'sold' the War through propaganda, Hörbiger began tirelessly propagandising for his own cause. An imposing figure, with a fearsome long moustache and beard, Hörbiger gathered a band of faithful disciples who together poured out a never-ending line of leaflets, books, pamphlets and posters in his name. There were even WEL-themed movies and radio broadcasts, and an actual WEL newspaper was launched, grandly titled *The Key to World Events*. An explorer and WEL-believer named Edmund Kiß (1886–1960) wrote a best-selling trilogy of novels detailing the impact past ice moons had had upon our planet, becoming known as the 'poet of Atlantis', and was later recruited by none other than Heinrich Himmler into a special SS archaeology unit tasked with finding traces of prehistoric Aryan man's former residence in the Arctic, and his encounters with frozen moons. In 1936, a document known as the 'Pyrmont Protocol' was drawn up and signed, giving official SS patronage to Hörbiger's theory five years after the pseudo-astronomer's death.

As this shows, Hörbiger was able to ruthlessly exploit the fact that the general public, then as now, had little meaningful scientific education, winning more and more adherents – including some, like Himmler, who would go on to play a central role in Nazism. Hörbiger himself was not a Nazi (he died in 1931, two years before Hitler came to power) but took advantage of the growing tide of German anti-Semitism to promote the idea that the WEL was some kind of antithesis to so-called 'Jewish Science'. With Hörbiger loudly denouncing the theories of Jews like Albert Einstein (1879–1955) as being in direct contradiction to his own ideas, belief in the WEL became a quick and easy test of whether a person was a 'true German' or not. Companies with WEL fans sitting on their boards refused jobs to non-believers, and people began campaigns of intimidation against those foolish astronomers who held on to the old 'Jewish' model of the universe. Connections were drawn between the 'healthy' idea of cosmic ice rings, and the equally 'healthy' Nazi myth that the German race had originated from the North Pole. 'Our Nordic ancestors grew strong in ice and snow; belief in the WEL is consequently the natural heritage of Nordic Man,' was one such opinion expressed. The perpetual struggle in outer space between the fire of the sun and the ice of the stars was taken to be a mirror of the perpetual struggle between *übermensch* and *üntermensch* down here on Earth. The WEL became no mere scientific theory, but a *kosmotechnische Weltanschauung*, a 'cosmotechnical worldview' which embraced not just astrophysics, but an entire poetic, mythic and political outlook upon the universe – nothing less than the 'astronomy of the invisible'.

If you could live long enough to see them, though, then the slow descent of our ice-laden super-moon would eventually produce some very visible changes indeed upon the Earth; due to the increased gravitational pull brought about by the falling moon's approach, people would probably grow into giants, it was said, being stretched upwards towards the lunar surface, much as little lizards had grown into gigantic dinosaurs during the fall of a previous ice moon. Cosmic rays also gained increasing penetration through the Earth's moon-damaged atmosphere at such times, leading to bizarre mutations occurring, lending animals amazing new powers like flight and increased intelligence. Maybe this was how human beings evolved from earlier primates, by having their skulls stretched up towards the moon to make space for bigger brains? During certain other phases of their life spans, however, the Earth's moons could also cause negative mutations in humanity, permitting such supposedly hideous sub-human freaks as Jews to emerge from swamps and sully our planet.

## The Pale Face of the Moon

Some of these ideas were Hörbiger's own, and others those of his later followers, of whom he had many. Millions of Germans professed belief in Hörbiger's doctrines, a phenomenon which probably had more to do with people's desire to get on in life than anything else. So intertwined did the worlds of WEL and Nazism become that the Party felt compelled to issue a press release reassuring worried citizens that it was not *absolutely* necessary to believe in cosmic ice in order to be a 'good National Socialist' – merely an advantage. Nonetheless, various Nazi meteorologists, such as the SS's Hans Robert Scultetus (1904–76), tried to make use of the WEL for forecasting weather patterns, whilst Hitler himself planned to build a massive museum devoted to Hörbiger in Linz following German victory. It was to have three separate floors devoted to history's greatest astronomers, with Hörbiger, 'the Copernicus of the twentieth century', naturally having the top floor all to himself, symbolising the towering nature of his triumph. Seeing as its main source of appeal was as an expression of ultra-nationalist, pro-German feeling, you would have thought that, following Allied victory in 1945, the whole idea of the WEL would have melted away instantly, like a tiny ice cube falling helplessly into a massive super-sun. This was not the case. Even into the 1950s, the WEL still had over a million adherents – many not even in Germany. The new post-war centre of the cult was across the Channel in England, where a disciple of Hörbiger's named Hans Bellamy (1901–82) continued to fly the frosty flag. As late as 1953, Hörbiger's remaining followers were writing that final proof of their hero's ideas would come when man first set foot on 'the ice-coated surface' of the moon – but, come 1969, the *Apollo 11* crew didn't need any ice skates. Hörbiger and his disciples had been wrong; but then, seeing as the whole theory had apparently been based upon emotional need rather than any form of actual logic, this fact should not surprise us.[21] This, for example, is Hans Bellamy's account of how he came to believe in the WEL:

> When I was a boy, I often dreamt vividly of a large moon ... glaringly bright and so near that I believed I could almost touch its surface. It moved quickly through the heavens. Suddenly it would change its aspect and – almost explosively – burst into fragments, which, however, did not fall down immediately. Then the ground beneath me would begin to roll and pitch, helpless terror would fall upon me – and I would awake with the sick feeling one has after a nightmare.[22]

When he first read about Hörbiger's theory in 1921, Bellamy realised that this dream had been a kind of 'racial memory' of a previous moon-fall which had been written into the minds of his white European ancestors and then passed down to him, generation after generation, by their collective ethnic unconscious. When old folk tales talked of demons, dragons and devils, Bellamy realised, they were really unconscious representations of mankind's age-old battle with the icy moon.[23] When in 1952 another of Hörbiger's disciples, Elmar Brugg, wrote of how for his hero the universe was not 'a piece of dead machinery', but a kind of living being in which 'every part reacts on every other part and which hands down from generation to generation its burning force',[24] then this was what he meant. For the Nordic race, their sacred destiny to achieve eternal mastery over both cosmic ice and the Earth itself was quite literally written in the stars – every bit as indelibly as, to the Nation of Islam, the supremacy of the black man was visible within the very darkness of space.

## Guardian-Readers of the Universe

Naturally, most readers will scoff at such absurdities as Nazi ice moons, white Aryan frost universes and black Muslim atom-men from outer space, and rightly so – it seems somehow appropriate that it turns out space's true colour is actually beige, which stands mid-way between the two competing skin colours of black and white, or cosmic forces of melanin and Vril, without actually taking the side of either. Enlightened modern-day members of post-racial Western democracies will doubtless rejoice at the aptness of such a metaphor, thinking themselves to be wise, just and free of all bias. However, in truth anyone can become the victim of their own invisible cultural prejudices, hubris and unconscious powers of self-projection when it comes to such issues. In 1972, for example, NASA's *Pioneer* space probe was equipped with a metal plaque bearing an image of a naked man and woman, hoping to give any aliens who might come across it a vague idea of the beings who had created the device. This might seem a reasonable scheme, but during a 2015 conference at that hardy bastion of modern metropolitan 'rationalism' the London School of Economics, gathered academics decided loftily that such an image was both highly sexist and racist in nature. Dr Jill Stuart, who is apparently an 'expert on the politics of outer space', had noticed that the man in the image was 'raising his hand in a very manly fashion' and seemed to be in charge of affairs, whilst the woman looked much more 'submissive' in her pose. Furthermore, both persons appeared to be hideously white, something which made Dr Stuart feel 'uncomfortable', leading for her to call for more 'diversity' in any future pictures we send floating off out into the cosmos. Another academic at the conference did point out that, for all we know, aliens might not have any eyes so may fail to be offended by the supposedly un-PC nature of the plaque, but was doubtless charged with violating Dr Stuart's safe space and consequently ignored.[25]

I laughed at this, but maybe it is our right to have such silly ideas. Mankind has always been drawn towards looking up into outer space, with the heavens becoming a kind of gigantic, ever-spreading ink blot onto which people could project their own personal fantasies and those of the societies they have lived in. From tales of crumbling canals and lost civilisations on Mars, to the notion that the planets were regular masturbators who sang songs of celestial joy amidst sparkling orgasms of pure lemonade, mankind has had to travel down a lot of blind alleys before finally being ready to initiate the

Cape Kennedy blast-off that took him up towards our surprisingly iceless moon in 1969. Central to the imaginative appeal of space throughout most of history, of course, was its utter physical inaccessibility. For millennia, almost whatever you chose to say about it would remain conveniently unfalsifiable. The advent of the Space Age was to put a stop to the most extreme kind of space-based nuttiness gaining any kind of serious mainstream airing, sadly. But what about the lunatic fringe? Religious fanatics, UFO fans, back-shed theorisers, inveterate iconoclasts and the profoundly mentally ill were to develop all kinds of ingenious, and often amusing, techniques for denying blatant reality as the twentieth century progressed onwards into the twenty-first. The pristine black canvas of outer space, for so long the repository of some of mankind's oddest ever dreams, was not about to be surrendered without a fight, and this book tells the story of some of the most interesting celestial skirmishes of all. Keep watching the skies!

# I

# Sphere of Fear: Sputnik Spooks the World

This book is first being published in 2017, to mark the sixtieth anniversary of the Soviet Union's successful launch of the first ever Earth-orbiting satellite, *Sputnik I*, on 4 October 1957. *Sputnik*'s full name was actually *Sputnik Zemlyi*, meaning 'fellow traveller'[1] and it proved to be such a persistent and dogged escort that some people ended up carrying images of it around with them everywhere they went, within the confines of their own head. To a degree, *Sputnik* was actively designed by the Soviets to promote Western paranoia. The iconic shiny silver sphere, the size of a beach ball, was actually visible to the naked eye as it passed overhead, which proved an irresistible sight to 1950s Americans, who stayed outside by the thousands to crane their heads above them, hoping to see what their arch-enemies had just achieved. For those who did not lie beneath its flight path, however, there was always the possibility of tracking it via radio, with its characteristic 'beep-beep' transmission signal being easily intercepted on radio sets across the world. Such ease of access was deliberate. This was the height of the Cold War, and by making the satellite simple to track, the Russians helped ordinary people to make a direct observation of their miracle, thus hammering home Soviet scientific superiority. Indeed, prior to *Sputnik*'s launch, the Russians had publicly announced what radio frequencies people should tune into in order to hear it.[2]

*Sputnik*'s beeps actually did nothing other than allow the satellite's progress to be tracked easily from the ground, but a whole mythology grew up around the eerie sound nonetheless. Dubbed the 'Red Moon', *Sputnik* was credited by the gullible with the ability to spy down on US citizens on the ground, listening to their conversations or messing with their minds; October 1957 proved the very height of tinfoil hat fashion for some. Maybe the beeps were some kind of coded message, or else an aerial equivalent of submarine sonar devices which could map the US via echolocation from outer space, allowing the Russkies to compile an inventory of American warplanes and tanks. Some thought it best such vehicles be kept inside hangars permanently from now on, so the eye in the sky wouldn't be able to count them. In such an atmosphere, pranksters inevitably took the opportunity to send out fake beeps themselves, leading some to conclude there were whole fleets of *Sputniks* orbiting overhead. Another panic was caused by the 'Spooknik' phenomenon, whereby *Sputnik*'s signal found itself reflected off the Earth's ionosphere, leading to a 'ghost signal' being picked up when the real satellite should have been on the other side of the planet; some asked themselves if the Soviets had two of those things whizzing about up there, and if so then why? (In fact a genuine second satellite, *Sputnik II*, was launched on 3 November, containing the famous but doomed 'space dog', Laika.)[3] Some radio hams became so addicted to listening out for *Sputnik* beeps that they continued surfing the static for years afterwards, claiming to have intercepted secret transmissions from a 'hidden' Soviet space programme whose feats were too suicidal to make public, with a pair of Italian brothers professing to have genuine sound footage of Russian Cosmonauts nobody was ever supposed to know about dying alone out in the icy depths of space.[4]

*Sputnik* was even blamed for an outbreak of self-opening garage doors. The first to report such an outrage was Thomas Rinaldi, a doctor from Schenectady, New York State, who was puzzled each morning to find his radio-controlled garage door standing wide open. Noticing these events coincided with *Sputnik* passing overhead, he got up early one day and stood watching his garage at the time the satellite was scheduled to whirl by. Sure enough, at 6.23 a.m. the door opened right on cue, and the connection seemed clear. When this was announced in the papers, similar stories spread like wildfire, some true, some false; but when events so trivial begin to be reported as if they are major news, you know you are at the centre of a social panic.[5] Rumours even began to multiply that Russia's next plan was to fire a rocket at the moon containing a giant canister full of crimson paint to be splurged across the lunar surface, thereby claiming visible ownership of a real Red Moon for Communism. It was the fortieth anniversary of the Russian Revolution on 7 November, and this was floated as a likely date for the outrage to occur, though it never did.[6]

## Going Nuclear

Of course, the real fear about *Sputnik* was far from trivial; namely, that it represented some new means for the Russians to drop nuclear bombs on US soil. American airspace, protected by the Atlantic and Pacific oceans, had never been penetrated by enemy aircraft, so the sight of *Sputnik* orbiting overhead proved a great shock. The USSR let it be known their satellite had been launched via an ICBM, or Inter-Continental Ballistic Missile, a new kind of projectile with a range great enough for them to hit targets almost anywhere upon the planet with, a disaster if such weapons were to be equipped with miniaturised nuclear warheads. 'Soon they will be dropping bombs on us from space like kids dropping rocks onto cars from freeway overpasses,' predicted future President Lyndon B. Johnson (1908–73), whilst *Newsweek* suggested a fleet of *Sputniks* could be fitted with an early form of 'dirty bombs', to spread radioactive fall-out over entire nations, poisoning millions.[7] The fear of falling further behind the Soviets sparked off the Space Race in America, which was ultimately won that fateful day in 1969 when *Homo Americanus* made his first giant leap onto the lunar surface, but in the meantime the country's nascent military-industrial complex was about to develop a few very peculiar counter-measures of its own to try and nullify *Sputnik*'s propaganda success. Another rumour about Soviet intentions was that they wished to fire an H-bomb at the moon,[8] an idea which certain US scientists decided was so good, they might try and copy it.

The idea had first been proposed prior to *Sputnik*'s launch by the physicist Edward Teller (1908–2003), often known as 'the father of the H-bomb'. A real-life *Dr Strangelove*, Teller advocated using nuclear weapons on a routine basis to perform tasks like dredging harbours and destroying inconveniently placed mountain ranges; listening to him, you could easily get the impression he would be quite happy to use small thermo-nuclear devices to unblock his drains. Teller's aim with sending a nuclear bomb to the moon was to subject the subsequent dust cloud following the huge explosion to spectrum analysis, thereby determining the moon's chemical and geological composition the easy way, without having to send astronauts up there to collect samples.[9] The idea was initially dismissed as insane, but in light of *Sputnik*'s success, Teller's scheme was reconsidered. 'Project A119', as it was known, had purposes other than spectrum analysis in mind, however. Causing a giant flash on the moon would have sent out a message about US military strength, and re-established Western prestige. There was also talk of stationing missile silos on the moon, militarising it as a further act of deterrence against Moscow. The project was eventually cancelled in January 1959 for fear of negative public reaction to the feat, and also after consideration of the dire consequences for life down below had the launch gone wrong.[10] (Incidentally, a genuine proposal to nuke the moon was once made by a Mr P. Norcott of Broadstairs in Kent, who in 1971 published a pamphlet demanding that scientists should split the moon into two equally sized hemispheres with H-bombs; if each half then orbited the Earth at such a distance that they were directly opposed to one another, Norcott reasoned this would act as a kind of gravitational 'brake', somehow preventing earthquakes.[11])

## Goodness Gracious, Great Balls of Fire!

American religious figures, too, responded to *Sputnik*'s launch, with the evangelist Billy Graham (b. 1918) using it as an excuse to harangue President

Dwight D. Eisenhower (1890–1969) about the need for moral renewal in the country to go hand in hand with compulsory scientific training in schools in order to combat growing post-war materialism. Some worshippers were outraged that an officially atheistic nation like Russia had been the first to penetrate God's Heaven, and demanded the government make quick redress. American atheists were pleased that *Sputnik* didn't seem to have encountered God or Jesus up there beyond our atmosphere, and crowed accordingly; contrariwise, some hell-fire preachers told their flocks that *Sputnik*'s launch heralded the imminent Second Coming of Christ. The Vatican seemed in two minds what to think about it all; one announcement stated that God placed no limit on space exploration, whilst another condemned the Soviets as 'child-like men who are without religion or morals'. The US National Council of Churches, meanwhile, called for prayers to be sent up to God, so that 'His will may be done ... on Earth and in outer space.'[12]

Perhaps the strangest individual spiritual reaction to *Sputnik*'s launch was that of the rock-'n'-roll legend Little Richard (b. 1932). His family were Seventh Day Adventists, which caused quite some conflict when Richard discovered he was bisexual and drawn towards gender-bending. Richard left home to live a wild life devoted to sex and music, subjects which came together in his most famous hit *Tutti Frutti* which, in its original form, was an ode (and in places practical guide) to anal sex. His religious upbringing left its mark, however, and during a tour of Australia in October 1957, Richard underwent a surprising *Sputnik*-based epiphany. During a flight from Melbourne, the glow of his plane's jet engines had led him to picture angels holding the vehicle up, and one of his stage performances in Sydney happened to coincide with the launch of *Sputnik* over in Russia. Things started off normally enough, with Richard entering the stage dressed in a bright yellow suit, emerald-green turban and glittery ruby-red jacket before climbing on top of a piano and beginning to dance. Looking up into the sky, however, he was incredibly disturbed to see what looked like 'a big ball of fire' flying directly over the concert stadium, 200–300 feet above the crowd. Richard wasn't to know that this was *Sputnik* being launched, and that the flames were a sight of the trailing ICBM rocket casing which had helped blast it into orbit, and concluded it was a fiery sign from God for him to quit singing the Devil's music and re-embrace religion. In the middle of his set, he announced he was leaving showbusiness with immediate effect to go away and study theology. He left his Australian tour ten days early, leaving half a million dollars' worth of cancellations behind. When Richard later discovered one of the planes he had been scheduled to catch on his tour had crashed into the Pacific, he decided that God had indeed shown him the path to true righteousness that day. He developed some kind of theory that *Sputnik* was intended to crash into his plane if he hadn't heeded the Lord's Word, and became an evangelical preacher, denouncing his previous style of music as 'demonic'. Little Richard later returned to singing gospel music, and did have the occasional minor moral lapse as regarded orgies, but nonetheless continued to maintain his immortal soul had been saved that day by a satellite.[13]

## Sputnik and the Saucers

Those who view ufology as being a new form of religion, tailor-made for the Space Age, may take heart from the fact that the launch of *Sputnik* also heralded an upsurge in 'saucer sightings', as such events were then still called. Of the 1,178 recorded UFO sightings in America in 1957, 701, or 60 per cent, occurred during the three months following *Sputnik*'s 4 October launch.[14] *Sputnik II*'s flaming death on 14 April 1958 later led to thousands of reports of a strange light being seen in the sky all along America's East Coast; witnesses did not realise they were seeing the flying coffin of Laika the sizzling space dog burning up on re-entry. Especially odd was the idea put forward by the leading UFO journal of the day, *Flying Saucer Review*, which disputed that it was possible to see *Sputniks* with the naked eye at all, meaning that the strange 'satellites' people were viewing above were actually alien spacecraft. The idea went that, now mankind had successfully launched an object into outer space, curious extraterrestrials were stopping by to see just what we were up to.[15] Other people speculated that perhaps *Sputnik*'s famous 'beep-beep' had been mistaken for some kind of mating call; as cultural historians David Clarke and Andy Roberts put it in their 2007 book *Flying Saucerers*: 'Perhaps, in some strange, metaphorical way, the penetration of space by man had an equal and opposite reaction in the penetration of Earth by entities from space.'[16]

Some of these space entities were then to find themselves being re-penetrated in turn by sex-hungry Earthlings. The most famous such case is supposed to have occurred in the immediate aftermath of *Sputnik*, although it didn't fully emerge until several years later. Now one of the most famous tales in all ufology, the strange experiences of the young Brazilian farmer Antônio Villas-Boas (1934–91) began on 5 October 1957, the very day after *Sputnik*'s launch, when a strange glow, 'brighter than moonlight', appeared in the fields around his farmhouse. The light returned on the night of 14 October when Villas-Boas was out in his tractor, and again the next night. To cut a long and oft-told tale short, Villas-Boas ended up aboard a curiously egg-shaped spaceship being forcibly undressed by a gang of large-helmeted alien semi-dwarfs who barked to one another like dogs, before being introduced to a beautiful, pale-skinned naked female entity with slanted blue eyes and fair, almost white hair (except around her genitals and armpits, where it was bright red 'like blood', apparently). Prior to her arrival, the male aliens had rubbed Villas-Boas' body down with special aphrodisiac gel, which led to him quickly gaining an erection. The ET woman took advantage of this fact to rape him, an experience the human found 'agreeable', apart from when she started woofing every now and then. Once they had finished, the woman pointed at her belly and then up towards the sky, implying that she was now going to fly away and have his baby in outer space. His sperm having been successfully stolen, Villas-Boas was then told to get dressed and kicked out of the space egg, leaving him feeling used and dirty.[17]

Villas-Boas' barking aliens did not specifically mention *Sputnik*, but things were different for a twenty-seven-year-old Birmingham housewife named Cynthia Appleton, who claimed that she was performing her household duties shortly after lunchtime on 19 November 1957 when a blonde gentleman from Venus (or 'Gharnasvarn', in native speak) materialised in her lounge wearing a dome-like glass helmet over his head, and telepathically warned her that humanity had best 'stop all wars' if it ever wanted to live in harmony with alien beings. The Venusian also informed Mrs Appleton that 'the bearers of the hammer-and-sickle' were on the point of inventing a matter-disintegrating ray, and told her the aliens were well aware of *Sputnik*'s successful launch. They were happy to leave it be as a simple scientific instrument, but cautioned her that, should the Russians send any actual weapon up there into orbit, the men from Gharnasvarn would 'confiscate it' immediately, like a naughty schoolboy's catapult.[18] In the weeks after the launch of *Sputnik II*, meanwhile, there was a small rash of stories about aliens landing and trying to make away with people's dogs, almost as if, having seen Laika floating around up there in Earth's orbit, the space people wanted to obtain some similar furry friends of their own.[19]

Whilst this isn't a book about UFOs and aliens as such, both subjects will unavoidably appear repeatedly throughout, and I shall largely be interpreting them as objects of human fantasy which represent some kind of psychological attempt to grapple with the twin topics of events taking place within wider society and the impact this then has upon our changing view of the heavens. Whether they are physically real or not (and I am not a complete sceptic upon the issue, by any means) many people's encounters with UFOs and ETs are certainly real *as experiences*, and *Sputnik*'s impact on ufology was clearly very genuine, even if most of the saucers themselves which appeared in its wake probably were not. However, UFOs do actually predate *Sputnik*. The saucer age itself is generally said to have begun on the afternoon of 24 June 1947, when a private pilot and fire-safety device salesman named Kenneth Arnold (1915–84) saw a group of 'nine peculiar-looking aircraft' flying towards him at great speed, in formation, above the Cascade Mountains in Washington State; this book's first publication is meant to mark the seventieth anniversary of this seminal event too, as well as sixty years since *Sputnik* being launched. Whilst these 'aircraft' were actually semi-circular or bat-winged in shape, Arnold described their motion as being like that of 'a saucer skipping on water'. This phrase was widely misunderstood, and so the phrase 'flying saucer', and the stereotypical image we have now of the disc-like shape of the average UFO, were born; it is interesting to speculate whether, if it were not for the media misinterpretation, the strange things people soon began seeing in the skies would have been described by them in the precise way they were. Arnold himself initially thought there was nothing alien about these things at all, presuming they were some kind of new secret military planes, but he was later to change his mind, as did many others.[20] The arrival of *Sputnik*, after all, proved that it was possible to launch artificial objects into space. Given this, why should mankind have been the only species in the universe to have achieved such a feat?

## Hollow Laughter

*Sputnik* continues to make waves amongst ufologists even today, with a Renaissance painting apparently depicting Jesus Christ and God sitting next to the iconic satellite in the clouds and fiddling with its trailing aerials being brought to public attention online. At first glance the image, *Detail of the Trinity* by Italian artist Ventura di Salimbeni (1568–1613), does indeed look as if it contains a clear image of *Sputnik*. However, closer examination shows that the bluish sphere is actually the 'Celestial Sphere' or *Sphaera Mundi*, a traditional Christian representation of the universe as a whole, not a metallic ball, and the antennae which appear to be sticking out of it are thin wands being held by God and Christ, representing the control they have over all Creation.[21] Salimbeni's picture now stands as the best of a whole genre of Renaissance images which have been combed through by ET-believers in search of spaceships. Some of them are quite amusing; I particularly like the picture of a little man apparently sitting in the elongated nosecone of another *Sputnik*-like object to be found on a 1350 representation of the Crucifixion which hangs above the altar in a Kosovo monastery, and the man and his dog looking up into the sky at a disc with multicoloured rays pouring out of it, to be found in the background of a painting of the Madonna by Domenico Ghirlandaio (1449–94). These things do *look* like spaceships, but are perfectly explicable in terms of standard (though now obscure) religious iconography of the time.[22]

An even stranger modern-day appropriation of *Sputnik* for ufology comes in the teachings of the British conspiracy king David Icke (b. 1952), who reckons our moon is itself another giant version of *Sputnik*, launched uncountable aeons ago by some alien race. 'If you are going to launch an artificial *Sputnik* it is advisable to make it hollow,' Icke has explained to his many followers, claiming that when NASA landed lunar modules on the thing in 1969, it 'rang like a gong', thus proving that the moon is not solid. Instead, it is a 'superbly constructed' *Sputnik*-shaped spaceship made of metals like brass, and intended to act as 'a Noah's Ark of intelligence' within which some gypsy-like extraterrestrial race had once stored away all the essential elements of their civilisation as they wandered throughout space in search of a new home. Scientists 'have no bloody clue where the moon came from and it shouldn't by physics be there', Icke has stated, with his basic theory since having expanded to accommodate notions of the satellite being full of either giant humanoid lizards or else special broadcasting equipment intended to surround us humans with a kind of fake, holographic 'reality', thereby turning us into the lizard-men's unwitting dupes and slaves.[23]

Bizarrely, Icke's speculations seem mostly to be derived from an obscure Soviet disinformation programme, in which speculative scientists and historians were allowed to develop and disseminate wild conjectures of the so-called 'ancient astronaut' type; ideas that Egypt's pyramids were really built by aliens, Sodom and Gomorrah destroyed by an atom bomb from space, that sort of thing. The general idea was to undermine Western faith in the accepted versions of both history and religion, thus lessening capitalist morale. Key figures within the Soviet space programme were happy to point out that, once they had penetrated Earth's atmosphere, there was no sign of the Christian Heaven anywhere to be seen. Cosmonauts joked about not being able to see God or any of His angels out in space, whilst President Nikita Khrushchev (1894–1971) himself facetiously claimed his space explorers had been given the specific but, as it transpired, impossible task of mapping out Paradise, remarks which were widely reproduced in the Western Press.[24] In terms of disinformation, one of the ideas put out by Russia was that one of Mars' moons, Phobos, was unnaturally light and therefore a hollow, artificially made satellite just like *Sputnik*. In 1969, more accurate data about Phobos was obtained, proving this could not be the case, but by 1967 an article had already appeared in the Russian equivalent of *Readers' Digest*, containing a new theory that Earth's moon too might be a hollow spacecraft. The basic idea was that the moon's craters, presumed to have been made via meteorite impact, were too shallow and therefore the falling space rocks must have penetrated only the soft moon surface and come up against super-thick alien metal, blocking their path. This text made a direct impact in the West, as was intended, and some specific details of Icke's claims appear to have been taken straight from it, although I would guess at second-hand. However, the Russian black-ops scheme backfired. Ancient astronaut theories became wildly popular in the West, most famously in the form of the Swiss hotelier Erich von Däniken's (b. 1935) 1968 best-seller *Chariots of the Gods?*, but many readers found pseudo-scientific claims about things like angels really being spacemen actually helped *reinforce* their religious beliefs, implying as they did that the events of the Bible genuinely did happen, albeit with an ET twist. Once they realised what had happened, the USSR adopted a new tone, condemning all ancient astronaut theories as credulous nonsense, their popularity being yet another manifestation of the childish decadence of the West.[25] They could hardly have anticipated that, sixty years later, a man like David Icke would still be putting them out!

# 2

# Flat-out Lies: Flat Earths and Unvisited Moons

A further unusual type of reaction to *Sputnik*'s launch was simply to deny it had occurred at all. Some commentators refused flat out to accept that the dirty Commies could have performed such a feat, whilst others guessed that the claims made for *Sputnik* were rather exaggerated. American physicist Charles E. Bartley (1921–96) disputed whether it had really been launched using an ICBM, or had instead been floated into the atmosphere using special balloons. The idea that Russia had missiles capable of reaching US soil, he implied, was just 'superb' propaganda, the launch thus functioning as a 'fake stunt' intended to imply the Soviets were stronger than they really were.[1]

Another man who refused to accept *Sputnik*'s launch was truly what it purported to be was Samuel Shenton (1903–71), the Devon-based leader of the International Flat Earth Research Society (IFERS). Shenton had only founded IFERS in 1956, and didn't want to give up on his big project so soon just because of some silly silver sphere flying around and beeping all over the place. Instead, he took the event as an opportunity to gain publicity for his cause, telling amused newspaper-men that such pathetic objects as *Sputnik* were 'just like marbles spinning round a saucer', with the saucer in question being Shenton's beloved Flat Earth, which was best imagined as a kind of giant Rich Tea biscuit surrounded by a big circular wall of ice cubes. Yes, *Sputnik* was up above us, admitted Shenton, circling around, but so what? 'Would sailing round the Isle of Wight prove that it were spherical? It is just the same for those *Sputniks*.'[2] In 1962, Shenton used a Press interview to condemn all of the 'Round Earthy' experts on satellites employed by the US and Russia, taking particular exception to a statement from the astronomer and mathematician Dr R. A. Lyttleton (1911–95) to the effect that 'When a satellite reaches required height, it has to be levelled off horizontally – or parallel with the Earth [in order to enter orbit].' Shenton, misunderstanding Lyttleton's words, took this as an admission that objects like *Sputnik* were not really floating around in 'space' (the inverted commas around that word were habitually his own) at all. Only IFERS had the right to use terms like 'levelled off', 'horizontal' and 'parallel with the Earth', he huffed, explaining that in his view, the *Sputniks* 'go up, flatten out above the Flat Earth, and then go round in a circle … The whole thing is a gigantic hoax.'[3]

Shenton was a most entertaining man, whose commitment to his doomed cause was admirably total – so much so that it ended up destroying his health, with the constant burden of correspondence and campaigning leading to numerous strokes and the collapse of his business as a sign-writer. Always keen to spot slights, Shenton objected strongly to the BBC's old spinning globe logo, and in 1967 had a comical letter read out on the viewer-complaints show *Points of View*, in which he berated the broadcast of a recent 'God-insulting TV programme ignorantly styled *Our World*', which had the temerity to include footage of model globes, intended 'to deceive and indoctrinate millions' with the wicked belief that the world was round. (Shenton had a theory that the entire fad of 'globularity' was being encouraged by globe-manufacturers to maintain their otherwise pointless business.)[4] In 1964, he made headlines after condemning the great Conservative free-marketeer Enoch Powell (1912–98) for comparing advocates of a planned economy to Flat-Earthers during a Commons debate, leading to Enoch receiving an angry letter in which Shenton made pointed mention of the fact that even the Science and Education Secretary Quintin Hogg (1907–2001) 'fondly imagines that he lives on a whirling planet', thus indicating that today's MPs were a sorry bunch indeed.[5] Shenton also occasionally speculated that missing persons who leave their homes one day and then never return may simply have walked too far by accident and then fallen off the edge of the world whilst trying to navigate the ice barrier.[6] In 1962, Shenton explained the basics of his theory thus:

> You see, the world is flat, like a plate. It is steady, and doesn't move or revolve. It is surrounded by a solid ice barrier, and the whole lot is at the bottom of a [cylindrical] pit in what we call "Mother Earth", which is a flat plain so large that it is endless. When you travel around the world and arrive back at the same place, it is like walking around the edge of a plate.[7]

The much-loved TV astronomer Sir Patrick Moore (1923–2012), who attended the first (and apparently only) meeting of IFERS in London in 1956,[8] managed to secure more details about Shenton's theory. The North Pole stood at the centre of the world biscuit, and the 60,000-mile edge, surrounded by a vast wall of ice, was what we called the South Pole. Beyond this, and above the confines of our pit, 'Mother Earth' stretched on forever. It wasn't quite clear what precisely this Mother Earth was, but it may have contained other cylindrical pits within which further inhabited world biscuits were also contained. As for what lay beneath Mother Earth,

Shenton said he did not know, but 'there is no reason to doubt that it is highly complicated'. There must have been lots of water under there, which seeped through our porous Rich Tea in the form of oceans and rivers, but other than that Shenton was at a loss. He also hypothesised that the Earth had a roof of some sort, but this was so far up it still let the rain through, so was clearly of poor design.[9]

## Falling Flat

Shenton's journey into flatness began when, as a young man in the 1920s, he had a marvellous idea for a new kind of airship. Still believing like most fools that the Earth was round and spun upon its axis at 1,040mph, Shenton decided it should be possible to make some kind of craft which would hover motionless in the air and then simply wait for the location to which it wished to travel to roll around beneath it. He didn't realise that the Earth's atmosphere spins alongside it, thus making his invention useless, and could not understand why the British Government were uninterested in what he had to offer. Seeing a conspiracy afoot, Shenton spent long hours in the library until he discovered writings by a certain Archbishop Stevens (1835–1917) who had proposed a similar flying machine of his own some years earlier. Surprised to find that Stevens had been associated with an earlier, now-defunct Flat Earth Society, Shenton decided to look into the organisation's publications, and quickly found himself a convert.[10] Inspired by what he read, Shenton became a 'zetetic' (from the Greek '*zeteo*', meaning 'I find out for myself') a word used by earlier Flat-Earthers to indicate their rejection of any reliance upon inherited wisdom in favour of the evidence of their own reasoning.[11]

Inconsistently for a true zetetic, the religious Shenton based his own world model upon one of the oldest authorities of all, the Holy Bible, despite the fact he refused to attend Church on the grounds that he couldn't find a single one whose vicar was willing to preach from the pulpit that the Earth was a pancake.[12] Even though it is debatable whether or not the Bible really does teach that the world is flat (it mentions 'the four corners of the Earth', but this could just be metaphor), Shenton became obsessed with the obscure Biblical figure of Peleg, in whose time Genesis 10:25 tells us that 'the Earth was divided'. This probably means it was divided up between different tribes, but Shenton preferred to believe it referred to the mythical lost continent of Atlantis disappearing beneath the North Pole following a major crack in the world biscuit, a crack through which Atlantean flying saucers still sometimes whizz up to greet us.[13] Although this narrative doesn't seem notably Christian in nature, Shenton was nonetheless convinced that belief in a round globe was part of a Satanic plot to encourage atheism. Christ Himself had warned us about false prophets, Shenton reminded humanity – although no one had ever previously thought such evil-doers would take on the form of astronauts.[14] His 1966/7 IFERS leaflet *The Plane Truth* – 'A modest pamphlet with a great aim' – was full of religious complaints, from moans about the lack of Archbishops who had bothered to join his club, to invective concerning the 'Godless thought and devilish mode of indoctrination which forces unproven THEORIES into the minds of the young' in schools. He did include testimony from a lone Norwegian Geography teacher to the effect that 'I have never really believed what I have to teach my pupils', but this was not enough to counter the tide of spherical filth being pumped into the minds of the defenceless young. Fortunately, Shenton told his readers, God was now fighting back. The death of three NASA astronauts during a ground test in January 1967 was attributed to their rocket being named after Apollo, 'one of the great divinities of the Greeks'. 'Greater care should be taken when applying names to such vehicles,' he warned. 'We cannot affront The Creator just as we please.'[15]

Once satellites began being equipped with cameras and taking photographs of what was manifestly a round globe, Shenton refused to believe it. When in 1966 the US *Lunar Orbiter* provided clear proof of a spherical world, Shenton proposed that the photos showed some other unknown object, which 'is not really the Earth at all'.[16] He later settled upon claiming such images were merely 'a fraud, a fake, a piece of trickery or deceit.'[17] When astronauts themselves began going into space and witnessing a globular Earth, he claimed it was just 'an optical illusion, quite common in these high-flying days'.[18] Or maybe they had gone up 'in an egg-shaped orbit', and their photographs of Earth become 'distorted' from flat to round, due to being taken at a funny angle?[19] Worse, Shenton said some astronauts were openly lying about their 'achievements', having faked their spacewalks down here on Earth – images of men in space were just 'studio-shots, probably'.[20] By 'using a wide-angle lens to give an impression of curvature', such a thing could be easily done.[21] After astronaut John Glenn (1921–2016) had successfully flown NASA's first manned orbital mission in February 1962, Shenton sent him a free membership card for IFERS together with a message reading 'OK, wise-guy!' After all, Glenn knew perfectly well that he had simply 'circled over a flat surface like a toy aeroplane on the end of a piece of string'.[22] Eventually, Shenton began refuting other standard truths of astrophysics, too. 'Late in his life,' Shenton falsely announced in 1967, 'Sir Isaac Newton himself denied his own theories of gravitation' and it was time for us to do the same. If gravity was real, then any astronauts out in so-called 'space' would simply 'fly off and never be seen again', whilst high-altitude parachutists would fall from their aeroplanes and miss the world biscuit altogether, landing upon the endless flat plains of Mother Earth.[23]

On 24 December 1968, Shenton observed NASA perpetrate the ultimate lie. To a live TV audience of 500 million, the *Apollo 8* astronauts turned their camera down upon Earth itself, displaying disturbing evidence it was round, before taking turns to read out the opening passage of the Book of

Genesis, starting with 'In the beginning God created the Heaven and the Earth.' Seeing as Shenton believed that other passages in Genesis proved Earth was flat, he reasoned this Bible reading was a duplicitous double-bluff or 'deceptive cloak' on NASA's behalf.[24] By April 1969, Shenton was putting the *Apollo 8* crew right in a shocking interview with the *Birmingham Evening Echo*. Above our Earth, he said, was a giant mass of water, which we were currently protected from by a big air bubble. When the astronauts had uttered the line from Genesis reading 'And God made the firmament, and divided the waters which were under the firmament from the waters which were above the firmament', they were secretly sending Shenton out a coded message that this was so. An atomic explosion or giant earthquake could make this air bubble pop any day now, flooding the world once more as in the days of Noah, when the world was square. The shock of a second Great Flood, predicted Shenton, could be so huge that it would make the disc-shaped Flat Earth revert back to its hitherto unknown original form, transforming the Rich Tea world biscuit into a giant Jacob's Cream Cracker.[25] Shenton reckoned the world's governments knew this and were secretly amalgamating into the UN for when the bubble finally burst. Interestingly, other Flat-Earthers have since pointed to the UN badge, which shows a bird's-eye view of the world's continents arranged on a flat disc (or 'a polar azimuthal equidistant projection', to be pedantic), as being a perfect illustration of the way our planet really is, so perhaps Shenton was onto something after all![26]

## The Eagle Has NOT Landed

Samuel Shenton was one of the first men on record to deny that the *Apollo 11* moon landing of 20 July 1969 had actually taken place. Nowadays, this is one of the most enduring of conspiracy theories, with reams of speculation available online purporting to show that NASA staged the whole thing in a Hollywood studio, so as to appear to win the Space Race. A bewildering array of photographic commentary upon US flags fluttering in a non-existent lunar breeze, or reflections of equally non-existent lunar light sources within moon-walking astronauts' helmets has arisen, purporting to prove the whole exercise was an audacious fake. Some of these analyses have been performed by professional photographic experts, and contain technical details and arguments it is difficult for a layman to refute, though common sense tells you this should be possible.[27] One event which helped spread such ideas was the release of the 1978 Hollywood thriller *Capricorn One*, in which the US fakes a Mars landing. One of the actors in the film was the former American football star O. J. Simpson (b. 1947), who played a false astronaut whom shadowy agents set out to kill to keep the fraud a secret. Bizarrely, this film was taken by some Flat-Earthers to be a kind of docudrama and tantamount to a public admission of guilt upon behalf of NASA. When in 1994 Simpson famously found himself accused of murdering his ex-wife, a ridiculous rumour arose that he was being framed by Washington so as to keep his knowledge of the 'truth' locked away forever behind bars.[28] So widespread have such ideas now become that, in 2002, astronaut Buzz Aldrin (b. 1930), the second man on the moon, faced the prospect of prison after he lost his cool and punched a documentary maker named Bart Sibrel (b. 1964) who had lured him to a restaurant in Beverley Hills on false pretences purely in order to call him a big moon-based liar to his face.[29]

The strangest attempt to explain away the moon landings came from John Bradbury, a chiropodist from Ashton-under-Lyne, whose development of a new kind of telescope filled with an amazing 15 lenses led him also to develop a new picture of the universe. According to Bradbury, the more lenses in your telescope, the better the view of the heavens you would get, something which allowed him to see the very edge of the universe itself, which was rectangular, made of metal, and magnetic. His special 'scope also somehow allowed Bradbury to discern that the Earth was not spherical at all, but flat on the top, where mankind lived, and hemispherical on the bottom, like a grapefruit cut in half. The North Pole was located within the centre of this oddly shaped world, meaning that every direction a person travelled was actually south. As for the moon, Bradbury determined it was constructed of a thin shell of carbon one or two inches thick, and slightly convex. As it travelled through the sky, it accumulated large amounts of phosphorescent plasticine from some unknown source, more and more each day, until it was completely covered in the stuff, leading to what we call a full moon. Then, as the weight of this plasticine grew too heavy, it all started to drop back off again, until we were left with no visible moon at all. One night in 1953, Bradbury claimed to have seen a giant finger made of plasticine emerging from the top of the moon, a remarkable sight indeed, but one which nobody else was able to confirm because no other astronomers possessed one of his multi-lensed telescopes. The reason nobody possessed one of these devices was because, by filling it up with so many pieces of glass, Bradbury ended up with massively distorted images bearing no resemblance whatsoever to reality. When news of 1969's moon landing reached his ears, Mr Bradbury had an explanation ready and waiting. If they had landed on the thin carbon surface of the moon, then the astronauts would simply have fallen through to the other side. Furthermore, they didn't find the place was covered in luminous plasticine. Therefore, Bradbury reasoned, they had not landed on the moon at all. Instead, when *Apollo 11* flew up into the air, the metallic craft was pulled sideways by the huge magnet which surrounds the universe, meaning that it landed in Tibet by mistake. When the astronauts emerged, they found themselves on a barren, rocky Himalayan mountain range at night, which they simply mistook for the moon because it was so bleak and inhospitable.[30]

## Flatly Refusing Orthodoxy

An especially militant moon landing denier was Charles K. Johnson (1924–2001), a California-based Flat Earth fanatic who inherited a good part of Samuel Shenton's archives when he died, allowing him to become President of a successor organisation, the International Flat Earth Research Society of America (IFERSA) in 1972. From his secret base in the Mojave Desert, Johnson and his wife Marjory spent three decades putting out their bizarre quarterly journal *Flat-Earth News*, the front page of which bore the exact same item of 'news' every issue: namely, that the Earth was still flat. Admittedly, this revelation was often expressed in an inventive way – 'AUSTRALIA NOT DOWN UNDER' and 'GALILEO WAS A LIAR!' being their best headlines – but could nonetheless come across as being a bit one-note.[31] Fortunately, the newspaper featured articles written in Charles' weird, stream-of-consciousness style, which were often highly readable in spite of their borderline illiteracy. Take the following extract, in which Johnson berates the ancient Greeks (from 'Grease') for creating the mad idea of the world being round, and thus a kind of slippery 'Grease-Ball' from which we would all one day fall, if the idea were really true:

THIS IS THE WORD OF THE LARDS
SCIENCE IS: Greasy Spoon Superstition. Now all know what we call the "Grease ball" world or delusion comes from the fact every school tells us, and is known by most, the "Ball-Planet" idea came from Grease. So, when we say the Grease ball, we mean and is obvious, is the idea or superstition from "Grease" ... Now, to anyone who has checked into this whole thing ... [will] know that the "great ones" of old Grease had a custom of eating "lard"! Now they didn't just dip it out with their hands, but used a spoon. So the glorious "founders" of present education-science-religion system sat around eating lard from a bucket with a spoon and hearing and telling tall tales. Many of these "lards" vied with each other to take the most absurd idea and concoct a way to tell it and make it seem reasonable or at least be believed ... [The idea of a round Earth is] ONLY for barbaric heathen mindless creatures too DUMB, too STUPID, too DEVOID OF REASON, LOGIC or COMMEN SENSE to know better, than to waste their time and money, marbles and chauk, on endless lunatic "explanations of the universe" and all therein![32]

I hope that's clear. The myth of globularity began with the Greeks, theorised Johnson, and was resurrected by England to facilitate the divorce of Henry VIII (1491–1547). When he established the Church of England, Henry also declared war upon the Bible, and spread a rumour that the Earth was round, not flat, hoping to undermine the Pope. Other Protestant nations were also in on the plan, and even today the British Government would not allow the world's top scientists, who knew the ways things really were, to reveal our planet's flatness. George Washington (1732–99) became so annoyed by this that he led a Revolution to free America from the influence of British globularity, thus making the USA itself one gigantic Flat Earth Society during its early years, but the American population had since been sold down the river during the Space Race. Everyone from the Disney Corporation to the Nazis were involved in spreading the lie of rotundity, but nobody more than NASA, which was stuffed full of ex-Nazi rocket scientists anyway. The moon landings were Hollywood fakes from start to finish, filmed in Arizona with scripts written by the evil British sci-fi novelist Arthur C. Clarke (1917–2008). The so-called 'Space Race' was simply a behind-closed-doors bidding session between the US and USSR to see who could claim to have set foot on the moon first. America won, but had to give the Russians Cuba in return. Johnson pointed out that a number of Hollywood films set on the moon had been made down the years, meaning there was no reason they couldn't make another; *Star Wars* was made 'in a man's garage', he said, and there were loads of fake round moons and planets in that. The space programme generated jobs and profits for big corporations, so continued to be funded by Washington to fend off unemployment. It turned out that the Mormons, in a touching if highly illegal gesture of public-mindedness, had kidnapped the billionaire Howard Hughes (1905–76) and assumed secret control of his companies in order to divert money towards NASA, most of whose employees were junkies with expensive drug habits in dire need of subsidising. Johnson advised the first man (not) on the moon Neil Armstrong (1930–2012) to 'confess and beg for mercy' if he knew what was good for him, but he never did.[33]

Johnson's favourite President, Ronald Reagan (1911–2004), wanted to tell the world the truth, but was gagged by Shadowy Forces. When in 1986 the space shuttle *Challenger* exploded, Reagan went on TV to give a speech which, according to Johnson, contained a series of secret messages in which the President admitted through code that he knew the Earth was flat and men had never landed on the moon at all. This 'code', however, consisted simply of Reagan mentioning God and looking up into the sky for obvious reasons. To Johnson, this was reinterpreted as Reagan indicating that he knew Heaven was above Earth in all directions at once (after all, the speech could have been watched simultaneously on Australian TV, too), thereby meaning that the world must indeed have been flat. Reagan felt obliged by circumstance to pay tribute to 'the phoney ANTICHRISTS, the slobbering foul degenerate dogs', the so-called 'astronauts' who had died in the explosion, said Johnson, but really what he meant to say was this: 'EARTH IS FLAT, GOD EXISTS. HE IS IN HEAVEN, A PLACE, THAT IS UP ABOVE EARTH, ABOVE THE USA.'[34] Having died in 2001, perhaps Charles K. Johnson's soul is also now to be found hovering somewhere above the USA. His body, however, is buried a mere six feet under the Flat Earth; any deeper and he might have fallen through the bottom.

## By the Light of the Silvery Non-Moon

Wrongly believing that 1969 would be 'a great year' for IFERS,[35] Samuel Shenton's own chosen mode of attack against the reality of the moon landings was especially inventive; namely, to claim that they couldn't possibly have occurred because the moon itself *doesn't exist*! Rather than a solid, three-dimensional sphere, Shenton taught that what we called 'the moon' was really a non-corporeal, self-luminous disc with translucent qualities; a kind of moving spotlight. 'The moon's transparent, you know,' he told a US newspaper in 1961, meaning that 'there isn't much to land on.'[36] The testimony of Genesis 1:14–16, which calls the sun and moon 'great lights' was part of Shenton's inspiration here, but he also derived support from the work of an obscure French theoriser named Gabrielle Henriet, who spent the years 1938–1956 writing her lunatic masterpiece *Heaven and Earth*, which was full of ideas every bit as incredible as those of Shenton, most notably her ridiculous but admirably unfashionable contention that the sky was solid.

Naturally, Henriet subscribed to various other Shenton-friendly beliefs, such as denying that the Earth spun on an axis on the grounds that otherwise tall buildings like the Eiffel Tower would lean over heavily to one side like giant metal drunkards. She maintained that the Earth did not move in any way at all, with ill-defined 'cosmic breath streams' being responsible for the existence of seasons, fluctuations of temperature and the day-night cycle. Henriet also claimed that all other heavenly bodies rotated around the Earth, not the sun, and that gravity did not exist, as 'any well-balanced mind' would admit. Her belief in a Flat Earth was justified by haughtily refusing to believe that Australians walked upon the bottom of the globe upside-down, 'in the manner of insects crawling on a ceiling'.[37] Henriet's basic idea was that the Flat Earth lay at the bottom of a vast dome of metallic rock which covered our pancake-planet like a silver platter over a dish of food. For Henriet, it was possible to view this 'vault of Heaven' with the naked eye, when briefly illuminated by lightning. She even claimed to have seen the dome herself once, during a particularly fierce thunderstorm:

> The aspect of this vault was that of a rather steep, slightly sloping dome of pyramidal shape, and it appeared to be composed of a bright metallic dark grey matter, uniformly showing small regular inequalities like lead which has been beaten or chiselled. The larger details, particularly the craters, were clearly visible against the background; but the most impressive circumstance yet was the incredible nearness of the vault,

the highest point of which did not appear to be, at most, any more than sixty kilometres from the Earth.[38]

At the bottom of this dome, surrounded by a vast circular wall, lay the Flat Earth, which came under periodic bombardment from 'meteorites' which were actually fragments of the heavenly vault above that had become detached and fallen down. Analysis of these rocks showed they contained high levels of metal, thus accounting for the brilliance of the sky during bright weather; the sky shone blue because the dome contained many blue-coloured metals like cobalt.[39] Sometimes lightning caused huge electrical vibrations in the dome, meanwhile, splitting it and causing booming thunderclaps, with water dripping through new cracks in the form of rain.[40] Sometimes, large stone fragments also fell down to Earth and were worshipped by primitive humans. Henriet seriously suggested that prehistoric structures like Stonehenge simply fell to ground in the basic shape and locations we can see them in now, thereby solving the age-old mystery of how they were constructed. Apparently, the Easter Island Heads and even Egypt's pyramids shared a similar origin. So did mountains; if you looked up at the sky dome above the Andes, then you would see a big dent corresponding in shape to those very same mountains, as snugly fitting as a hand and glove.[41] Probably the entire Earth itself had fallen down from the 'Heaviside Layer', as Henriet dubbed the dome, although doubtless bit by bit over a number of years, rather than all in one big go.[42]

## The World Is Your Oyster

Henriet supported her beliefs by pointing out they were in broad correspondence with certain of the earliest pre-scientific ideas about the cosmos held by ancient man – which to an extent they were. The Babylonians believed the universe was a sort of giant oyster with water beneath and above it, kept at bay by a solid sky dome. The sun, moon and stars passed through windows in this sky dome, whilst some water seeped through as rain, much as Henriet said. The ancient Egyptians likewise viewed the universe as a kind of rectangular box, covered either by a huge metal lid or the underside of a gigantic cow, whose hoofs stood upon the four corners of the Earth. On a kind of elevated platform far above the ground flowed a raised river, along which the sun, moon and stars flowed in their special boats, entering and exiting the world box through stage doors. The fixed stars were lamps hanging from the ceiling lid.[43] James Bowdoin (1726–90), a one-time Governor of Massachusetts, had also once proposed that a giant solid shell surrounded the universe, holding the stars in position, and even acting as a possible residence for 'vast numbers' of alien life forms.[44] Helpfully for the modern Flat-Earthers, therefore, it could be claimed that their ideas had actual historical precedents.

Most useful for Shenton personally was Henriet's idea that all moons and planets seen in the sky were not really 'solid, opaque masses of matter' but 'immaterial, luminous and transparent discs' which it was utterly impossible to ever land upon. When these huge searchlights made their way across the sky, they illuminated bits of the dome above them, throwing the craters, dents and cracks of the Heaviside Layer into sharp relief. Any valleys or mountains seen on the moon, therefore, were not really *on* the moon at all, but *behind* it, illuminated by its light.[45] Thus, 'when poetical reference is made to the silver mirror of the moon, it is the metallic surface of the dome appearing under the transparent disc which, in reality, may be described as a silver mirror.'[46] All other celestial bodies are also simply incorporeal lights like the moon in Henriet's model. The sun is so bright a light that when it passes beneath the dome it melts the rock and metal behind it, turning it into glass. This glass then becomes a lens, focusing the sun's rays back down onto the Flat Earth below, like a child turning his magnifying glass upon ants.[47] Mars, meanwhile, only *appeared* to be red because the portion of dome behind it was rusty, and the very word 'planet' itself now had to be abandoned immediately, as there was no such thing. If the 'planets' really were solid, then wouldn't they make a tremendous racket when flying around everywhere at top speed? To Henriet, their actual silence spoke volumes.[48] What were these lights, then? Henriet was unsure, but said they probably shone in through certain open windows and doors in the sky dome, as the Babylonians had said.[49] Furthermore, seeing as UFOs were often described as luminous metallic discs, Henriet considered they were just smaller 'erratic' versions of the moon, with some dome windows only being left open temporarily, accounting for their irregular appearances; alternatively, maybe UFOs were simply baby moons which had not yet settled into their proper habits.[50] The stars, too, were imposters, being nothing but 'latent sparks' of radioactivity which floated up towards the dome and condensed upon the ends of 'a network of ethereal cords', hanging down like lit chandeliers. This network of cords then rotated across the sky dome, making it look to the untutored eye as if the Earth was floating and spinning in interstellar space – which it wasn't, because as Henriet had expertly proved, there was no such thing as space at all.[51] And, if there was no such thing as space, then there could also be no such thing as a genuine space programme – just as Samuel Shenton had been arguing.

Seeing as Henriet finally completed her book only a year prior to *Sputnik*'s launch, Shenton was a natural ally in the doomed quest to keep pretending that nothing much had really happened, and the two became correspondents, although their alliance was an unequal one in the sense that Shenton kept on grabbing all the major PR opportunities. After all, Henriet habitually turned down all potential TV appearances on the grounds that her false teeth were too loose and might fly out of her mouth at any minute.[52] However, your false teeth falling out is nothing compared to the effect the moon landing had upon Samuel Shenton's health. He had tried to head off any future criticism just prior to the moon landings by claiming that 'What happens up there on the moon doesn't really concern us', but this gambit did not work.[53] Weirdly, the initial impact of lunar touchdown led to an upturn in positive correspondence to IFERS, with a number of people around the world simply refusing to accept that the pale face of Mistress Moon had been sullied by mankind's muddy footprints, and wondering what other lies science might have been telling them down the years too. Ultimately, however, the moon landing proved the final straw for Shenton who, exhausted and discouraged, died of heart disease in 1971, aged 68.[54] If he could see how the foul disease of globularism has continued its spread unabated everywhere throughout the civilised world in the years since, I'm tempted to say Samuel Shenton would be spinning in his grave – except, of course, this could be taken as evidence that the world is really round, orbits the sun and spins upon an axis of 23.5 degrees at a speed of 1,040mph, all of which he flatly denied.

# 3

# Diana's Dust: The Accidental Murder of a Moon-goddess

Why were fringe thinkers so keen to deny that mankind had been to the moon? Perhaps because, since the dawn of time, the moon had stood as one of the ultimate imaginative blank canvases, a place where anything might exist or take place. Most, if not all, cultures have developed their own myths about the moon, many of them rather delightful and magical in nature. In ancient Greece, the moon was thought to be a kind of gigantic heavenly recycling plant for human souls. The third-century author Porphyry (233–305), in line with the widespread Classical belief that bees were vehicles for human spirits, taught that, when such souls were ready to be reincarnated, the moon-goddess Artemis would send them back down from her lunar home to Earth in the form of special ghost bees called *melissae*. Nowadays the name 'Melissa' seems like a simple everyday girl's name, but it could be taken as originally meaning something like 'moon-bee soul waiting to be reborn in the form of a human baby'.[1] Equally as charmingly, there was a time when the tropical Bird of Paradise was known to European naturalists only from dead specimens. Therefore, it was theorised by the French physician Pierre Borel (1620–71) that such animals actually lived on the moon, only dropping down to Earth after their natural spans had passed.[2] Other moon-related myths were just plain weird, with some primitive cultures believing that exposing yourself to the moon could lengthen your penis. The Tumupasa Indians of Bolivia, for example, believed that the tapir has a particularly long penis (which apparently it does) because, according to their ancient myths, he had sex with his wife on a night when she happened to have swallowed the waning moon – according to the Tumupasa, on nights when there is no moon, it has been swallowed by

Mrs Tapir, who later burps it back out again when the time comes for it to begin to wax large once more.[3]

Both ordinary folk and the well born have believed strange things about the moon, crediting it with quasi-supernatural powers. During the reign of France's Louis XIV (1643–1715), a Royal Order was issued to the effect that any trees used to make timber for French naval vessels should only be felled during a waning moon, when they would possess more vigour than at other times, due to having absorbed more lunar energy from the recent full moon. However, around 1734, Duhamel du Monceau (1700–82), General Inspector of the French Navy, declared this belief in the powers of so-called 'Moonwood' was nonsense, and ordered a comparative study be made of the quality of timber produced during different stages of the lunar cycle. As expected, he found that King Louis' belief was pure superstition. Instead, he concluded that trees should be felled during a *waxing* moon for optimum naval efficiency![4] Amongst commoners, the moon was also often associated with magic and the uncanny. Right into the twentieth century, it was a popular British custom to bow to the new moon upon first sight, perhaps whilst reciting a rhyme or doffing your cap, to guarantee being given a gift or secret knowledge of some kind. On the other hand, you could climb a stile to get closer to the moon, and whisper a spell to it, requesting the shining object to send down a dream of your future wife or husband to you that night. If you preferred, you could try and see the moon reflected in a pail of water; the number of lunar reflections you counted on the liquid equated to the number of years you must wait to be married.[5]

Most notably, the moon was believed by many ancient cultures to be a kind of sky-goddess. This 'White Goddess', as the moon-worshipping English poet Robert Graves (1895–1985) was to call her in his celebrated 1948 book of that title, had many names, amongst the most famous being Diana, Artemis, Hecate, Luna, Cynthia, Astarte, Selene and Phoebe, the sister of the sun-god Phoebus. Often personified as a tripartite goddess, whose three female forms of virgin, mother and crone corresponded to the growing, full and dying moon as observed in the night sky, many famous legends attached themselves to this most worshipped of all celestial objects. There was the tale of Endymion, for example, the sleeping Greek shepherd who appeared so beautiful to Selene in the light of her own shining that she descended down to the hillside where he slept and seduced him, or of Orion the hunter, the desired lover of Artemis, who was shot with an arrow by the goddess after she had been tricked into doing so by her jealous brother Apollo, who had earlier sent out a giant scorpion to try and kill the man. Aghast, Artemis transformed Orion's body into the constellation of that same name, and the scorpion's corpse into that of Scorpio, as an eternal starry reminder of the tragedy.[6] But what use are such tales of timeless poetry nowadays? 'The function of poetry,' wrote Graves, is the 'religious invocation of the Muse',[7] that is to say of the moon imagined as being a kind of goddess, an uncanny representation of Woman as a whole. As Graves says:

> The test of a poet's vision ... is the accuracy of his portrayal of the White Goddess ... The reason why the hairs stand on end, the eyes water, the throat is constricted, the skin crawls and a shiver runs down the spine when one writes or reads a true poem is that a true poem is necessarily an invocation of the White Goddess, or Muse, the Mother of All Living, the ancient power of fright and lust – the female spider or the queen-bee whose embrace is death.[8]

## First Men on the Moon

Who could go on truly believing or writing about stuff like that after the momentous – and, to some, momentously disappointing – events of 20 July 1969? Certainly not the Union of Persian Storytellers, who complained in *Apollo 11*'s wake that the irresponsible antics of NASA had fatally damaged their members' livelihoods.[9] Prior to the moon landings, the only possible trips to the moon were performed using the imagination. What is arguably the first sci-fi story on record, the Greco-Roman satirist Lucian of Samosata's (*c.* 125–*c.* 180) *True History*, which tells the tale of a traveller whisked up to the moon by a freak whirlwind at sea, was first published in English in 1634 and seems to have been at least partly responsible for a rash of subsequent lunar-travel fantasies penned within these isles.[10] The Bishop of Hereford Francis Godwin's (1562–1633) *Man in the Moone*, published posthumously in 1638, better deserves the label of sci-fi than Lucian's wild yarn, containing as it does a pre-Newtonian, if confused, description of what we might now term the effects of gravity, or at least weightlessness. Admittedly, the book is far-fetched, telling as it does the tale of a Spaniard named Domingo Gonsales who attempts to escape from being stranded on a desert island by training a troop of swans to carry him into the air. Sadly, he does not realise that it will soon be time for the swans to migrate – towards the moon! It was not then known precisely where such birds did migrate to, so this notion is not actually as ludicrous as it now seems. From up above in space – not then realised to be an airless vacuum, a deduction first made in 1643 by Evangelista Toricelli (1608–47)[11] – Gonsales is able to see proof of the Earth's rotation upon its axis, thereby solving another long-standing scientific conundrum. When he actually lands on the moon, however, such attempts at scientific speculation virtually cease. Godwin thinks that the so-called lunar 'seas' really are bodies of water, and so his Spanish protagonist encounters a world filled with life, primarily a race of human-like beings some 28 feet tall who communicate via musical utterances, not words. However, it later transpires that there are in fact three races of men on the moon: the aforementioned giants, who could stand the full glare of the sun; an intermediate race some 10–12 feet in height; and a third race of human-sized 'bastard-men', who could only stand to go outside in the weak light when the sun was being reflected off the Earth onto the lunar surface.[12]

Godwin's fantastic voyage proved influential upon later, more serious speculations about the nature of life on the moon. The very year that Godwin's book was first published also saw the first appearance of *Discovery of a New World* by the early proto-scientist and later Bishop of Chester John Wilkins (1614–72). In the 1640 edition of this book Wilkins added a section upon Godwin's fable, assessing it from a scientific perspective – his learned conclusion being that it would be 'easily conceivable' to fly to the moon using swans, provided you could train them properly![13] The main problem was how to keep oneself alive once beyond the Earth's orbit. Seeing as space travellers 'should scarce find any lodging by the way' in the shape of space inns, Wilkins proposed that, under weightless conditions, the human body would have no cause to move around whatsoever, and so would not need to eat to replenish energy. Instead, you could just laze your way through the journey in a kind of hibernation like a bear in winter, feeding off nothing but thin air and the nourishing smell of outer space itself: 'We cannot desire a softer bed than the air, where we may repose ourselves firmly and safely as in our chambers.'[14] Alas, not everyone at the time took Wilkins seriously, with one wag making the satirical suggestion that he be appointed as Britain's official ambassador to the Man in the Moon immediately, 'in regard he hath the greatest knowledge' of that newly discovered land.[15] An eighteenth-century Italian children's book paid similarly double-edged tribute to Wilkins by making him the chief protagonist of its journey to the moon – a most ridiculously depicted region where people lived within gigantic pumpkins growing out of lunar trees and fought off the attentions of giant anteaters whilst riding around on the back of flying snakes.[16]

## The Eagle Has Been Branded

When the Eagle landed in 1969, all such fantastic possibilities were closed off forever – or so you might have thought. The moon landings came as a great disappointment for some, with those more sensitive souls who shared Graves' outlook on the world coming to feel that Divine Cynthia's face had essentially just been violated by a giant phallic rocket. Some people really didn't want to know what was up there on the moon. 'Can't we be free *not* to know some things, to keep the timeless romance of the White Goddess intact?' they asked. No. Once something is known there is no way to un-know it, and so we are left to cope with the disappointment of all those NASA images of a dead and dusty moon, shorn of its former magic. If Friedrich Nietzsche (1844–1900) had announced the death of God during the nineteenth century, then Neil Armstrong had followed this up in the twentieth by very publicly killing off the Goddess too.

Even Armstrong can't have imagined that one day Diana's corpse would be used to flog consumer goods, however. The idea of constructing giant adverts on the moon so that people down below can't escape them is a hideous one indeed; nobody wants to look up into the night sky and see 'The Moon Sponsored by PEPSI™'. In 1993, when the American company Space Marketing Inc. came up with the idea for an illuminated 1 km² 'space billboard' to enter orbit in such a way that it would be 'roughly the same apparent size and brightness as the moon', there was such outrage that a Bill was passed through the US Congress banning any such scheme from ever taking place.[17] However, this hasn't stopped a new outfit named Moon Publicity from proposing in 2015 that special robots should be sent to the moon and programmed to carve away masses of dust from certain areas of the lunar surface, thereby creating a number of ridges which, via a process termed 'shadow-shaping', would spell out the logos and web addresses of certain high-paying sponsors. Their objectionable promise, 'Twelve billion eyeballs looking at your logo in the sky for several days every month for the next several thousand years', was outdone in crassness only by the company's claim that, by flying robots to the moon, they would really be helping develop space science to such a degree that it would help us settle other planets in case of major disaster down here on Earth. According to Moon Publicity, we should be thankful to them for making this benign and selfless proposal, as turning the moon into giant billboard would be 'a small price to pay for saving mankind'.[18] I disagree.

A more plausible attempt at moon advertising is imminent. The Otsuka Pharmaceutical Co. of Japan may be no household name in the West, but domestically speaking their tasty-sounding energy drink 'Pocari Sweat' is big business, being 'a beverage very close to you', so they say. The latest plan to grow the brand further involves buying space on a robotic lander being sent up to the moon by a private company, the Pittsburgh-based Astrobiotic Technology, sometime soon. On-board the lander will be a time capsule shaped like a huge can of Pocari Sweat, containing several plates engraved with descriptions of Japanese children's dreams about the future. Also inside will be a serving of powdered Sweat – just add water, and you will get a refreshing serving of ion-revitalising energy drink on the moon! The company's hope is that, one day, one of the Japanese children whose dream involved them growing up to be an astronaut will fly to the moon for real, discover the giant can, open it, and, as the official PR release puts it, 'drink SWEAT mixed with water', something which would be a 'first in the world's beverage history'. But where are future Japanese astronauts ever going to find a viable source of water on the moon? My only guess is that the powdered Sweat will be brought to life by the thirsty space travellers bottling up poor Cynthia's tears.[19]

# Heavens Above!

It is easy to forget, but the heavens above us once really were thought to be just that – the abode of God and His angels. In his book *The Discarded Image*, the writer and academic C. S. Lewis (1898–1963) paints a clear picture of the old medieval and Renaissance religious concept of what was still not yet referred to as 'outer space'. It is partly an inheritance from the Classical world of Greece and Rome, and partly a construct of the Christian faith. Men like Galileo and Kepler were soon to wreck it completely, with their radical ideas about a heliocentric (sun-centred) rather than geocentric (Earth-centred) cosmos, but it has a certain beauty of its own nonetheless. Firstly, we have to forget about the very idea of physics-based 'laws'. The medieval universe was governed by something called 'sympathy', a kind of natural inclination of everything in existence to find its correct place. The sea, for example, was said to 'desire' to follow the moon, causing the fluctuating tides, whereas today we might say that the water was 'forced' by the laws of physics to obey the pull of the moon. This shouldn't be taken as implying that medieval man thought everything around him, from pebble to ocean, was literally alive; instead it reflected the idea that, in a God-centred universe, there was a happy natural order of things which man, beast and mineral should be glad to embody through sympathy, not forced to obey by impersonal physical laws.[20]

The moon played an absolutely central role in the old Christian cosmos, for below it lay the so-called 'sublunary world' of decay, growth and change, but above it was a more perfect 'translunary world' of the unchanging celestial bodies, an idea derived from Aristotle (384 BC–322 BC) and his treatise *On the Heavens*.[21] This was a reasonable assumption because the stars and planets had appeared unchanging. Across the course of a lifetime, you might live to see the tallest tree shrivel and die, but should you live forever, the Pleiades and Sirius would still be there – or so it was thought. On 11 November 1572, astronomers were astonished by the appearance of a new star in the sky; what would become known as a 'nova'. It appeared in the constellation of Cassiopeia, and was so bright that it could be seen in broad daylight, through cloud cover. By 1573 the nova began to fade from sight, however, and by 1574 had disappeared completely. Despite some valiant attempts to maintain the old model of the universe by, for example, claiming that the nova represented the miraculous reappearance of the Star of Bethlehem, the stage was set for a scientific revolution.[22] A particularly desperate attempt to deny reality was proposed by the German painter Georg Busch; he said the nova was really a comet made from the condensed vapours of human sins, set aflame by God, with its smoky vapours then drifting down to Earth to blast us all with such dread afflictions as 'bad weather, pestilence and Frenchmen'.[23] Few agreed, and the slow realisation that stars might be born, live and die over the course of unimaginable aeons was one which would have grave implications for the classic Aristotelian model of the unchanging world above the moon.

The greatest astronomer of the Classical world had been the Greco-Egyptian Ptolemy (*c.* 100–*c.* 170), whose model of the universe, later adopted and adapted by Christian scholars, placed the spherical Earth at the centre of the cosmos, surrounded by a series of hollow concentric globes enclosing one another like globular Russian dolls. Known as the 'spheres', they were crystalline and transparent in nature, and fixed in each of the first seven spheres was a luminous body – the moon, Mercury, Venus, the sun, Mars, Jupiter and Saturn – an idea derived from another ancient Greek, Anaximenes (*c.* 585 BC–*c.* 528 BC).[24] Beyond this was the *Stellatum*, where the fixed stars resided, and beyond this the *Primum Mobile*, or 'Prime Mover', which caused the whole lot to orbit around Earth. Outside of this, taught the Christian Church, lay Heaven.[25] What we would now call 'space' was not even black in this model; the dark night sky was caused by the rotating shadow of our own planet, cast outwards towards Venus like a giant black finger of gloom. Beyond, all was suffused with a godly light which would have made our own terrestrial daylight appear dull indeed.[26] All the celestial objects emitted a kind of glorious natural melody known as the 'music of the spheres', meanwhile.[27] This idea had its origins with the mystic Greek philosopher Pythagoras (*c.* 570BC–*c.* 495BC), founder of a cult which worshipped numbers as divine. He reasoned that the revolutions of the planets in their celestial spheres must naturally have produced a harmonious hum, with each planet emitting sounds of a different pitch, thus transforming the heavens into a gigantic musical instrument or lyre, forever hymning the eternal glory of God and Maths.[28] To sing you have to be alive, of course, and so it was with the translunary world. The *Primum Mobile* felt real love for God and, so moved, imparted its own movement towards the rest of the universe.[29] As for why the *Primum Mobile* moved in a circular fashion, this would have been obvious to the medieval mind. God was perfect, and the most perfect of shapes was a circle which, like the Deity, had no beginning and no end. Thus, by moving in a circular fashion, the universe was demonstrating its love for God through imitation. The crystalline spheres were also alive, with each occupied by angelic masters known as 'Intelligences'. These too loved God, so were only too happy to manifest as spherical entities called planets, which orbited in circles around another sphere called the Earth, in worship of the ultimate circle of all, Jehovah.[30] All in all, it was an image unlikely to impress any early members of the Flat Earth Society.

## It's All Greek to Me

The extent to which this model was a modified inheritance from the Classical world can be seen in the fact that in ancient Greece, too, the translunary universe was felt to be alive. To the Greeks, the power of independent movement was a sign of animation, as with animals. It was the same with planets, which also appeared to move by themselves. Although planetary movements were more regular than those of beasts, this was simply a sign the celestial bodies were more perfect in nature, and thus gods. Some philosophers like Aristotle developed this idea into something more subtle, speaking of a force called '*physis*' causing orbital movements, a word often translated as 'nature', though also somewhat equivalent to the medieval idea of 'sympathy'. It was in the nature or *physis* of things to achieve their natural ends; for an acorn to grow into an oak, or a planet to follow its path. This *physis* was ultimately ordained by some supreme Creator-God meaning that, in effect, the planets moved from inclination to follow the will of God, much as in the later Christian model. God is thus a so-called 'Unmoved Mover'; He may have set the planets in motion, but as the First Cause behind everything, was never actually set in motion by anything Himself.[31] Anyone caught denying the popular view that the heavenly bodies were living gods could expect to be given a hard time, however. Ancient Greece even had its own version of the trial of Galileo, when the philosopher Anaxagoras (*c.* 510BC–*c.* 428BC) was impeached for blasphemy due to his alleged teaching that the moon was made of ordinary rock and soil, and the sun simply a gigantic, red-hot stone. Forced to leave Athens, Anaxagoras lived the rest of his life in exile.[32] Aristarchus of Samos (*c.* 310BC–*c.* 320BC), who advanced a version of the later truth that the planets all orbited around the sun, not the Earth, as early as the 3rd century BC, had his astounding insight denounced for similar reasons, picking up only one open disciple, the astronomer Seleucus (b.*c.* 190BC).[33]

Evidently the Classical mind was not yet ready to accept such an idea. It was far easier to believe the cosmology of a man like Plato (428BC–348BC) who, along with Aristotle and Ptolemy, was one of the three main Classical influences upon the medieval Christian cosmos. Like many contemporaries, Plato taught that the celestial bodies were alive, calling the fixed stars 'divine and immortal animals', whose spherical 'bodies' were made of fire.[34] Such beliefs were contained in his *Timaeus*, which for a long time was the only text of the great philosopher available across medieval Europe. In it, Plato taught of how the entire universe itself was one gigantic life form, a 'perfect animal' which could be known as the macrocosm, or 'big world'. However, all things smaller than the cosmos still shared within its nature, no matter how imperfectly, and so could each be labelled as being a kind of 'microcosm', or 'little world'. For example, the macrocosm or world animal, being a truly perfect creature, was described as being a perfect sphere by Plato. The planets, being smaller and slightly less perfect, were spheres of a very high quality, though not entirely faultless in shape, and orbited around the heavens in circles, as was only right and proper. Circles and spheres were deemed the most perfect of shapes by Plato seeing as each point upon one's surface or circumference would by definition be equidistant from the centre, where he thought the shape's soul lay. Thus, the soul of a spherical planet was diffused equally throughout its entire rotund body, making it intrinsically good throughout its whole being. Because of the near-perfection of the heavenly bodies' movements, this then made them the most appropriate means by which to measure the passage of time, lending us the definitions of things such as years and months, and giving an idea that time itself was somehow circular, which survives today in the shape of most clock faces.

However, there was also something called a 'Great Chain of Being' operating throughout the universe, linking everything together within a kind of gigantic ladder leading up towards God, the perfect sphere. The higher up this ladder you were, the closer your nature to the macrocosm, and the lower down, the further away. Mankind was on the middle rung, being the most perfect of all animals, but an animal nonetheless. Unlike the planets, mankind was not spherical but, being a microcosmic version of the world animal, still possessed various relatively spherical or circular aspects to his being. Human heads, for example, were broadly round, and so were the most perfect part of our bodies, fit to act as the seats of our minds. The internal circulation of our breath and blood, too, were essentially circular or cyclical in nature, leading Plato to one grand conclusion; if you wanted to better know the nature of man, you should look not at actual men down here on Earth, but at the greater and more archetypal macrocosmic man embodied within the heavens above. To see yourself a little better, it is always useful to employ a mirror – and the best and most telling such device mankind has ever known is the night sky.[35]

## Breaking the Circle

This age-old picture of a living universe, within which God and man alike were both as intimately connected as a mirror and the thing it reflects, was ultimately smashed into tiny shards and replaced by the inanimate image of a so-called 'Clockwork Universe' as the late Renaissance faded away into the early Enlightenment. The irony was that most of the main slayers of Plato's venerable world animal were not themselves disbelievers. The chief murderer was Sir Isaac Newton (1642–1727), whose theory of universal gravitation finally destroyed any last lingering belief in Greek-style notions of planetary movements being caused by some in-dwelling life force. Newton's model demonstrated that, once set in motion, objects like planets would keep on moving forever, unless their momentum were checked by some external force; there was now no need for any supernatural intervention from God to keep the clockwork cosmos ticking along nicely. For example, once the astronomer Edmund Halley (1656–1742) had used Newton's work to correctly calculate the next visible appearance of a certain comet in the sky, the old view of these fiery objects as portents from Heaven disappeared as it became clear they were simply ordinary celestial objects winding their way through space in perfect accordance with the laws of physics. However, as a committed religious fanatic who spent years combing through the Bible in search of hidden codes, Newton had no desire to kill off the Deity, always maintaining that, whatever the case today, the planets had originally been thrown into motion by the hand of God. Had he known his discoveries would later be made great use of by atheists, Sir Isaac would have been aghast![36]

If Newton dealt the cosmos of Plato and Ptolemy the final blow, then there were several earlier men of genius who had already weakened their victim with prior slashes from logic's razor. The first major wound was inflicted by Nicolaus Copernicus (1473–1543), the first man since the days of Aristarchus to have seriously proposed the Earth might orbit around the sun. A Polish ecclesiastic, Copernicus' seminal *The Revolutions of the Heavenly Spheres* was only published in the year of his death, with his heliocentric idea cautiously presented as an interesting theory, not an assertion of fact. Largely ignored initially, the book's mental apparatus is hardly modern, with Copernicus still claiming that planetary orbits are all precisely circular upon the grounds of that shape's evident godly perfection, for instance.[37] Another seemingly reluctant assassin was Johannes Kepler (1571–1630), who made the important deduction that, far from planets enjoying perfectly circular orbits, they may actually orbit our sun in an elliptical fashion. Such had been the previous commitment to the notion of circular orbits that astronomers, imitating Ptolemy, had made use of something called 'epicycles', a series of fictional 'circles within circles' which allowed them to maintain that planetary paths were indeed as perfect as the flattened shape of God.[38] In his strange 1618 text *The Harmony of the World*, Kepler appeared to regret what he had done, engaging in an elaborate Pythagorean quest to restore that sense of harmony to the universe that he himself had done so much to destroy. Dealing initially with the concept of harmony in mathematics and geometry, Kepler goes on to apply similar ideas to the worlds of music, astrology and astronomy too, finding certain key mathematical and geometrical ratios common to each field, and concluding that these were the pure harmonies and proportions which God made use of when constructing His universe. The end effect is to present the entire cosmos as being one big holy mathematical song, with the sense of beauty we feel when listening to harmonic music being due to an unconscious reconnection occurring between the microcosm of man's mind and the macrocosm of God. 'The heavenly motions are nothing but a continuous song for several voices', said Kepler, these voices being the planets. These were the elaborate – and beautiful – lengths Kepler ultimately went to in order to repair his self-broken circle.[39] Concern with aesthetics can also be seen in the Catholic Church's reaction to the discoveries of our final murderer of the macrocosm, Galileo Galiliei (1564–1642). One of the first men to make use of a telescope, Galileo discovered many curious things in the night sky, including the four moons of Jupiter. This was awkward, as hitherto there had been thought to be only seven celestial objects in existence (the sun, moon and five known planets). Seven was a sacred number, with the Holy Sabbath being the seventh day of the week – whereas eleven, created by adding Jupiter's four moons to this sum, was not. Many churchmen consequently denounced the telescope as a tool of the Devil, fit only for producing delusions; it was not only for maintaining a heliocentric doctrine that Galileo was tried by the Inquisition in 1633, but for upsetting the very numerological balance of the universe.[40]

One simple way of fighting back against the deadness of the Clockwork Universe was to try and transform the solar system's most famous points into locations of holy import. For example, in 1714 Tobias Swinden (1659–1715), an English vicar, published his *Enquiry Into the Nature and Place of Hell*, which logically enough located Satan's home squarely within the flaming ball of the sun.[41] The religious writer Edward King (1735–1807) disagreed, proposing that the sun was not in fact hot at all, but filled with the glorious light of God, and was therefore Heaven.[42] The Dominican friar Tommaso Campanella (1568–1639) wrote a response to some of Galileo's ideas in the form of 1622's *Apologia pro Galileo*, proposing that the Garden of Eden was located on the moon, on account of it being high enough to escape the effects of the later Great Flood. This lunar paradise, he further argued, must have had a high temperature, seeing as its most famous citizens were always wandering around there in the nude.[43] Perhaps the oddest attempt to locate something of religious significance within outer space came in the proposal that the rings of Saturn were really the jettisoned foreskin of Jesus Christ. This was apparently the view of Leo Allatius (*c.* 1586–1669), one-time Keeper of the Vatican Library, who responded to the puzzlement occasioned by Galileo's discovery of Saturn's rings through his new-fangled telescope by proposing that they were an item known to religious scholars as the 'Holy Prepuce'. Allatius' unpublished (and now unfindable) essay, *Discourse on the Foreskin of Our Lord Jesus Christ*, apparently sought to settle a contemporary theological dispute concerning the fact that Jesus, as a Jew, had been circumcised at birth. Seeing as after His death Christ's entire body had ascended to Heaven, leaving His tomb empty, people wondered what had happened to the severed bit of his penis. Around AD 800, the Emperor Charlemagne (742–814) presented a bit of dried flesh to Pope Leo III (750–816), claiming it to be the disputed foreskin in question which he had been gifted by an angel, but before long this original Holy Prepuce had acquired twenty imitators housed within holy sites across Europe – meaning that, if they were all real, then Jesus must have had no fewer than twenty-one penises. Allatius attempted to resolve the situation by saying that, at the exact moment Christ had ascended to meet His Father, His abandoned foreskin must have also floated up from wherever it had been discarded, before expanding to gigantic size and wrapping itself around Saturn as an infallible sign of the presence of the Creator within His universe.[44]

## The Little World of Man

During the early twentieth century there was a surprising upsurge in attempts to restore the connection between macrocosm and microcosm, this time from the scientific and political worlds. This was an era when, following the successful invention of the aeroplane, the idea of humanity one day managing to touch the stars suddenly became much more thinkable. Russia was a particular centre of such thought, producing revolutionary left-wing fantasies about downtrodden peasants rising up and overthrowing not only their corrupt tsarist rulers, but also the force of gravity itself. The Marxist terrorist N. A. Morozov (1854–1946), for example, as well as writing 1880's hugely influential *The Terrorist Struggle*, also penned a number of mystical reveries based upon the idea of man conquering Newton's laws and floating away into outer space, where he might meet alien beings and become awed into realising that, in the end, mankind and the universe were as one. Happy to discredit Christianity in his writings, the militant atheist Morozov instead proposed a new universal religion of panpsychism, in which human beings became little more than agglomerations of 'feeling atoms', the temporary manifestations of the same chemical elements which had once made up the stars. His 1910 poetry collection *Star Songs* landed him in prison for a year as it was considered seditious and blasphemous, provoking men to rebel against the authority of both Church and State and reach instead for the moon ... by violent means if necessary.[45]

Another Russian trying to reunite man with the cosmos around this time was Aleksandr Chizhevsky (1897–1964), whose big idea held that there was an inherent link between cycles of sunspot activity and major historical events taking place on Earth; the basic theory was that mysterious cosmic rays were beamed down to our planet during periods of extreme solar activity, interfering with our brains, causing 'excitability of the masses' and fomenting war and revolt. Our heart beats with 'the pulse of the cosmos', said Chizhevsky, a statement which would have won Plato's approval. Seeing as one of the theories of the day was that sunspots were caused by the gravitational influence of planets upon the sun, Chizhevsky's idea neatly implied that there were inherent links between the revolutions of the heavenly bodies and Communist revolutions down here on Earth. This sounds like something which would have been lapped up by the Kremlin, but in fact the idea that 1917's revolution may have actually been caused by the proletariat's brains being frazzled by space rays won Chizhevsky little favour, and he was accused of being a 'sun-worshipper' in the Soviet Press.

One 1930 sci-fi story even satirised him by imagining a future war between America and Russia in which Yankee scientists destroyed Communism at a single stroke, simply by firing space missiles into all the spots then visible upon the sun, bringing the population of the Eastern Bloc suddenly to their senses. In 1942, Chizhevsky was arrested as an 'Enemy Under the Mask of a Scientist' and sent to rot in a gulag for sixteen years.[46]

Russia's most significant scientific lover of the macrocosm was Konstantin Tsiolkovsky (1857–1935), a giant of early rocketry to whom we owe a great deal; some of his calculations are still used by NASA today. Rarely, however, is any use made of his more esoteric formulations, such as his belief that the universe was filled with invisible 'space angels' who read our minds before broadcasting their thoughts down to certain chosen genius figures here on Earth, inspiring them with ideas for poems and inventions; such an angel could 'take possession of some fly and make it Newton' he once wrote. He claimed to have twice witnessed mysterious signs in the heavens being sent down from these spirits himself – today, might he have labelled them UFOs? A believer in the Great Chain of Being, Tsiolkovsky spoke of the cosmos as a *zhivotnoe*, or 'animal being', with planets and solar systems having their own gigantic souls. Possibly, some of these living planets and moons might have been hiding in plain sight, like dark matter – 'Do invisible suns, planets and creatures surround us, like bacteria?' he asked. Tsiolkovsky considered each rung on the cosmic ladder to have its own 'President of Various Merit', and taught that 'The minds and might of the leading planets are guiding the universe into a state of perfection'. Seeing as everything within this Platonic world animal was made up of 'lively and happy atoms', the reincarnation of dead souls was a given. 'Death does not exist', he said, being simply 'one of the illusions of the weak human mind'. 'Who is an immortal citizen of space? It is an atom!' he once proclaimed, which meant that *we* were all immortal, too. For Tsiolkovsky, the point of space travel was to cause mankind to be exposed to zero-gravity conditions and bizarre cosmic rays which would then, in an echo of Hanns Hörbiger, cause him to mutate into strange shapes, before eventually becoming an immortal mass of intelligent atomic radiation floating in space – at which point, microcosm and macrocosm would have become one. In space, Tsiolkovsky predicted, first we would shed our possessions, then our clothes, and finally our material bodies themselves, becoming 'free children of the ether' or 'angels in human form', or possibly just living crystalline points of light filled with pure love. Possibly, humanity would metamorphose into a whole new set of moons and planets, orbiting the sun. If you didn't want to join in with this programme you would have to be killed of course, but that was just a minor matter, seeing as one day you might well wake up again in the shape of an invisible super-moon.[47] Tsiolkovsky lived in the obscure provincial town of Kaluga, as did Chizhevsky for a while, and it is no accident that the place – which was once hit by a meteorite – is also a main centre of Russia's Theosophical Society and the home of its associated Lotos publishing house.[48]

## Rockets of the Reich

Germany, one of the main centres of early rocket research, was also one of the last major world centres of Neoplatonism, at least to judge by the attitude of some early pioneers in the field. There was a profound nationalist element to the well-known German 'rocket craze' of the 1920s and early '30s, with members of organisations such as the famous *Verein fur Raumschiffart* rocket club eager to make up for defeat in the First World War by demonstrating Germany's superior technological prowess, but the words 'nationalism' and 'rationalism' don't always go together.[49] Due to his many scientific achievements, Hermann Oberth (1894–1989) was a hero to many members of the *Verein*, but his peculiar occult philosophy held a place every bit as dear in his own heart as did rocket science. Like Tsiolkovsky, he believed that all atoms were somehow alive, meaning that the universe had a soul, and that mankind had amazing capacities for telepathy, transmitting ideas and feelings from one living atom to another. These ideas also led Oberth to believe in reincarnation, with his ultimate aim being to push the science of rocketry onwards in the hope that, one day, his own living atoms would be reincarnated in the shape of a starship captain like James T. Kirk. Wisely, the later UFO fan Oberth kept such speculations separate from his more technical books about rocket propulsion, lest his religious speculations got in the way of people being able to accept his groundbreaking science.[50]

Another German who felt that the universe was a 'living and besouled organism'[51] was Oberth's colleague Max Valier (1895–1930), who was to become rocketry's first ever casualty when, on 17 May 1930, his own atoms were forced into a premature bout of reunion with the macrocosm when a fuel explosion sent a big chunk of metal careering into his head.[52] Interestingly, Valier was a follower of Hanns Hörbiger, giving public lectures with titles like 'The Ice-Haze Horn of the Milky Way' to support himself during his own astronomical researches.[53] Even Hörbiger thought Valier's more spiritual works were an embarrassment, which he feared critics would latch onto to discredit his own WEL theory, but his student dismissed such warnings and books like *The Transcendental Vision*, *Things of the Beyond*, *Trinity of the Original Being* and *An Occult Theory of the Universe* poured from Valier's pen with alarming regularity throughout the 1920s.[54] Valier also wrote a mainstream introduction to astronomy aimed at the general reader, called *Orbit and Nature of the Stars*. Hörbiger was doubtless delighted that Valier included much laudatory material about the WEL within it, but would have been less pleased that he also used the older man's theory to attempt yet another reunion of microcosm with macrocosm by arguing that astrology was a valid science, with weather, crop growth and flooding all being affected by the regular passage of zones of cosmic ice near to our sun, altering its emission of solar rays and thereby affecting life down here on Earth. Sounding not unlike a German Chizhevsky, Valier was rash enough to proclaim that 'astrology happens to be an empirical science – perhaps the oldest, at that'.[55]

Maybe such men derived their leanings from the work of the naturalist writer Max Wilhelm Mayer (1853–1910). Using the familiar model of the subatomic world resembling a small solar system to argue that microcosm and macrocosm reflected one another perfectly, Mayer proposed that a quasi-erotic force of 'universal love' permeated the universe, binding it all together, a collective synthesis which could be demonstrated even by the rotation of the Earth.[56] One particular German whose life proved that love did not really make the world go round, however, was Wernher von Braun (1912–77), the Nazi rocket engineer who rained death down upon London with his V-2 ballistic missiles near the end of the Second World War. Following German defeat, von Braun and his men were spirited away to Texas by the Americans under the infamous 'Operation Paperclip', which aimed to exploit Nazi rocket expertise to win the looming Cold War. Here, von Braun found religion, taking every opportunity to spout off about how it was 'God's purpose' to send 'His Son to the other worlds to bring the Gospel to them'.[57] Given that in 1959 he seemingly tried to explain the deflection of a US test missile from its expected flight path by blaming aliens for the fact – 'Far stronger powers face us than we had thought originally, operating from an unknown base', he said[58] – there is a horrible possibility that von Braun held this intention of Bible-bashing towards the inhabitants of other planets seriously. On the other hand, perhaps von Braun simply meant that humanity had a holy responsibility to set up Christian communes beyond our Earth: 'It is profoundly important for religious reasons that [mankind] travels to other worlds, other galaxies; for it may be man's destiny to assure immortality not only of his race, but of the life spark itself.'[59] This was why America's first attempt to put a man into space was code-named 'Adam' – because, for von Braun, engineering the whole possibility was really a means to transplant spiritual life into an otherwise dead universe.[60] Possibly von Braun had been listening to the speculations of Oberth and Valier, and maybe even looking up Tsiolkovsky upon the prospect of atomic reincarnation as well. 'Nature does not know extinction; all it knows is transformation,' von Braun once said. 'Everything science has taught me, and continues to teach me, strengthens my belief in the continuity of our spiritual existence after death.'[61] Von Braun's innovation was to combine this essentially Neoplatonist doctrine with that of Christianity. Christ's resurrection, he implied, was a covenant between mankind and God, demonstrating how our atoms, too, would one day be resurrected after death ... maybe even in the shape of a starship captain, as Hermann Oberth had hoped. 'It is inconceivable there should not be something else for us after our Earthly voyage,' von Braun wrote in 1962.[62]

It is difficult to see how this former SS officer could have reconciled his earlier actions with his later evangelism, but as the old Tom Lehrer (b. 1928) song explained, von Braun had little problem in dividing his abstract science and its concrete consequences into two wholly separate compartments. 'Once the rockets are up, who cares where they come down,' sang Lehrer. 'That's not *my* department, says Wernher von Braun.' During his youth, von Braun had studied at the boarding school of Weimar's Ettersberg Castle, where Goethe had written his *Faust*; the feeling would persist for many that this modern-day seeker after knowledge of the heavens had also sold his soul to Mephisto rather than to God.[63] The irony about mankind needing to settle other planets in order to ensure the 'immortality ... of the life spark itself' was that the most plausible reason why such a move might prove necessary was that nuclear warheads fixed to von Braun's own rockets would one day wipe out all life on Earth. Von Braun was far from alone in his quest to seed the universe with human life for millenarian ends, however. One recent student of early rocketry, Michael G. Smith, has summed up the hopes of many of the field's pioneers, especially those from Russia, by pointing out that the parabolas traced by the first rockets in their flight were hoped to become metaphors for the forthcoming ascent of the human race to evolutionary and spiritual greatness in outer space, too:

> These parabolas ... were stories: parables. As arcs, they positioned the human being "side by side" (*parabolē*) with the rocket, both in motion through the universe. The spacecraft's expanding arcs to the planets became the human form writ large, fulfilling the human form writ small, that speck of dust in the vacuum of the cosmos, now tracing new forms across its vast expanses. By way of the rocket we became the vector pointed beyond the planet, the microcosm seeding the macrocosm, the pathogen humanising outer space ... Human history was turning biology into technology, the human being becoming a star-traveller ... a "spaceship" all of its own.[64]

So how come, when we finally did fly up into space, we ended up killing the living universe, not seeding it?

# 4

# Heavenly Bodies: NASA, Alien Sex and Marrying the Universe

You would think walking on the moon would have been an uplifting experience; but in fact some astronauts found their grand day out to have been really boring, a complete and utter let-down. Buzz Aldrin grew so depressed by the sense of disappointment that in later life he turned to drink to numb the pain. Apparently, he had anticipated undergoing some kind of major epiphany up there on the lunar surface, but when it failed to arrive he felt something of a failure. A man who had trod the heavens ought to return back with something profound to tell the world, he thought, and he had not.[1] Just after *Apollo 11*'s successful landing, Aldrin had served himself some blessed wine and a Communion wafer – the first food and drink ever consumed upon the moon – but this act had not produced the overwhelming effect within his soul he had hoped for.[2] Prior to blast-off, Aldrin had commissioned his Presbyterian Minister, Dean Woodruff, to write a paper entitled *The Myth of Apollo 11: The Effects of the Lunar Landing on the Mythic Dimension of Man*, in which Woodruff had predicted that the moon landing would lead to an upsurge in global spirituality, restoring an age-old sense of connection between man and the heavens which had now been lost.[3] He was wrong. Far from reviving his faith in God, one *Apollo 8* astronaut, Bill Anders (b. 1933), actually *lost* his faith whilst orbiting above Earth, but kept the fact quiet, worrying about negative PR if word ever got around that travelling in space might make men turn atheist.[4] His fellow *Apollo 8* astronaut Jim Lovell (b. 1928) was less guarded in his statements, openly denigrating the moon as an ugly, worthless thing upon his return to Earth. 'I'm kind of curious,' he told the American media, 'how all the songwriters can refer to it in such romantic terms.'[5]

Space flight affected different people in different ways. Probably the most extreme example of an astronaut doing the direct opposite of Buzz Aldrin and undergoing a transformative religious experience on the moon is that of Jim Irwin (1930–91), of *Apollo 15*. Wandering the lunar surface in search of ancient rock samples, Irwin felt the presence of God surround him, before his attention was guided towards a 'strange, light-coloured rock sitting on a base of grey stone', like a trophy on a pedestal. Irwin's companion David Scott (b. 1932) wiped away the dust from its surface and found it was composed of large, white crystals, a sign of great age. When analysed it turned out to be 4 billion years old, being therefore dubbed the 'Genesis Rock'. Convinced that God had shown him towards the item as a sign of His presence on the moon, Irwin whipped out a Bible and read a passage aloud whilst looking over the alien hills prior to leaving back for Earth. At that moment, Irwin felt 'the beginning of some sort of deep change taking place inside me', he said, indicating he had just been re-born in Christ. Back home, Irwin founded his own missionary-type organisation, High Flight, planning to build a spiritual retreat fit for the Space Age out in the mountains of Colorado. By 1978, Irwin was publicly proclaiming his belief that Earth and moon alike, just like the Bible said, were no more than a few thousand years old, in complete contradiction to the evidence of the Genesis Rock. In later years, Irwin sought out the remains of Noah's Ark on Turkey's Mount Ararat; by proving Noah was real, he hoped to show that the dinosaurs had been wiped out by the Great Flood, not a super-comet, thereby aiding his case as to the relative youth of the planet. 'Jesus Christ walking on the Earth is more important than man walking on the moon,' he declared, claiming that *Apollo 15* had been propelled not only with rocket fuel, but by 'the power of God and Jesus Christ'.[6]

For men like Irwin, it might be claimed that the very fact of being in outer space can prime their minds somehow to be on the lookout for odd things. For example, during their orbit around the Dark Side of the moon in 1969, the crew of *Apollo 10* once heard what they described as 'weird music' of an 'outer spacey' kind whistling through their ears like an eerie 'Whoooooo!'[7] But what could it have been? Aliens? Space angels? Moon ghosts? The uncanny event still stands as unexplained – although that does not, of course, necessarily mean that there was anything genuinely abnormal about it. Some scientists point out that certain planets, like the gas giant Jupiter, often emit large amounts of radiation which may not be heard on Earth, but would be easier for human ears to detect on the moon, with its lack of shielding atmosphere. The human brain, with its innate love of recognising patterns, could then have 'filled in the gaps' for the astronauts, transforming what was in fact a series of wholly random sounds into an apparently recognisable melody.[8] So, it turns out that Pythagoras and Kepler were right after all – there really is a music of the spheres!

## Stars and Gripes

Given the range of reactions of the early astronauts towards their lunar experiences, it would be useful to ask what the corporate attitudes of NASA itself were. Did the organisation think that its work had any kind of spiritual dimension to it? Two differing opinions can be seen in the work of a pair of American academics. Kendrick Oliver (b. 1971), in his 2013 book *To Touch the Face of God*, argues that NASA was in no sense an inherently religious organisation, it was simply that, with its glory years coming in the 1960s and early '70s, NASA was operating within a fairly religious society, and so there was a good chance that a number of its employees, including astronauts, would be inclined to possess a somewhat religious worldview. The main purpose of NASA was always scientific, and Mission Control actually specifically tried to select astronauts who would be unlikely to be religiously overcome by the experience of space flight; enjoying a sense of union with the Godhead is all very well down on Earth, but not when placed in control of a spacecraft. Rather than finding God in outer space, said *Gemini 10*'s Michael Collins (b. 1930), 'I didn't even have time to look for Him.'[9]

In any case, from the early 1960s onwards public organisations in the US had become hyper-sensitive about being accused of promoting religion in the course of their main duties, since the Supreme Court had recently ruled such activities to be unconstitutional. Whilst as many as eight million ordinary Americans wrote to NASA offering support for the right of astronauts to express their religious beliefs on the moon,[10] other less tolerant souls disagreed. Most notably, a tedious right-on busybody named Madalyn Murray O'Hair (1919–95) demonstrated her lasting commitment towards unfree speech by filing a lawsuit against NASA for having allowed *Apollo 8*'s astronauts to read some lines from the Book of Genesis on live TV, thereby supposedly 'attempt[ing] to establish the Christian religion of the US Government before the world'. Whilst in narrow terms O'Hair didn't succeed in her petition, in a wider sense she did indeed triumph; despite her pathetic whinings being thrown out of court, NASA still felt obliged to tread more carefully from hereon in, abandoning their previous plans to construct a Chapel of the Astronauts on public land adjacent to the Kennedy Space Centre. O'Hair's lawsuit even included a complaint against NASA for allowing Buzz Aldrin to privately serve himself Communion on the lunar surface; no wonder he turned to drink.[11]

So, by law, NASA would have been unable to operate as an openly Christian organisation. Not even Madalyn Murray O'Hair possessed a window into men's souls, however, so it was perfectly possible for NASA employees to fulfil their duties out of a private sense of devotion towards God, if they so wished – and many did. This is the view of David F. Noble (1945–2010), whose book *The Religion of Technology* provides much evidence that a surprisingly large number of NASA employees were what might be termed 'highly religious'. General John B. Medaris (1902–90), the Head of NASA's immediate predecessor ABMA, for example, became an ordained Episcopal Minister in the 1960s, and openly preached that America would never have got into outer space 'without God's help'.[12] Another NASA Head, James Fletcher (1919–91), was a committed Mormon, a religion which teaches that there is life on other planets, something which might help explain why, under his control, NASA put an unusually large amount of effort into searching for signs of extraterrestrial life.[13] The Episcopalian Thomas O. Paine (1921–92), Fletcher's predecessor, went so far as to call the Bible 'the best operating-manual ever written'.[14] Much odder were a number of NASA employees who felt that, by putting man into space, they were helping hasten the Day of Judgement. Following his retirement, an electrical engineer who worked on the moon programme called Edwin Whisenant (1932–2001) wrote several books predicting the coming Rapture, for example, whilst NASA systems engineer Jerry Klumas was obsessed by a prophecy from the Book of Daniel which implied that, as human knowledge about things like space flight increased, the End Times would come ever nearer.[15] Before Madalyn Murray O'Hair became involved, NASA's references to their employees' faith could be much more overt. In 1958, after a number of failures had befallen their *Vanguard* rocket prototype, the device's designers strapped a St Christopher's medal to its gyroscope, something which was recorded on official design documents with the medal's function listed as 'Addition of Divine Guidance'.[16] Just remember that those rockets were the later basis of US ICBMs designed to hold nukes on them!

## Fly Me to the Moon

Surely the most spiritual of all NASA's employees was *Apollo 14*'s Edgar Mitchell (1930–2016). The sixth man on the moon, he was also by far the weirdest. Raised as a Southern Baptist, Mitchell found his faith waning by the early 1960s, and was actively searching for some new source of spirituality to replace the Christianity of his youth.[17] To this end, whilst travelling between Earth and the moon, he secretly conducted a series of experiments in telepathy, trying to beam down information to four colleagues below in America, to see if they could receive the same mental images of shapes stamped upon cards which he was thinking of. Mitchell claimed the results were statistically significant, although others have preferred to disagree.[18] Mitchell's conclusion was that, with telepathy, distance was no object, with thoughts instantly traversing the distance between moon and Earth as if nothing were there (seeing as space is a vacuum, in a sense nothing *is* there …). It was as though, thought Mitchell, man and the universe were inherently and intimately connected, something confirmed for him during his return voyage to Earth when he experienced a sudden 'ecstasy of unity' between himself and the cosmos. Everything in existence, he realised, was linked, with the elements that made up his own body having been born aeons ago within 'the furnace of one of the ancient stars that burned in the heavens about me'. Seeing as microcosm and macrocosm were but twin aspects of one another, Mitchell theorised that mind could extend outwards into apparently 'dead' matter, with thought energy being transmitted from brain to brain because, in the end, no two minds were truly separate from one another, or from the universe surrounding them.[19]

Re-examining his experiences at NASA, Mitchell came to believe that the organisation worked so smoothly because its employees collectively constituted some kind of 'common mind', as if every worker had been absorbed into 'the anatomy of a larger animal', just as we were all a part of Plato's macrocosmic world animal.[20] Despite his upbringing, Mitchell was really much more mentally primed to participate within the contemporary New Age movement than traditional Christianity, as can be seen from the fact that he labelled his epiphany in space an experience not of rebirth in Christ, but of '*savikalpa samdhi*', a yoga term indicating a state of heightened consciousness.[21] Mitchell resigned from NASA in 1972, bored with 'flying a desk' instead of a rocket, and founded a parapsychological research establishment, the Institute of Noetic Sciences, in the New Age (and, later, new-tech) hotbed of Palo Alto, California. Funding was a struggle, however, and in 1982 disputes about the Institute's direction led to Mitchell being ejected from his own organisation's board.[22] Perhaps Mitchell's most well-known work at the Institute centred around the celebrated spoon-bending Israeli psychic Uri Geller (b. 1946), who claimed to have gained his seemingly miraculous abilities after seeing a UFO as a child.[23] Mitchell was impressed with Geller's apparent mastery of mind over matter, declaring that the Institute's investigations into him 'could be as important as *Sputnik*'.[24] Certainly, Mitchell wasted no time in introducing Geller to his old NASA colleagues – upon one memorable occasion the Israeli impresario greatly impressed none other than Wernher von Braun by magically fixing his calculator for him.[25] Hopefully the conversation at no point strayed onto the more awkward matter of what the old man had helped do to six million of Geller's ethnic compatriots back in Germany some thirty or so years previously.

Considering himself an 'interplanetary citizen', Mitchell was convinced of the reality of alien life, hoping to find evidence of it during one of his trips into space. He had grown up on a cattle ranch near Roswell, New Mexico, scene of the famous alleged UFO crash, and firmly believed that flying saucers had been visiting Earth for decades, something which 'a cabal of insiders' in Washington had been covering up, together with the dead bodies of their occupants. Mitchell was sure, from talking to those in the know, that ETs had intervened in some way to prevent America and Russia going nuclear during the early days of the Cold War, with the White Sands Missile Range in New Mexico (where von Braun had claimed one of his rockets was intercepted by an alien craft in 1959) being the epicentre of the unknown visitors' attentions. Questioned about these statements, NASA replied 'Dr Mitchell is a great American, but we do not share his opinions on this issue.' In 2004 Mitchell claimed to have beaten cancer with the aid of a faith healer, but in February 2016 'Mr Spock', as his fellow astronauts affectionately called him, finally passed away, having lived long and prospered.[26]

## The Marriage of Heaven and Earth

There are other ways of reconnecting oneself with the universe than those pursued by Edgar Mitchell. A more physical union between Earth and outer space can be seen in all those strange stories about people having sex with aliens that have become a common staple of the tabloids; Mitchell may have done some odd things up there in orbit, but he never tried that. One of the first meaningful attempts at interplanetary intercourse on record was surprisingly chaste in nature – but not for want of trying, at least by the human half of the couple. The tragic tale of American construction worker Truman Bethurum (1898–1969) and his doomed love for the beautiful maiden from beyond the moon, Aura Rhanes, is the ET equivalent of *Romeo and Juliet*, and one which has moved some to resort to poetry – namely, Mr Bethurum himself. Consider these haunting lines:

> *Her flesh was real and plenty firm,*
> *Her shape was like an expensive urn.*
> *She was just over four feet tall,*
> *And certainly entrancing, all in all.*[27]

So entrancing was the pint-sized Ms Rhanes that she became the first alien ever to be cited in a divorce petition, with Bethurum's wife Mary claiming to have become sick of her materialising within the marital bedroom. A resident of Santa Barbara, California, by summer 1952 Bethurum had been married to Mary for seven years, and was feeling the proverbial itch. It was fortunate, then, that late one night out at Mormon Mesa, a large flat-topped hill in the Nevada Desert, Truman Bethurum awoke from a snooze in his car to find a giant circular saucer hovering atop the mound. Invited aboard by a group of midget males wearing space uniforms of some kind, Bethurum was soon introduced to the craft's female captain, Aura Rhanes. It was love at first sight – although not for Aura, who spoke in rhyming couplets and came from a world where strong women were in charge, not puny menfolk. Nonetheless, the captain agreed to further dates with Bethurum, telling him that she hailed from the planet Clarion, which was hidden from Earth's view as it sat directly behind the sun and moon, and also because of 'moisture, clouds and light-reflectors making an impenetrable screen' for astronomers. Bethurum had been working away from home whilst all this was going on, with Mary quite happily minding the fort back in California, at least until he made the mistake of telling her about the lovely Ms Rhanes, who, despite being a dwarf, was nonetheless 'tops in shapeliness and beauty', a 'queen of women' with a 'fully developed' figure and rosy olive-coloured skin. So obsessed with the lovely alien did he become that Bethurum kept on approaching random females in the street, thinking they might have been Aura in disguise. After Rhanes offered Bethurum a trip up to Clarion, where the climate was 'mild enough for you to sleep in the nude if you wish', Mary blew her top and filed for divorce. In 1954, Bethurum quit his job and set up a New Age commune, the Sanctuary of Thought, hiring a dwarfish female secretary to remind him of his lost love, who had heartlessly returned back to her hidden planet without him after all.[28]

More successful was Howard Menger (1922–2009), a New Jersey sign-writer who not only had sex with an alien, he married one – or so he said. To those who met her, Connie Menger just looked like an ordinary blue-eyed, blonde-haired Earth woman, but maybe she looked different naked. Menger's first encounter with an alien blonde had occurred in 1932 when he was ten and a space lady wearing a see-through dress had approached him in a wood, apparently causing him to undergo some sort of sexual awakening. 'As you grow older, you will come to know your purpose', the woman told him, in what sounds like a clear case of grooming. Certainly, the age difference between the pair was unacceptable by modern standards; whilst she appeared to be in her twenties, the alien temptress was in fact 500 years old. According to the adult Menger, saucers were always landing on his shaded property and their long-haired, Scandinavian-looking occupants, clad in tight ski suits, dropping in for a quick coffee before taking him up on trips to the moon where he was given gifts of lunar potatoes, which greatly resembled ordinary Earth potatoes, but with five times the nutritional value. Menger performed many mundane but helpful tasks for his alien visitors, cutting the long hippy-style hair of the menfolk and buying Earth underwear for the women – although not bras which, they told him, they never wore. With all this talk of bra-less blondes from another world, Menger started getting media attention and at one of his radio talks a young woman called Connie Weber turned up, affording him a good look at her own space potatoes. It later transpired that this woman's name was really Marla, that she hailed from Venus, and that Howard himself was actually from Saturn – Connie explained it all in her highly romantic (in both senses of the term) book, *My Saturnian Lover*. In the early 1960s, Menger admitted he had made all this up, supposedly at the instigation of the CIA, who had wanted to test out public reaction to the possibility of alien contact; presumably, few secret agents had expected that one possible human reaction to meeting an alien was to try and sleep with it![29]

The next step following sex and marriage is to fall pregnant by an alien, a claim first made by a white South African lady named Elizabeth Klarer

(1910–94). Klarer had her first encounter with a UFO in 1917, when a saucer swept down from the sky and destroyed a huge meteor which was headed for the family farm in the Drakensberg Mountains, an event only a local native named Ladam believed had actually occurred. Rather than thinking it was a spaceship, Ladam called the silvery machine a herald of the Sky-Gods, guessing Klarer's long blonde hair had caught the attention of the 'white men from the sky' who would one day return to mate with her – was she too one of the *Vril-Damen*? In 1954, the first stage in this heavenly courtship began when a craft flew down towards Klarer when she was standing on top of a local prominence later christened Flying Saucer Hill. Looking out from one of the portholes was a white-haired man with 'the most wonderful face I had ever seen' – Klarer's future lover, Akon from the planet Meton. In 1956, Akon returned, letting Klarer aboard his ship and giving her 'electric' kisses. He said he wished to strengthen his race with an infusion of pure new blood, which makes him sound like a Space Nazi from Aldebaran, but Elizabeth appeared unconcerned. In 1957, the year of *Sputnik*'s launch, Klarer was taken on board again, and told to strip off. She removed her tartan kilt, and enjoyed a nice bath full of bubbly green water. Then she was given a ring set with a stone of pure light, and taken to Akon's room to enjoy 'the magic of his love-making', during which she felt their 'two bodies merging in magnetic union as the divine essence of our spirits became one'. Finally, they ate some tasty post-coital vegetables and Klarer was returned home, pregnant with Akon's child. Following a close shave in which some Communists had fired a death-ray at Klarer and tried to kidnap her in their rocket, Akon returned to Earth in 1959, taking his lover and her car (whose engine he fixed!) up to Meton so she could give birth safely, an absence of several months which, inexplicably, neither of Klarer's human children remember having ever occurred. Her half-breed child Klarer called Ayling, and, whilst she was able to keep in contact with him via holographic projection, his dad rarely allowed him to visit Earth in the flesh because the natives were so dangerous – although, in Klarer's case, hardly unfriendly.[30]

## Mars Fills That Gap!

The natural response to such stories is to say that Klarer, Menger and Bethurum were all mad. Maybe they were, but if so then the particular course their folly ran is still an interesting one. The native Ladam called Elizabeth Klarer *Hlangabeza* or 'one who brings together',[31] and it would seem that what she 'brought together' most was Heaven and Earth – or, to put it another way, macrocosm and microcosm. Consider Klarer's following description of her inter-species love-making:

> As our bodies became one, the fusion of the electric essence of life
> was attained, and the ensuing ecstasy and balance of electrical forces

transcended all things experienced in life. To love and be loved, encompassed within the magnetic emotion of mind and body in perfect union of affinities ... I found the true meaning of love in mating with a man from another planet ... The eternal magic of wholeness bonded our love with the everlasting light of the universe, and ... as I lay in Akon's arms ... I sensed the life and continuous movement within each tiny particle of air, a thrilling awareness and knowledge of the whole, of magnetism, the essence and stuff of life. To become whole oneself is to find that magic lease we sense as life. The pulse of life throbbed through the air ... Throughout intergalactic space, on the surface of other planets, other Earths, it is the same. All are relative, all have the magnetic stuff of life, and all are within the whole.[32]

Such ecstatic unions with humanoid representatives of the heavens are no invention of the Space Age; consider the age-old claim of certain Roman Catholic nuns to have experienced a so-called 'Mystical Marriage' with Christ in their cells at night. Some such nuns have even, like Elizabeth Klarer, been given a wedding ring by their new husband; stigmatic ones, with red bands of raised and thickened skin becoming visible around their ring fingers as a sign of their betrothal to Christ.[33] The most famous such Catholic mystic, St Catherine of Siena (1347–80), even claimed that her heavenly wedding ring constituted Jesus' severed foreskin; no longer did it float around Saturn.

Whilst clearly not normal, I would suggest that persons such as Elizabeth Klarer and Truman Bethurum should be regarded more as modern-day Space Age mystics, as successors to the likes of St Catherine, rather than insane as such. Whilst their interplanetary romances did not literally happen, it is nonetheless reasonable to suppose that they really did *experience* them occurring, inside their own heads. Their fantasies seem to have satisfied some kind of deep-seated emotional needs, which were every bit as spiritual as they were sexual. Like the Mystical Marriage of the nun and Christ, physical opposites such as Heaven and Earth, spirit and flesh, are joined together as one during such accounts. After all, many of the early participants within the nascent UFO scene of the 1950s did not actually consider the aliens they met with to be physical, flesh-and-blood beings, but instead to possess so-called 'ethereal' or 'astral' bodies of some kind, having evolved into something not unlike the disembodied 'space angels' conceived of by Konstantin Tsiolkovsky. Take the Borderland Sciences Research Association (BSRA), one of the first-ever saucer-investigation clubs. Based in California, the BSRA liked to explain UFOs and aliens along supernatural lines, arguing they were a '4-D' phenomenon, existing within 'other dimensions of matter' – BSRA lingo called aliens 'etherians' and their craft 'ether ships'. Such theories later helped account for the fact that the first space probes found Mars and Venus to be uninhabited, even though many 1950s aliens had claimed to hail from just these planets. This, it could be maintained, was because the Martians

and Venusians lived upon the *etheric equivalents* of these worl
primitive physical ones NASA was probing, hence accounting
apparent non-existence.[34]

## Sweet Child of Mine

Many of these stories seem to emerge from California – as this bc
on, places like Los Angeles, Hollywood and the Mojave Desert will
every bit as central to our story as the plains of Mars and moons of Ju
and there is good reason for this. California being the world centre c
Age thinking, it makes sense that persons of a mystical bent who have
sublimated wish to merge their souls back with that of the universe s
feel themselves drawn there. One modern-day group of alien-fanciers k
as the 'Hybrid Baby Community' (HBC), for example, seem to have a nu
of members in the area. The group's leading light, a former marke
executive named Bridget Nielsen, actually hails from neighbouring Ariz
but her mind-set is pure downtown LA. In 2016, Nielsen, then twenty-se
decided to spread the word about her group through the Press, leading
some very odd articles appearing in which she spoke of her 'incredil
super-primal sexual experiences' with a loving alien reptile-man, which ha
led to her giving birth to no fewer than ten half-human, half-lizard 'hybri
babies, with big reptoid eyes and scales. Another member of the group, a the
twenty-three-year-old videogame designer from Los Angeles called Alun
Verse, gave the following account of her first amorous alien encounter:

> I was in a classroom setting with other humans. All of a sudden I'm
> next to this green reptilian creature and immediately I'm so sexually
> turned-on by looking at this being. I was very surprised. We're making
> love in the classroom in front of everyone. Everyone turned their
> attention to us. It sounds crazy … But this is really happening![35]

Is it? Visiting the HBC website, it becomes apparent that most of these
encounters take place within the women's dreams – or at least that is what
it *sounds* like. The group bill themselves as being 'a family, a tribe and a
community', who are 'excited to live by the highest vibrational values',
something which apparently 'aids in shifting us to higher dimensions' –
which I get the impression often happens during sleep or meditation. In
language the BSRA would have recognised, the HBC say that their hybrid
children 'operate on a very different frequency than we're used to', seeing
as they 'exist in a very etheric, dream-like, wondrous reality with infinite
possibility', as do we all when we close our eyes at night. Apparently, the
HBC's mission is to 'create a high-frequency cocoon', with them hoping
thereby to 'shift actual physical reality' to enable their half-lizard children to
land here on Earth for real.[36] Obviously, the HBC want to make their dreams

ome true. But is having wet dreams about aliens normal? Being sexually
tracted towards lizards, explains Nielsen, is actually a sign of high spiritual
evelopment. In fact, only liking humans in this way is akin to racism. 'If
e're still looking through filters of prejudice or Brad Pitt standards, how
an we openly embrace our star-brothers and sisters?' she asks. Rather
han looking at an alien's hideous scaly body, Nielsen advises, the women
of tomorrow will instead have to learn to find what is *inside* the lizards
attractive. Nielsen claims to have 'majorly redefined' the kind of lover she
herself finds appealing, seeking only those who are 'vibrationally aligned'
with her, even if they possess claws. During her 'many loving experiences'
with 'SUPER-positive and smart reptilians', Nielsen has discovered that
'the reptilians you encounter are, a majority of the time – YOU', or at least
your 'cross-connecting incarnational counterparts'. If so, then these alien
sex acts would appear to be some form of weird spiritual self-impregnation;
following which you will become 'telempathically connected' to your hybrid
child living on some other etheric plane in outer space, who will turn out
to be 'the sweetest thing', with 'SOOOOOO much love' and be 'totally
physically cute as well!'[37]

A typical tale of a loving HBC alien encounter comes from a woman named
Miesha Johnston, who began having regular liaisons with a humanoid
reptile called Iyano after moving to Los Angeles in 1989. Awakening in bed
one night, 'with a feeling of my kundalini rising', Johnston was surprised
to find a towering, muscle-bound, scaled creature (who, to judge by an
accompanying picture, looked like Julian Assange, but green) standing next
to her bed. Apparently, he never actually touched her physically, but still
made Miesha pregnant no fewer than thirteen times, via an 'unconditionally
loving' transfer of 'spiritual energy' into her body.[38] The ambiguity about
whether or not such alleged couplings are truly spiritual or physical in
nature is well summed up in a series of messages channelled from outer
space by Bridget Nielsen, in which she suggests the hybrid children are made
from pure female orgasms. Apparently, space babies are constructed from
'sixth-density' etheric matter, which Nielsen says is exactly the same kind
of energy released by a woman during climax. This sixth-density matter
appears really to be some kind of divine light, and the children demonstrate
to humanity what we will one day evolve into – living, light-based orgasms.
According to the adult aliens, the hybrid kids' role is 'to be an example of an
ignited flame' of the sort that one day 'humanity will live as'. Apparently, 'in
moments of orgasm, the human [body's] sixth-density energy is captured to
create the vibrational and archetypal template for the hybrids', which leads
to a 'meeting between Heaven and Earth'. The aliens' final message to the
Hybrid Baby Community, then, is one of hope: 'Your children were created
from ephemeral bliss, to live as unceasing ecstasy'.[39] This, it seems, is what
the old myth of Endymion, seduced by the moon-goddess during a hillside
dream, has now come to.

Some people clearly have a deep psychic need to feel themselves seeded with the fruits of the cosmos and vice versa, reuniting microcosm and macrocosm in loving unity. And yet in the end this whole wild fantasy is no more than one aspect of yourself having sex with another aspect of yourself – or, in other words, masturbation. One woman whose sexual encounters with a reptile sound very telling in this respect is Pamela Stonebrooke, a Los Angeles-based jazz singer who in 1999 made headlines after allegedly being offered $100,000 to tell her story in book form. Stonebrooke's tale began in 1998, when she awoke to find herself being raped in her own bed. Whoever her rapist was, he 'was so much larger than most men' Stonebrooke explained, which made sense when she saw she was 'making love to what appeared to be a Greek god' in what was simply 'an exceptionally lucid dream'. Had this been ancient Greece, then that would have been that. However, this being 1990s America, when Stonebrooke next opened her eyes she saw that the Greek god had transformed into 'a reptilian entity with scaly, snake-like skin'. 'We've always been together, we love each other', the alien whispered, and Stonehouse climaxed several times – 'the orgasms were intense'. However, these revelations later caused Stonebrooke grief, on the grounds that 'reptilians are not [regarded as] a very politically correct species in the UFO community, and to admit to having sex with one … is beyond the pale.' Other members of the UFO community may also have had problems with Stonebrooke's admission that perhaps her various alien rapists were 'simply different aspects of myself' and not real ETs anyway. Stonebrooke later admitted to having been an astral traveller for around twenty years, with her soul 'floating in a sea of knowingness beyond thought' and enjoying sex sessions with various non-human entities on the astral plane, which constituted 'a complete merging of energy and spirit'. Explaining that humans were 'multidimensional beings', Stonebrooke speculated that, by producing hybrid babies, we were creating a new, future species for our own souls to one day inhabit.[40] Never has the term 'daydream believer' seemed so apt.

## Written in the Stars

Another way of reconnecting mankind back with the universe was through astrology, which has its origins not in its modern-day hub of California, but amongst the ancient Babylonians and Chaldeans, in that land now known as Iraq. The basic idea held that the moving stars and planets had some kind of innate influence over humanity down below, with the night sky being divided up into twelve bands of 30 degrees, each of which was then named after a nearby fixed constellation of stars – the signs of the Zodiac. From the perspective of Earth, the sun seems to pass through each of these 30 degree sections in turn through the course of the year. As the sun passed by, each Zodiac constellation in turn was thought to cast its influence down upon Earth below, altering human behaviour and fate.[41] Whilst the idea has no true validity, one modern theoriser who chose to believe in the astral influence of the starry macrocosm was Gabrielle Henriet, that strange Frenchwoman we met earlier, claiming the sky was solid. Henriet's belief in astrology may seem odd, as we will recall she thought that neither stars nor planets actually existed, being mere incorporeal lights emanating from open windows in the sky. Her alternative explanation for astrology's efficacy was that, when such luminous discs passed overhead, they stimulated the vault upon which they shone into emitting certain cosmic rays down onto the Earth below. These different areas of the geo-metallic vault had different mineral and magnetic properties, affecting humans in different ways, and these striations of heavenly rock were what had actually been mapped out by the ancient Babylonians when splitting the night sky up into the signs of the Zodiac. So, for instance, when the sun disc passed over one particular 30 degree section of dome, it would emit cosmic waves which made humans angry and aggressive, leading this section of sky to be associated with Ares, god of war. Thus, there was an inherent connection between man and the universe. God, Henriet said, could be defined as being all the different parts of the dome added together and considered as one, and all other minor gods like Ares were simply partial manifestations of Him.[42]

Like Plato, Henriet also felt that the Earth itself was a kind of living animal, contained within another, larger, animal called the universe. Whilst she generally defined 'the universe' as meaning 'everything within the sky dome', Henriet sometimes spoke of the possibility of other similar universes being present outside of our own particular metallic dish-cover. Each dome universe then corresponded to a particular bodily organ contained within the frame of a gigantic cosmic human – with our own Earth as Cosmic Man's beating heart. Seeing as Henriet refused to believe the human heart was really a pump, or that blood circulated around the body at all, preferring to think that it stood still inside our veins 'in the same way that marrow is enclosed in the bones',[43] her explanation of how the Earth-Heart works is somewhat obscure. It seems to centre upon the idea that earthquakes 'represent cardiac phenomena' whilst volcanoes 'represent the apertures through which blood is assumed to leave the heart and pass into the blood vessels'. When volcanoes erupted, this corresponded to 'the expulsion of the blood', represented by lava. Basically, her idea was probably that, when volcanoes explode, the Earth-Heart expands; when earthquakes strike, the Earth-Heart contracts, and the alternation of the two keeps Cosmic Man alive.[44] Given that the size of the average heart is approximately one-thirteenth of the size of an average human being (so she says, anyway), Henriet was able to extrapolate outwards from this, calculating that the 'gigantic Man-World possessing the characteristics of the human body which is the universal prototype' in which all galactic citizens unknowingly lived was approximately 517,920 km in terms of circumference, a size which, she said, 'does not seem so absolutely inconceivable'.[45]

## Periodic Bloodshed

Now let us return to our original discussion of the moon, for it turns out that, in light of what has been discussed above, Phoebe is not completely dead after all. If, like Ms Henriet, you still believe in astrology and read your daily horoscope in the newspaper, then little mention seems to be made of the moon's alleged astrological influence over your day ahead. This is odd, because traditionally this was not the case. The moon, as a wanderer through the night sky, was supposed to induce a certain *wanderlust* in those born under its sign – either that, or cause a person to develop 'wandering wits', that is to say madness.[46] The very word 'lunacy' is derived from the word 'lunar', with man and the moon once being thought so intimately connected to one another that man's mind was known as 'the microcosmic moon'.[47] There must be something archetypal about such an idea, because many vestiges of the notion that the cycles of the moon have some kind of impact upon life down here on Earth have survived even into the present day. To some degree, this is in fact true; the reproductive cycles of certain species of seaweed are linked to the lunar cycle, for instance.[48] However, what about the bizarre claim, made by the travel writer Laurens van der Post (1906–96) in his 1986 book *A Walk with a White Bushman*, that more atrocities occurred within Second World War Japanese POW camps during times of the waning moon, seeing as more darkness then began to enter into the guards' souls, leading to 'a rearrangement, a reversal of the poles within the Japanese spirit'?[49] Or what about the infamous study which supposedly proved that male Japanese bus and taxi drivers had secret hidden 'periods' linked to the lunar cycle, being more susceptible to accidents during 'their time of the month'? The men had their work rotas rearranged, leading to a large reduction in the number of reported road accidents, allegedly.[50] Can such tales really be true?

The idea that the lunar cycle has some physiological influence over humans is most commonly seen in relation to menstruation. This is because the moon is often said to pass from nothing to fullness and then back again within the space of exactly twenty-eight days (actually the correct figure is twenty-nine-and-a-half days). Multiply this conveniently simplified figure of twenty-eight by thirteen, and you get 364 days, the approximate length of a solar year (the time it takes the Earth to revolve around the sun). The term 'month' is therefore derived from the word 'moon', with a month being originally defined as the time it took for a full lunar cycle to complete. Seeing as the average length of the female menstrual cycle has been measured at something between twenty-seven-and-a-half days to twenty-nine-and-a-half days, the lunar and menstrual cycles seemed to fit well with one another, thus leading to the general conclusion that the moon must have been a female goddess. Charles Darwin himself speculated that this might all have been a relic of the evolutionary past of *Homo sapiens*' fish-like ancestors; reproduction amongst marine life is often dependent upon the tides, which are governed by the phases of the moon. Maybe women's twenty-eight-day lunar period cycles are some kind of reminder to even the most elegant of ladies that they are all, at root, nothing but modified fish-wives?[51] However, experiments on apes have shown this cannot be true. Our simian siblings do have periods, but scientists given the unpleasant job of watching them do so and then taking notes have determined that female orang-utans go on the blob every twenty-nine to thirty-two days, gorillas every forty-three to forty-nine days, and so on. Other mammals have periods too, and their cycles are also different from those of humans. Curiously, the only animals to share the approximate twenty-eight-day cycle with human women are opossums! Clearly, the whole 'link' between periods and the moon is simply down to coincidence.[52]

Nonetheless, some women still insist upon the reality of the 'empowering' connection between their bodies and the allegedly 'inherent' femininity of the moon. It has been known for feminist American New Agers to gather together in so-called 'Moon-Huts' during their monthly 'Moon-Time', in order to worship the White Goddess by going outside, sitting on the bare soil and unleashing their fluids, allowing their red fertility to mingle with the moonlight before seeping down into the Earth, whose holy pulse they can feel beating against their bare but bloodied genitalia.[53] Hillary Clinton voters, evidently. Clearly, such persons are fulfilling their emotional, not their logical, needs here, as can be demonstrated by a classic statistical study performed by the American biologist Lamont Cole in 1957. Cole had read a report about oysters which, transported from their coastal homes to an inland lab in Illinois, had supposedly re-set their internal rhythms to correspond with the lunar cycle in their new environment. His paper, *Biological Clock in the Unicorn*, proposed a thought experiment in which the legendary creature of fable did indeed exist, with fictional data about its breathing rate being produced from a random-number generator. Although this data was arbitrary nonsense, Cole analysed it using the same mathematical methods used by others in relation to the real-life oysters, and still managed to come up with a means of interpretation 'proving' that respiration rates amongst unicorns were directly linked to the appearance of the nocturnal moon. What he had actually proved, of course, was that such methods of analysis were profoundly flawed and that, as is commonly said, there are lies, damned lies, and statistics.[54] In truth, the connection between universal macrocosm and human microcosm is much more of a mental one than a physical one; looking up at the beauty of the night sky, you would have to be hard-hearted indeed not to feel any link. But that does not necessarily mean that any link is actually there ...

# 5

# Planetary Pareidolia, Part I: Seeing Ourselves through a Telescope, via the Moon

If the macrocosmic moon was once thought to cause lunacy within the microcosmic human brain, then the most pervasive form of moon-related madness down the years has been the habit people have had of seeing things on it which were not really there. The English poet Samuel Butler (1613–80) mocked this trend expertly in his c.1670 poem *The Elephant in the Moon*, a satirical examination of the astronomical activities of London's Royal Society.[1] Founded in 1660, the Royal Society was England's most prestigious scientific institution, the first of its kind, but the sometimes bizarre experiments conducted under its auspices frequently attracted censure for their oddness, cruelty, or sheer wrong-headedness. Whilst such criticisms were often rather unfair, Butler's poem stands in a long tradition of mocking those who dare peer too closely into the true nature of God's Creation. The phase 'to find an elephant in the moon', meaning to make a mistaken discovery of some kind, was once proverbial, though I am unsure whether the proverb predates the poem, or was inspired by it.[2] Either way, Butler's verse involves the clever conceit of a group of Royal Society men gathered around one of their new-fangled telescopes and mistaking a number of animals which have crept in under the glass for distant yet gigantic alien creatures living on the moon. Initially, the scientists and their task are described in dignified, colonial-era terms:

> *A virtuous, learn'd Society, of late*
> *The pride and glory of a foreign state,*
> *Made an agreement on a summer's night*
> *To search the Moon at full, by her own light;*
> *To take a perfect invent'ry of all*

*Her real fortunes, or her personal,*
*And make a geometrical survey*
*Of all her lands, and how her country lay,*
*As accurate as that of Ireland.*
*T' observe her country's climate, how 'twas planted,*
*And what she most abounded with or wanted;*
*And draw maps of her prop'rest situations,*
*For settling and erecting new plantations.*

The idea that the moon might one day become a new colony – or 'plantation' – in Britain's Empire was once a popular literary fancy, but Butler's moon already seems to be experiencing conflict over its territorial possession. At first, the Royal Society's astronomers are amazed to see two warring armies engaged in battle on the lunar surface. This seems incredible enough, but then:

*A stranger sight appears*
*Than ever yet was seen in all the spheres;*
*A greater wonder, more unparallel'd*
*Than ever mortal tube [telescope] or eye beheld;*
*A mighty Elephant from one of those*
*Two fighting armies is at length broke loose,*
*And with the desp'rate horror of the fight*
*Appears amazed, and in a dreadful fright.*
*It is a large one, and appears more great*
*Than ever was produced in Afric yet;*
*From this we confidently may infer*
*The moon appears to be the fruitfuller [world than Earth].*

The astronomers should be able to work out that this 'elephant' is not entirely what it seems, seeing as it appears able to run across the whole lunar globe within a matter of seconds:

*A member peeping in the tube by chance*
*Beheld the Elephant begin t'advance,*
*That from the west-by-north side of the Moon*
*To th' east-by-south was in a moment gone.*
*This being related gave a sudden stop*
*To all their grandees had been drawing up;*
*And ev'ry person was amazed anew*
*How such a strange surprisal should be true,*
*Or any beast perform so great a race,*
*So swift and rapid, in so short a space.*

Nonetheless, the Royal Society men refuse to admit they may have been mistaken in their observations. They would 'rather choose their own eyes to condemn/Than question what was beheld with them'. Instead of admitting their obvious error, the astronomers prefer to construct an elaborate theory, saying that the elephant only *appeared* to move so quickly due to the way the moon and Earth were constantly revolving on their respective axes away from one another, a notion instantly acclaimed as a 'solid mathematic demonstration/Upon a full and perfect calculation':

> *As th' Earth and Moon*
> *Do constantly move contrary upon*
> *Their sev'ral axes, the rapidity*
> *Of both their motions cannot fail to be*
> *So violent, and naturally fast,*
> *That larger distances may well be passed*
> *In less time than the Elephant has gone,*
> *Altho' he had no motion of his own,*
> *Which we on Earth can take no measure of.*

The astronomers then all leave the room and it is left to one of the Royal Society's footboys, made curious by all the commotion, to discover the obvious:

> *He no sooner had apply'd his eye*
> *To th' optic engine, but immediately*
> *He found a small field-mouse was gotten in*
> *The hollow telescope, and shut between*
> *The two glass windows, closely in restraint,*
> *Was magnify'd into an Elephant.*

When the footboy tells his employers this, they are reluctant to believe him, and begin to re-examine the 'elephant' one by one. It quickly becomes apparent that each scientist has a completely different way of seeing than his fellows:

> *In no one thing they gazed upon agreeing,*
> *As if they'd different principles of seeing.*
> *Some boldly swore, upon a second view,*
> *That all they held before was true,*
> *And damn'd themselves, they never would recant*
> *One syllable they'd seen, of th' Elephant;*
> *Avow'd his shape and snout could be no Mouse's,*
> *But a true nat'ral Elephant's proboscis.*

*Others began to doubt as much, and waver,*
*Uncertain which to disallow or favour.*

To settle the matter, the telescope is dismantled – whereupon not only a mouse, but a collection of fleas and gnats jump out, these insects having been the two warring armies of moon people. The field mouse, it transpires, has 'catch'd both himself, *and them*, in th' optic trap' – although surely the truth is that the astronomers trapped themselves. The moral of Butler's story is thus:

*That learned men, who greedily pursue*
*Things that are rather wonderful than true,*
*And, in their nicest speculations, choose*
*To make their own discoveries strange news,*
*And nat'ral hist'ry rather a Gazette*
*Of rarities stupendous and far-fetch't,*
*Believe no truths are worthy to be known*
*That are not strongly vast and overgrown,*
*And strive to explicate appearances*
*Not as they're probable, but as they please,*
*In vain endeavour Nature to suborn;*
*And, for their pains, are justly paid with scorn.*

The real elephant in the Royal Society's room was the inherent fallibility of human perception when faced with the apparent presence of something people wish desperately to believe in.

## Too Many Men in the Moon

There is a name for seeing things which aren't really there. It is called pareidolia, and is something we are all familiar with. The human brain is inherently hardwired to recognise visual patterns where none exist; everyone has seen a strange shadow or bush they have mistaken for a lurking human figure at night. Imaginary faces are one of the most common such illusions, as endless news stories about coffee stains which resemble Jesus show. The most famous lunar example is that enigmatic-looking fellow known as the 'Man in the Moon'. 'There liveth none under the sunne that knows what to make of the Man in the Moone'[3] said the English poet John Lyly (1553–1606), and he was right. Rather than one man in the moon, there are actually several to be seen lazing around up there – and quite a few women and animals, too. Most people in the West see one particular version of the Man in the Moon, but as you go further down towards the South Pole, it becomes less and less possible to spot this particular optical illusion, until eventually he stops appearing at all.[4] In China, Korea and Japan, people prefer to see a rabbit in the moon; to their eyes, the Moon-Rabbit looks as if he is pounding something inside a small container with a stick. Some say it is the magical elixir of ever-lasting life; others that it is simply rice cakes. As a PR exercise, just prior to the *Apollo 11* landing NASA told Buzz Aldrin about this Moon-Rabbit, adding that legends said he was kept company on the moon by a 4,000-year-old Chinese girl named Chang-o who had been banished there after stealing an immortality pill. 'OK, we'll keep a close eye for the bunny-girl,' joked Aldrin, but they never seem to have found her.[5] The ancient Aztecs also saw a rabbit on the moon; their story is that, once upon a time, the moon and sun were of equal brightness, making it hard to sleep at night. Annoyed, a god threw a rabbit up into the moon, with the creature creating all the dark spots we can see across its face today, thus dimming its light and enabling a better night's rest for all.[6]

Once a person or creature has been spotted on our lunar neighbour, the human mind quickly gets to work constructing a narrative to explain the phantom's presence. Western fables about the Man in the Moon include that he was banished there for sinfully gathering firewood on the Sabbath, or that he is Judas Iscariot, entombed upon the silvery moon for all eternity for the crime of betraying Christ for a handful of silvery coins. On the other hand, there may be a Woman in the Moon, namely Mary Magdalene, with the moon's spots being the tears she famously shed for the life of Jesus.[7] Naturally, everyone knows that there isn't really a man – or a giant crab, as some have also claimed – on the moon for real. As early as the Greco-Roman philosopher Plutarch's (*c.* 46–120) dialogue *The Face which Appears on the Orb of the Moon*, people have debated how such an illusion might be created. Plutarch's guess was that the moonman was a reflection of Earth's oceans on the moon's white disc, but the ancient Roman was wrong.[8] In fact, the lunar surface is made up of dark areas of dried-up basaltic lava known as seas, or *maria*, and whiter, paler sections called highlands. The interplay between these areas gives rise to innumerable optical illusions; even the designation of the lava flows as seas can be traced back to early astronomers like Galileo discovering them through telescopes and thinking they were actual bodies of water, which they are not.[9] All illusory lunar images are caused by the interplay of such geographical features with the human eye and brain. The right eye of the Western moonman, for example, is simply the dark *Mare Imbrium*. *Apollo 11*, meanwhile, actually landed within part of the left eye, the *Mare Tranquilitatis*.[10] Nobody ever believed this act might have blinded the poor fellow. Some Shia Muslims, however, really do believe that the name of Muhammad's son-in-law Hazrat Ali (599–661) appears on the moon, and invest this illusion with actual religious significance; *Apollo 11* landing on that instead might have led to a much earlier call for anti-American *jihad*.[11] Other people claim to see things on the moon whose significance is more obscure. Go online and innumerable posts can easily be found in which buildings, roads, bridges, and even animals are pointed out by excitable conspiracy theorists. A good random example is a video created by someone calling themselves 'Cavorite Rising' (sci-fi fans may recognise the reference), in which the narrator claims he can spy things resembling a truck, a big pipe, a square structure 'about the size of a Los Angeles hotel' and even a brontosaurus sitting on the lunar surface.[12] A Californian organisation called the Orion Observatory of Santa Monica, meanwhile, once produced an academic paper purporting to demonstrate that, by joining up lines between various rocks on the moon, it was possible to see the entire place was covered with dead dinosaurs.[13]

You might think such bizarre speculations would have died out following the *Apollo 11* landings, but actually the golden era of the Space Race coincided with a real upsurge in cases of lunar pareidolia. After all, the moon was big news back in the '60s, so more people were looking at it, with Russia and America each launching probes to scope out the lunar surface prior to any potential landings. In fact, one set of images taken by Russia's *Luna-9* probe appeared to show an open invitation for just such a landing to occur! Bird's-eye photos of the Ocean of Storms revealed, in the description of the journalist, businessman and paranormal investigator Ivan T. Sanderson (1911–73), 'two straight lines of equidistant stones that look like the markers along an airport runway.' These stones, he said, were 'all identical' and 'positioned at an angle that produces a strong reflection from the sun, which would render them visible to descending aircraft.' Certain images taken by NASA's *Orbiter-2* probe, meanwhile, seemed to show a series of eight pointed spires, massive in size, and shaped like obelisks. The Russians, continuing their early disinformation programme of promoting belief in aliens to undermine belief in God, publicised the speculations of the Soviet space engineer Alexander Abramov to the effect that these obelisks formed a specific kind of geometrical arrangement known as an *abaka*, or 'Egyptian Triangle'. The centres of some of the spires, Abramov said, were aligned in precisely the same way as the apexes of the Great Pyramids at Giza. Might this mean the Pharaohs had been in contact with aliens?[14] Not everyone saw these strange 'structures' through the same imaginative lens, however; some paranoid Westerners viewed the apparent spires as Soviet ICBMs, stationed on the moon and aimed right at America. In fact, the things were neither lunar monuments nor missiles. They were simply low hills which cast long shadows when the sun was positioned down near the lunar horizon.[15] Panic over – such wonders were available to be seen only through the mind's eye.

## Mining the Moon

A further misunderstanding involving *Luna-9* centred upon the fact that the Soviets didn't release the photos it had captured immediately. Transmission of these images back down to Earth was intercepted by Britain, however.

These pictures were then compressed to facilitate their distribution to the Press across wire services – the problem being that this fact was never actually announced. The distorted photos were thus misinterpreted as showing that the moon had gigantic mountains running all across it, apparent evidence of volcanic activity. One excited geologist speculated that such activity could have led to the formation of precious metals on the moon, something reported by the media as indicating that *Luna-9* had discovered a new lunar gold mine! The Russians then released the original, non-compressed photos, and laughed at the degenerate capitalists and their greedy obsession with gold.[16]

Another man who saw lunar mines was an American writer named George H. Leonard (b. 1921), whose 1976 book *Somebody Else is On the Moon* used a combination of official NASA photographs and his own sketches to make the case that the moon was occupied by at least one underground-dwelling alien race, who were either mining it for useful resources, or else engaging in large-scale repairs of its surface, with the whole moon really being nothing more than a giant 'macro-spaceship' in disguise.[17] Thinking there was no way NASA would have spent billions flying to the moon purely for scientific purposes ('Not while our cities decay'), Leonard reasoned there must have been something strange lurking away up there to warrant all the effort, and set out to find it.[18] He didn't have to look far, as to his eye the entire moon was little more than a gigantic building site, littered with industrial machinery. For instance, in the crater Lubnicky A, he sees 'a motor as big as the Bronx', complete with 'the shaft of a gear sticking out' and with part of it having been 'ripped away by some cataclysm', exposing the inner teeth of a huge gear wheel. According to Leonard, this gigantic mega-motor would, if dropped from the sky over Manhattan, 'obliterate everything from Midtown to the Bowery', so let's hope this never happens.[19] Further items of colossal machinery on the moon include huge 'super-rigs', which are 'several miles long and capable of demolishing the rim of a 75-mile-wide crater'. Leonard speculates these super-rigs are used either to mine the lunar surface for minerals, or to dig out trapped moon-people from the after-effects of some titanic catastrophe of years ago.[20] Other moon machines Leonard has spotted include giant 'X-drones' resembling 'two crossed earthworms' which lie flat on the ground and 'vary in size from under a mile to three miles in any direction'. Having many possible functions, the 'arms' of these devices are detachable and replaceable with other giant earthworms, although their main task appears to be that of 'pulverising rock'.[21] Many crosses can also be seen carved into the moon, which Leonard speculates are alien foremen's marks meaning 'I've already dug here, don't bother'.[22] Leonard has carefully catalogued various other apparently meaningful markings all over the moon, from arrows to ancient runes, but admits he doesn't know what they mean.[23] Giant domed tents like yurts dot the lunar landscape, meanwhile, presumably for the space workers to rest in,[24] and the author even provides a drawing of an alien comms tower he has spotted, which looks strangely like the Loch Ness Monster.[25] Enormous u-bends and pieces of plumbing are visible upon Leonard's moon too,[26] alongside a number of alleged vehicles which strongly resemble living bagpipes or mutant pig bladders.[27] Leonard reckons the *Apollo* astronauts saw all this stuff for themselves, but disguised the fact by using female names like 'Barbara' as code-words so it made them sound like they were talking about girls instead.[28]

Leonard's most original piece of thinking is the notion that the moon's surface is artificial and spread out over some kind of framework in the same way that *papier mâché* might be strung out over a wire-frame grid to make a model globe.[29] To support his case, Leonard points to several chasms with small pairs of thin rectangles passing over them. Some people might think them lunar bridges, but Leonard prefers to hypothesise that they are huge surgical stitches used to prevent the gaping moon falling apart any further.[30] Leonard's best piece of evidence is found in the form of a giant screw which passes through two deeply cracked parts of the moon, 'to hold together parts of the skin' – although there are other possible explanations, the author does admit. For example, it might be a visual way of saying 'Screw you, Armstrong!' to any overly inquisitive Earth dwellers tempted to try and claim the moon as their own.[31] 'Have you ever kicked over an anthill and watched the mega-myriad creatures work feverishly in repair?' asks Leonard. 'Is this the activity which we are glimpsing on the moon?'[32] No, but this doesn't stop him further speculating that UFOs seen hovering over bodies of water on Earth and sucking up liquid through tubes might have been moonmen stocking up on supplies,[33] or that missing persons from our own planet may have been spirited away to the moon to supply the aliens with spare limbs.[34] Rarely have such elaborate castles in the sky been founded upon such incredibly weak foundations.

## Mirror, Mirror

One notable predecessor of Leonard was the German astronomer Johann Heironymous Schröter (1745–1816), whose 1791 *Selenotopographische Fragmente* provided easily the most detailed telescopic description of the moon's surface available up to that point; detailed, but also partly fictional. A firm believer in the existence of 'selenites' or 'lunarians', as he called his longed-for moonmen, Schröter speculated wildly about a race of 'calm' and 'rational' beings who 'give thanks for the fruit of the field' living idyllic lives up there, whose only worry was about the possible damage lunar volcanoes (or 'cratermountains') might wreak upon their 'many moon-cottages'. Meanwhile, Schröter accounted for certain areas of the moon changing colour during observation not by acknowledging that the Earth's atmosphere was interfering with the optics of his telescope, as was the truth, but by speculating that smogs and smoke created by heavy moon industries might be drifting over the lunar scene, obscuring it from view.[35]

Due to his many legitimate achievements, Schröter was sometimes known as 'the German Herschel', after Britain's celebrated Astronomer Royal, William Herschel (1738–1822), fêted for his sensational discovery of Uranus in 1781, the first new planet to have been sighted since ancient times. This was ironic, because Herschel was in fact German himself, not British; also ironic was that, for all his fabled powers of observation, Herschel too had once been involved in bizarre speculations about life on the moon. To Herschel, the moon was a pareidolic paradise; in a letter to a colleague he once claimed that 'For my part, were I to choose between the Earth and moon, I should not hesitate to fix upon the moon for my habitation' because to him it seemed so lush. Whilst Herschel was well aware that perfectly good evidence existed to say that the moon had no atmosphere and thus was unable to support any life as we knew it, he claimed to have used his telescope to see trees (possibly *giant* trees) arranged in vast forests, as well as canals and 'two small pyramids'. Even weirder, he felt it possible that the various roundish craters visible on the moon may well have been circular buildings, or even entire circular cities for moonmen to live in, structures he dubbed 'circuses', after the curved rows of Georgian townhouses in Bath where he had once lived.[36]

Interestingly, Herschel was amongst the first astronomers to make truly effective use of mirrors rather than simple glass lenses within his telescopes, thereby transforming each into a reflecting telescope rather than a refracting one. The advantage was that mirrors focused light more efficiently, giving clearer, brighter images with less prismatic distortion.[37] Materially speaking, with his mirrors Herschel saw more clearly what was lurking up there in space, in physical reality. Metaphorically speaking, however, the mirrors also gave him back a profoundly distorted reflection of his own mind, with his speculations about the existence of lunarians being based upon no real evidence at all, other than an apparent desire to repopulate a dead universe with life. One person who also felt such an idea attractive was the great man's son, John Herschel (1792–1871), another astronomer of note, and a big believer in life on other worlds. John was a fan of the astronomer Peter Andreas Hansen (1795–1874) and his idea that the moon's centre of gravity lay on its Dark Side, meaning that the satellite would actually be egg-shaped, with all of its water and atmospheric air being pulled away there out of Earth's view. If so, reasoned Herschel Jr, then the Dark Side may have been filled with lashings of unknown life.[38] Even odder, in 1861 John made the astonishing proposal that the surface of the sun was filled with gigantic luminous creatures which roamed across it day and night, and that *these* were the sources of its intense light and heat, not the sun itself![39]

## Pure Moonshine

Given these fantasies, it was perhaps no surprise that John Herschel was later to find himself at the centre of the greatest astronomical hoax in history – or, at least, something which is generally *thought* to have been a hoax, but was actually intended as a piece of satire which got out hand. The famous 'Great Moon Hoax' began on 25 August 1835, when a New York paper called *The Sun* printed a front-page notice, reading as follows:

GREAT ASTRONOMICAL DISCOVERIES
Lately Made
By Sir John Herschel, LL. D., F.R.S., &c
At the Cape of Good Hope

*The Sun*, the first successful penny tabloid, did a roaring trade on the back of this headline, with sales soaring over the next few days as a series of increasingly sensational descriptions of John Herschel's alleged findings began to be printed. Herschel, it was claimed, had built a gigantic new telescope, whose mirror had magnified the moon's surface to such an extent that various fantastic details of this distant world could for the first time be revealed. It sounded rather like fairy-land:

> A beach of brilliant white sand, girt with castellated rocks, apparently of green marble, varied … with grotesque blocks of chalk or gypsum and … clustering foliage of unknown trees … [left us] speechless with admiration. The water … was nearly as blue of that of the deep ocean, and broke in large white billows along the strand … [There was also] a lofty chain of obelisk-shaped, or very slender pyramids, standing in irregular groups, each composed of about thirty or forty spires, every one of which was perfectly square … They were of a faint lilac hue, and very resplendent … monstrous amethysts … glowing in the intensest light of the sun!

Even better, the moon contained animal life; there were miniature bison, with 'a remarkable fleshy appendage over the eyes, crossing the whole breadth of the forehead and united to the ears', moon sheep and moon zebras, lovely blue unicorns, and 'a strange amphibious creature of a spherical form, which rolled with great velocity across the pebbly beach'. More amazing yet was a species of tail-less bipedal beaver, which had developed its own basic form of civilisation: 'Its huts are constructed higher and better than those of many tribes of human savages, and from the appearance of smoke in nearly all of them, there is no doubt of its being acquainted with the use of fire.' Best of all, Herschel and his assistants described a species of man-bats, primitive-looking, hairy humanoids with huge wings and the power of speech:

They averaged four feet in height, were covered, except on the face, with short and glossy copper-coloured hair, and had wings composed of a thin membrane, without hair, lying snugly upon their backs ... The face ... was a slight improvement upon that of the orang-utan, being more open and intelligent in its expression, and having a much greater expansion of forehead. The mouth, however, was very prominent, though somewhat relieved by a thick beard upon the lower jaw, and by lips far more human ... The hair on the head was a darker colour than that of the body, closely curled, but apparently not woolly, and arranged in two curious semi-circles over the temples of the forehead ... Whenever we saw them, these creatures were engaged in conversation; their gesticulation ... appeared impassioned and emphatic. We hence inferred that they were rational beings ... [and later discovered] they were capable of creating works of art ... We scientifically denominated them as *Vespertilio-homo*, or 'man-bat'; they are doubtless innocent and happy creatures, notwithstanding that some of their amusements [like mating in the open-air] would but ill-comport with our terrestrial notions of decorum.[40]

Whilst people actually believed this twaddle – there were rumours of an American Minister raising funds to send Bibles to the intelligent but as-yet unredeemed bats and beavers on the moon[41] – it was soon enough discovered that the whole thing had been a skit perpetrated by a young *Sun* journalist named Richard Adams Locke (1800–71). Well educated in astronomy, Locke had intended his stories first and foremost as a satirical mockery of Schröter and the Herschels, taking their ideas about life on the moon to their illogical conclusion by presenting a bizarre, impossible world full of talking man-bats, blue unicorns and hut-building beavers.[42] Locke's other main target was a Scotsman, the Reverend Thomas Dick (1774–1857), whose books explaining the world of astronomy through the prism of Christianity were best-sellers, despite their numerous absurdities. To Dick, there was no point in God having created all those other planets if He wasn't going to populate them, and so, extrapolating from the average population density of England, he claimed to have worked out that there were precisely 21,894,974,404,480 aliens living within our solar system, some 4,200,000,000 of whom hailed from the moon.[43] Other totally unprovable opinions held by Dick included that 'THE THRONE OF GOD' was located at the very centre of the universe, and that there could be no volcanoes on the moon as this would imply the lunarians were being punished for their sins – which was impossible, seeing as Earth was the only planet upon which the Original Sin of Adam and Eve had ever occurred.[44] Eventually Locke admitted his reports were a satire, with Dick chastising his tormenter in print; Locke responded by saying that maybe Dick's own ideas were also a hoax, seeing as they were so utterly ridiculous.[45]

John Herschel himself took matters more calmly, finding Locke's articles entertaining, at least initially. Later, however, he was known to lament that 'the world ... should be brought to believe in my personal acquaintance with the Man in the Moon'.[46] He should not have been too ashamed. As this book shows, John Herschel was hardly the first person to be mocked for claiming to have seen impossible things on other worlds. Nor would he be anything like the last.

# 6

# Planetary Pareidolia, Part II: Seeing Ourselves through a Telescope, via Mars

Nowadays mass outbreaks of interplanetary pareidolia have largely switched from the moon towards Mars, whose rocky surface provides plenty of opportunity for simulacra to be spotted. Giant faces are a perennial favourite, both human and animal. Amongst the most notable in the Martian portrait gallery are a panda, an anteater, a screaming George Washington and a full-body side-profile outline of Kermit the Frog formed from old lava flows; Kermit's eye, placed in precisely the right place, is handily provided by a random impact crater.[1] Now that NASA's *Curiosity Rover* has landed on the Red Planet and taken close-up images from on the ground, sharp-eyed persons have also noted the presence on Mars of a set of fossilised traffic lights, a protoceratops dinosaur, a large elephant head, a fossilised human finger and a giant penis with attached testicles, this latter image having been formed accidentally by the *Curiosity Rover*'s tracks in the alien sand.[2] Other planets and moons have proved similarly favourable to producing pareidolic images. Stalin has been spotted on Venus, and Bugs Bunny seems now to reside on one of the moons of Uranus – must have taken a wrong turn at Albuquerque.[3] Two of Saturn's moons, Mimas and Tethys, provide clear images of 1980s arcade game sensation Pac-Man upon their surface when examined with special cameras used to map temperature variations. On both moons Pac-Man appears in the correct yellow colour, with his mouth spread wide as if about to pop a Power-Pill, whilst the lunar background is dark blue and purple, like the mazes through which he used to be chased endlessly by electronic 8-bit ghosts. The explanation is that the bluish sections are bombarded with more electrons than the yellow Pac-Man sections are, thereby compacting the lunar surface at these points, causing them to develop a hard, icy nature with a lower temperature. Furthermore, the moon

Mimas has a giant round impact crater on its right-hand side, making it greatly resemble the Death Star from *Star Wars*, thus bringing home the way such illusions are wholly dependent upon the culture of the person who sees them; prior to the '70s and '80s, such resemblances would have been by definition impossible to discern, as neither Pac-Man nor *Star Wars* existed back then.[4]

## Planetary Politics

The majority of the above optical illusions have been taken by most sensible people in a wholly jokey frame of mind. Others, however, have used the never-ending gift of pareidolia to re-enchant the disappointingly dead planets around Earth with new signs of life by claiming that they demonstrate there are secret inter-governmental conspiracies afoot up there in space, with Martian mimetoliths (rocks that look like things) eagerly accepted at face value as valuable 'evidence' that the US space programme is not all it seems. This particular sub-genre of conspiracy theory is often dubbed 'exopolitics', dealing as it does with alleged political and governmental conspiracies taking place within outer space, or else involving alien life forms interacting with shadowy statesmen down here on Earth. The basic exopolitical motifs come in three equally implausible strands:

- Governments covering up UFO crashes and back-engineering the advanced technology found within them for their own benefit, thus gaining access to things like fuel-less transport, anti-gravity devices, death-rays and invisibility cloaks.
- Presidents and PMs negotiating secret treaties with ETs, allowing them to do things like abduct innocent citizens from their beds and extract their sperm for use in interplanetary breeding programmes.
- The existence of a second hidden 'black' NASA (or Soviet) space programme lurking behind the standard 'white' one we all know of today, which has discovered alien bases scattered throughout the solar system but kept all evidence from us.

The frequently low standard of exopolitical thought can be well gauged by a recent speech given in the European Parliament following the Brexit vote of June 2016, in which EU Commission President Jean-Claude Juncker (b. 1954) appeared to accidentally admit that, as many had suspected all along, he spent a lot of time on a different planet from most voters:

It should be known that those who observe us from afar are very worried. I met and heard and listened to several of the leaders from other planets who are very concerned because they question the path the EU will [now] engage on. And so, a soothing is needed for both the Europeans and those who observe us from farther away.

Was Juncker now planning to throw open Europe's borders to millions of aliens as well as alleged 'refugees' from Africa and the Middle East in the name of increasing the continent's diversity? Not even he was that arrogant. Juncker had simply experienced a slip of the tongue, meaning to say that 'other planetary leaders' from lands like America and Japan were worried by Brexit, not 'leaders from other planets'. Nonetheless, the official EU transcript of Juncker's speech was subsequently altered by Eurocrats so as not to appear so weird. Was a cover-up at hand? During the same speech, after being justifiably derided by victorious Brexiteer Nigel Farage (b. 1964), Juncker had also stated that 'I'm not a robot, I'm not a machine. I'm a human being, I'm a European.' Why had Juncker been moved to say these words, asked certain 'experts'? Might it be because Juncker *really was* a robot, and trying to put people off his trail? Or was he secretly speaking *to* a hidden audience of robots, making his excuses to metal men from Mars for having failed to keep Britain chained within its EU shackles? Exopolitical opinion was split as to why Juncker's alien masters might have disapproved of Brexit. Maybe there was some kind of Lefty outer-space treaty in existence stating that no planet's wider population could openly be contacted by aliens until they had agreed to submit to a unifying One-World Government, of which the EU was a necessary precursor. Alternatively, perhaps the EU was secretly a Nazi 'Fourth Reich' which had been surreptitiously created in a deal between Hitler and a race of evil alien reptoids called the Draconians, and the fascist lizard-folk just didn't want the British to escape their oppressive clutches once again.[5] As a keen Eurosceptic myself, I can well believe this may indeed have been the case.

## Making a Mountain out of a Mars-Hill

Exponents of exopolitics seem determined to exploit the phenomenon of pareidolia to back up their theories. An excellent example occurred in 2004, when the NASA *Opportunity Rover* sent back an image of Mars showing what appeared to be a cute little bunny rabbit, with two big floppy ears sticking up in the Martian wilderness. Closer examination showed that the 'rabbit' was more of a random white blob, with no actual facial features or legs, but that didn't stop online speculators from concluding that there really was life on Mars. However, what happened next was truly shocking – the *Opportunity Rover* drove straight across to the rabbit and ran it over, seemingly deliberately! Had NASA killed the space bunny to cover up The Truth? No, because the white object was simply a bit of light, soft debris, probably an impact-softening airbag, which had become detached from the *Rover* and blown away. That there are people out there who genuinely think that NASA is secretly killing rabbits on Mars speaks volumes.[6] The advent of the Internet has led to an explosion of such silliness. You can even find online a NASA-released image from 1999 allegedly showing a giant translucent Martian earthworm … or a large gully-like channel with striated ridges running along its bottom, as some spoilsports have preferred to label it. You might have thought the fact that the 'worm' never moves between photographs would have been a dead giveaway that it was not a living creature, but apparently not! Other commentators, meanwhile, take a far more sensible position upon the feature; they say it is actually a huge glass tunnel constructed by Martians.[7]

The most famous saga of exopolitics involves another piece of planetary pareidolia *par excellence* known as the 'Face on Mars'. The existence of the 'Face' was first revealed in a NASA press release for 31 July 1976, in which it was announced that the *Viking 1 Orbiter* probe had photographed a number of 'mesa-like landforms' in the Cydonia region of Mars, one of which, NASA pointed out, 'resembles a human head', something they explained was caused by 'shadows giving the illusion of eyes, nose and mouth' on top of it.[8] A second image of the flattened hill-top taken from another angle, in which the mesa no longer resembled any kind of head whatsoever, satisfied most that NASA were telling the truth. However, others disagreed, and used image-enhancement software to conclude that the photo did show an artificially constructed face after all.[9] Was this proof that an alien civilisation had once existed on Mars? Not really, but some of the Face's greatest champions were credible figures like Dr Brian O'Leary (1940–2011), a former member of NASA's *Apollo* programme, and the astronaut chosen to be the first man on Mars before the plan for any such landing was shelved back in 1967.[10] The most notable populariser of the Face, however, is Richard C. Hoagland (b. 1945), a one-time employee of various American science museums, who in 1976 successfully lobbied to name NASA's new space shuttle *Enterprise*, after Captain Kirk's spaceship in his favourite TV show *Star Trek*. In 1987, Hoagland once again confused sci-fi with reality through publication of his exopolitical classic *The Monuments of Mars: A City on the Edge of Forever*, which contained speculation that, alongside the Face, Cydonia also contained an entire Martian metropolis filled with ancient pyramid-like structures, which apparently contained great secrets.[11]

A second book, 2007's *Dark Mission: The Secret History of NASA*, extends Hoagland's theories towards the idea that NASA, acting upon the advice of a specially commissioned 1960 document called the 'Brookings Report', has been deliberately concealing evidence for ETs having once lived upon the moon and Mars so as not to panic Earth's population. Effectively, this logic means that the total absence of evidence for the existence of advanced alien life within our solar system in fact equates to direct evidence of its existence; those many thousands of space-probe photos of the moon which do not show extraterrestrial cities, for example, might have been altered by NASA, meaning that they *do* show extraterrestrial cities after all, but hidden ones that you can't see. How, then, to account for the presence of things like the Face on Mars on some of NASA's photographs? In Hoagland's view, a carefully managed dribble of such evidence has been being released by NASA for decades now, in order to prepare us slowly for the revelation of the ultimate truth when humanity is finally deemed ready. NASA can't openly say that the Face is real, as that would ruin the whole exercise, but each (deliberately?) inept denial is to be taken with a pinch of salt. To speed up this process, NASA have also been seeding popular culture with references to alien life in films and TV shows like *The X-Files* and *2001: A Space Odyssey*, which, to those in the know, are basically documentaries.[12]

## Alternative Thinking

The idea that NASA's dark works might slowly be being revealed through fictional media products did not begin with Richard Hoagland. Consider the 1977 Anglia TV programme *Alternative 3*, whose intended transmission date of 1 April was unfortunately missed due to industrial action. The broadcast, a joke documentary, proposed that the US, Britain and USSR had all clandestinely been working together behind the scenes to create colonies on the moon and Mars to provide shelter for a chosen global elite for when our abused Mother Earth finally succumbed to fatal environmental disaster from pollution. All NASA footage of moon landings and a lifeless Mars were fakes; as with Hoagland, any official shots showing no cities on the moon were ironic proof-positive that there really *were* cities on the moon. Various scientists had been going missing for years, the show said, being beamed up to Mars to make things ready for the day of escape. Despite the show deliberately featuring numerous bit-part actors whose faces should have been familiar from popular dramas and sitcoms, many viewers failed to see through the spoof. The mother of a real missing person wrote in to the producers, thanking them for explaining where her son had got to these days, and expressing hope he would enjoy his new life of slavery on Mars. Even once the whole thing had been officially debunked, many refused to believe the denials. Some contemporary conspiracists, whilst admitting the broadcast was a hoax, still make the bizarre argument that the show itself was only faked in order to cover up a series of real-life events which were exactly the same in nature, thus undermining any public belief in them! As one man at the time said, he had checked out all the facts, names, dates, and locations mentioned on-air and found that not a single one was true; given this, he argued that the programme *must* have been genuine, because otherwise why would its makers have gone to so much trouble not to include any real facts in the show? Such is the admirable mental flexibility of the truly committed conspiracy theorist.[13]

The endless adaptability of Richard Hoagland's own mode of thought is well illustrated by the surprising turn his thinking took in light of the much more high-resolution images released of the Face on Mars by NASA in 2001 following another fly-by probing. These photos clearly revealed that the mesa really was just a mesa, but Hoagland didn't let this discourage him. In 1992 he had given a speech to the UN, telling delegates that the Face on Mars, upon closer inspection, might turn out to depict the head of a lion-human hybrid, not a normal person at all. On NASA's original 1976 photos, the right-hand side of the Face was almost entirely covered with shadow, allowing Hoagland to guess this. When the 2001 images emerged, Hoagland found to his delight that, if you squint at the newly revealed right-hand side of the Face, it could be interpreted as looking a bit like a lion's head. Seeing as the Sphinx in Egypt also represents a human-lion hybrid (though not facially!), and sits near some pyramids, just as the Face (supposedly) does on Mars, Hoagland concluded there must be some profound connection between Cydonia and Egypt; the lesson to be drawn being that '*WE* are the Martians!' Human civilisation once stretched across the whole solar system but was destroyed somehow, with life on Earth being its only large living remnant; the presence of pyramids and Sphinxes all across our neighbouring planets proves it, says Hoagland. When the Nazis spoke of themselves as being 'Aryan' supermen, meanwhile, they really meant 'Martian' supermen, because they also knew the truth. Thus, seeing as many of NASA's leading early employees were captured Nazi rocket scientists, one hidden aim of NASA's 'black' space programme was to allow men like Wernher von Braun one day to reclaim their ancestral Martian homeland in the name of Hitler.[14]

Committed Trekkie Hoagland has since helped construct a website, enterprisemission.com, which boldly goes where logic has never gone before. The best example of this involves Hoagland's elaborate recasting of the popular 1950s US children's TV show/media franchise *Tom Corbett: Space Cadet* as a sinister propaganda exercise created by Nazis with the aim of subconsciously preparing America's youth for the future revelation of the white race's former existence on Mars. Following his 1992 UN presentation, Hoagland was apparently approached by an unnamed architect who possessed a set of old View-Master slides from 1955, telling a new illustrated entry in the *Tom Corbett* saga. The tale begins with the discovery of a small but ancient pyramidal object with anti-gravity properties. Remembering that a similar larger pyramid exists on the Dark Side of the moon, the Space Cadet team fly there, finding that the tiny pyramid is the capstone of the larger one. Once put in place, the pyramid opens revealing a globe of Mars, marked with a metaphorical 'X'. Travelling to the corresponding spot on Mars, the Cadets then find a Sphinx-like statue, with a face a bit like the one Hoagland later discerned in the 2001 photos from Mars. The Sphinx clutches within its paws the remains of a shattered planet, which the Cadets deduce refers to a nearby asteroid field. Numbers carved on the

Sphinx provide them with co-ordinates leading to one particular asteroid which, they find, has an entrance leading inside its hollow core. Within, the asteroid is really an ancient alien 'time-tomb' built by the same beings that sculpted the Sphinx, and containing powerful anti-gravity technology for the Space Cadets to retrieve. The End – or was it? Not to Hoagland. To him, these slides proved that, before NASA was even founded in 1958, certain shadowy forces knew of the existence of the lion-man's face on Cydonia. So, was the asteroid in the story real too? The enterprisemission.com team think so; according to them the Martian treasure trove is located on asteroid 433 Eros, which in NASA images appears to contain a small item with straight-looking edges which just *must* be the door to the time-tomb! NASA say the object is only a broadly rectangular boulder, but as everyone knows boulders are not usually rectangular whereas doors very often are, so it must be a secret entrance – or so claim Hoagland's own band of brave Space Cadets.[15]

## Opening the Floodgates

Such pareidolic Martian delusions are no mere modern phenomenon. Everyone has heard of the fiasco of the alleged 'canals on Mars', a cautionary tale whose central character was the American astronomer Percival Lowell (1855–1916), who liked to claim he had superhuman powers of vision. During Lowell's day, even the best telescopes could only give stable visions of distant objects like Mars for brief periods at a time due to the ever-present interference of the Earth's atmosphere, but during those precious few seconds of clarity Lowell said he had the miraculous ability to see what was on the Martian surface with 'copper-plate distinctness' – and what he saw were canals, plenty of them.[16] Of course, as we now know, these 'canals' were really mere optical illusions. In his 1913 book *Are the Planets Inhabited?* the English astronomer E. W. Maunder (1851–1928) described making a map of Mars upon which he had drawn a series of dots and irregular markings, positioned very closely together, but not actually touching. Then, he had asked a class of schoolboys to sit at their desks and draw what they saw on the map, displayed at the front of the classroom. Those close to the image drew it accurately. Those farther away drew the dots and dashes as a series of straight interconnected lines – in other words, as something very like Lowell's canals. The human eye, it transpired, was an inadvertent liar.[17] However, Percival Lowell didn't want to hear such arguments; he was a man so obsessed with the idea of canals on Mars that he spent a good part of his honeymoon (although hopefully not the actual consummation) hovering above London's Hyde Park in a hot-air balloon, carefully comparing the layout of the paths below to those of the irrigation channels up above.[18] Lowell desperately wanted to see canals up on Mars, and so he did. But how did the idea arise in the first place?

The canals were first brought to the world's attention by the colour-blind Italian astronomer Giovanni Schiaparelli (1835–1910), for whom the Red Planet may not have been that red at all. During detailed observations of Mars in 1877, Schiaparelli saw a number of apparently straight features on the planet's surface which he dubbed *canali*, meaning simply 'channels'. At no point did Schiaparelli imply they were artificial features. However, *canali* was translated into English as 'canals' and some people got the wrong end of the stick completely. Down here on Earth the Suez Canal had been completed in 1869, demonstrating that such large-scale engineering works were possible. Perhaps the Martians were a few centuries ahead of us, and had managed to build an entire network of Suez-like structures?[19] This initial rash of breathless speculation over with, the *canali* simply dropped out of the news until 1894, when astronomers were anticipating another of Mars' periodic close approaches towards the Earth.[20] By this stage, however, Percival Lowell was ready to take up the cause. Before Lowell, the canals were generally thought to be natural features; after him, the waters grew more muddied. This is Schiaparelli, writing in 1893:

> The most natural and the most simple interpretation [of the canals] ... is of a great inundation produced by the melting of the [polar] snows ... It is not necessary to suppose them the work of intelligent beings, and notwithstanding the almost geometric appearance of all of their system, we are now inclined to believe them to be produced by the evolution of the planet, just as on Earth we have the English Channel.[21]

Compare that to the following purple passage, from Lowell's 1895 best-seller *Mars*:

> We may ... consider for a moment how different in its details existence on Mars must be from existence on the Earth ... Gravity on the surface of Mars is only a little more than one third what it is on the surface of the Earth ... If we were transported to Mars, we should be pleasingly surprised to find all our labour suddenly lightened threefold ... If Nature chose, she could afford there to build her inhabitants on three times the scale she does on Earth ... Now apply this principle to a possible inhabitant of Mars, and suppose him to be constructed three times as large as a human being in every dimension ... Consider the work he might be able to do. His muscles, having length, breadth and thickness, would all be twenty-seven times as effective as ours. He would prove twenty-seven times as strong as we, and could accomplish twenty-seven times as much ... Owing to decreased gravity [his labours would require] but one third the effort ... His effective force, therefore, would be eighty-one times as great as man's, whether in digging canals or in other occupation ... Mars being ... old, we know that

evolution on the surface must be similarly advanced ... The evidence of [the canals] points to a highly intelligent mind ... a mind certainly of considerably more comprehensiveness than that which presides over the various departments of our own public works. Party politics, at all events, have had no part in them; for the system is planet-wide. Quite possibly, such Martian folk are possessed of inventions of which we have not dreamed, and with them electrophones and kinetoscopes are things of a bygone past, preserved with veneration in museums as relics of the clumsy contrivances of the simple childhood of the race.[22]

So: super-strong, giant Martians, with amazing (though conveniently unspecified) futuristic inventions and a One-World Government? Lest you conclude from such fantasies that Lowell was a loon of the highest order, it should quickly be pointed out that this was not so. Amongst his other accomplishments, Lowell was smart enough to deduce the existence of an unknown 'Planet X' from mathematical observations made about perturbations in the orbits of Neptune and Uranus. In 1930 Clyde Tombaugh (1906–97), following Lowell's deductions and working in Lowell's own observatory complex, succeeded in finding this Planet X. He dubbed it Pluto, after the god of the underworld – and after Percival Lowell, whose initials make up the first two letters of Pluto's name.[23] Lowell was also to an extent responsible for spreading the radical idea that observatories should be built not in large population centres for the convenience of astronomers themselves, but far away from the light and smoke of the city, in dry and cloudless areas like deserts, or high up on mountains where the interfering atmosphere itself is thinner. The Lowell Observatory, built on Mars Hill near Flagstaff in the desert fastness of Arizona at Lowell's own expense, still stands today as a tribute to the man's clear-sightedness upon at least one issue of astronomical import.[24]

As this fact implies, Lowell was a rich man, coming as he did from a well-off Boston family, and had no real need to work at all. Having a long-standing interest in astronomy, when media attention began to focus upon Mars during the 1894 approach of the planet, Lowell dropped his previous literary ambitions and, as only rich men can, ordered the building of his new observatory.[25] His basic theory was as follows. Mars was an older world than our own, a dry and dying planet. On it lived super-advanced Martians, gigantic of limb and brain, who exploited the alternate annual melting of the polar ice-caps to irrigate their otherwise bone-dry equatorial regions. It wasn't the canals themselves we could see from Earth, but lush bands of crops which flourished on either side of them. He compared this to the way that Egyptians had used the land around the Nile to irrigate their own lands, making the African desert habitable; 'Seen from space, the Nile would not look otherwise,' he wrote.[26] The idea of Super-Egyptians living on Mars was an appealing one but, as we now know, completely wrong. Many scientists at the time had already realised this fact, with knowledgeable critics lining up to pan Lowell's ideas. The American astronomer W. W. Campbell (1862–1938) pointed out that, given how Lowell's network of canals on Mars were mapped out, it would be as difficult to irrigate the equatorial locations envisaged from Mars' poles as it would be to divert water from the South Pole to San Francisco, Chicago, New York, Rome and Tokyo down here on Earth.[27] Another detractor tried to estimate how much meltwater would be contained on Mars' poles and found that, spread out over Lowell's canals, each would gain less than two inches of liquid to benefit civilisation with across a whole year. Those Martian crops had better be hardy![28]

## Channels of Communication

Why did Lowell's ideas become so popular? Not all astronomers dismissed what he was saying, by any means, and he certainly had the newspaper-reading public firmly on his side. To look for answers, we should recall that Lowell once had literary ambitions. A well-cultivated man, his phrasing could often be humorous and pithy, and sometimes relied upon neatness of phrase for its impact more than its inherent strength of logic, as with the following statement about Mars' exceedingly thin atmosphere being no inevitable barrier to life:

> A fish doubtless imagines life out of water to be impossible; and similarly to argue that life of an order as high as our own, or higher, is impossible because of less air to breathe than that to which we are locally accustomed is … to argue not as a philosopher, but as a fish.[29]

Lowell actually stole that last line from someone else, but it doesn't matter – his point is made, and the average reader probably wouldn't stop to think how weak his actual argument here is. Naturally, many newspapers were happy to promote Lowell's wit and ideas, with some of their reports now making entertaining reading. 'Mars, it has been found, is like Holland,' announced the *New York World* for 24 February 1895, thinking of Amsterdam's canal network, whilst the *Chicago Evening American* for 14 February 1910 ran with 'New Canal Built by Martians!' as their breathless front-page headline.[30] Obligingly, many papers printed aerial pictures of railway tracks, the grid-like street plans of American cities, or actual canal networks here on Earth alongside the *canali* drawings of Lowell, with readers invited to draw the obvious conclusion.[31] Even Hanns Hörbiger got in on the act, speculating that Mars was covered by a 250-mile-thick layer of space ice, and the so-called 'canals' were really just deep cracks in it.[32] With such friendly publicity, it was no wonder Lowell preferred to court a large audience through the mainstream Press rather than through academic journals which were likely to question more closely just how he had drawn his conclusions.[33] Due to all the hype, Lowell quickly accumulated masses of fan mail. Various mediums wrote in to him, pronouncing their own astral travels on the Red Planet had confirmed his basic hypotheses, while a dealer in building materials from Delaware asked him for a scoop on the specific system of hydraulics and pumping the Martians were making use of to irrigate their land, information he hoped to turn a profit from.[34] Occasionally, sci-fi writers would send Lowell in their ideas for stories, and he would actually send back further suggestions and corrections.[35]

As time passed, even Schiaparelli threw caution to the wind and in 1895 began speculating that the canals may have been artificial after all, and that if so then the canal-builders must have been living in 'a socialist paradise ... a grand Federation of Humanity' where, seeing as the 'common enemy' was drought, all wars and social differences had been put aside in a state of 'universal solidarity' because 'the interests of each individual and those of all are not to be separated'. He even started talking about a 'Martian Minister of Agriculture', whose responsibility it was to decide at what point in the year to open the sluice gates, flooding the canals with water![36] Schiaparelli may have been speaking half in jest here, but others who put forward their own mad ideas about the canals were not. In 1895, an orientalist from Washington DC studied one of Lowell's maps and claimed the lines spelled out the name of God in Hebrew; in 1897, the Irish physicist J. Joly (1857–1933) declared the canals to really be raised ridges caused by asteroids flying over the Martian landscape in straight lines prior to impact, with their strong gravitational attraction causing the land to rise up after them like solid smoke trails.[37] Another proposal was that Mars was covered in water and floating seaweed, with the canals being areas where this seaweed had been pushed away by ocean currents.[38] The Anglo-American engineer Elihu Thompson (1853–1937) said the *canali* mapped out the broadly straight paths of seasonal migration by herds of Martian cattle, which carried seeds from their grazing lands along with them in their hoofs or excrement like a trail of breadcrumbs, leading to long strips of vegetation sprouting up behind them.[39] Whilst most astronomers and informed members of the public had dismissed the canals as being definitely illusory sometime prior to the outbreak of the First World War, not everyone got the message. Embarrassingly, in 1989 the US Vice-President Dan Quayle (b. 1947), discussing a possible manned NASA mission to Mars, expressed the opinion that such a feat would succeed because 'We have seen pictures where there are canals, we believe, and water. If there is water, there is oxygen. If oxygen, that means we can breathe.' Disturbingly, at the time he made these comments, Quayle sat as Chairman of the US National Space Council![40] Surely the oddest idea of all about the canals was put forward by E. H. Hankin (1865–1939) of the Aeronautical Society of Great Britain, who in a 1908 *Nature* article proposed that there was but a single life form on Mars, a gigantic alien vegetable which straddled the planet with its roots or branches 'like the arms of an octopus' so that it could suck water from the polar ice caps, with these green tendrils looking like canals to us down here on Earth, from far away.[41] Some people even started seeing canals elsewhere, too. In 1902 Lowell's one-time assistant W. H. Pickering (1858–1938) spotted canals on the moon, which he explained as long strings of vegetation. Even worse, he then tried to claim that certain shifting dark spots on the satellite were really huge swarms of lunar insects that periodically migrated.[42]

In 1965, a stop was finally put to all such conjecture (except in the mind of Dan Quayle) when NASA's *Mariner 4* space probe produced the first reliable close-up photographs of the Martian surface. They were grainy and low-definition, but showed not even the slightest traces of life, let alone evidence of a canal network.[43] Still, never mind; there were only another eleven years to wait before people could begin speculating about the presence of a gigantic alien Face on the planet instead. A final story about the canals sums up the whole affair quite nicely. In 1907, a Professor from New Jersey's Rutgers College had just finished giving a public lecture about Mars illustrated with slides of its famous *canali* when, as *The New York Times* put it, 'a monstrous thing was seen to walk upon the landscape and sit down beside a canal as though to take a drink.' The alien beast appeared just after the Professor had told his audience that Mars was definitely inhabited, enhancing the drama of its appearance. The monster had 'many legs', 'wicked eyes', a 'horrible head' and came complete with 'a flying-machine attachment' on its back. The 'Martian' was, of course, a horsefly, which had got into the slides somehow and become magnified up on the projection screen. Like the elephant on the moon, the creature then calmly 'walked around to the other side of the planet and was lost to view.'[44] Somewhere up above, I like to think that Samuel Butler was laughing.

# 7

# California Dreaming: Eccentric Orbits around the Desert Sun

Astronomy was not Percival Lowell's only love in life. During his gilded earlier years, the wealthy American had gone on an extended 'Grand Tour' to Japan, whose folklore he fell in love with. It was in Japan that Lowell had begun investigating the supernatural, a prior fascination of his, leading to the publication of his book *Occult Japan* in 1894 – the very same year he began making his Martian observations out at Flagstaff. Whilst in the Orient, he had taken a particular interest in the subject of spirit possession, sticking pins into Japanese people who had fallen into trances to test whether or not the 'spirits' had made them insensible to pain.[1] Lowell himself was to prove every bit as difficult to awaken from his own incurable possession by dreams of life on Mars; in a 1988 book, the writer William Sheehan (b. 1954) entitled a chapter about Lowell 'The Visions of Sir Percival', and the play on words is a good one.[2] Sheehan notes that, whilst in Japan, Lowell sought out old feudal ruins, sights he said took him back to his childhood, when he had played at exploring Arthurian fairylands in his mind. Maybe the dying, dried-up world of Mars was the adult Lowell's own personal version of the blasted wasteland of the original Arthurian legends, which could only be revived via a dose of refreshing water poured straight from the Holy Grail of his imagined canals?[3] *Occult Japan* opens with a description of Lowell walking up the lonely old mountainside of a half-extinct volcano called Mount Ontaké which sounds suspiciously like his later descriptions of Mars itself:

> Active once, it has been inactive now beyond the memory of man. Yet its form lets one divine what it must have been in its day. For upon its summit are the crumbling walls of eight successive craters, piled in

parapet up into the sky. It is not dead; it slumbers. For on its western face a single solfatara [steaming crater] sends heavenward long, slender filaments of vapour, faint breath of what now sleeps beneath; a volcano sunk in trance.[4]

This scene doesn't just sound like Lowell's dying Mars, but also like his Flagstaff observatory itself, whose desert location was actively acknowledged by Lowell to resemble the Red Planet upon which he loved to gaze – the sheer lifeless dryness of the place, and the preciousness of water there, made Lowell appreciate just how harsh and unforgiving an environment his Martians must have inhabited.[5] However, Flagstaff had certain mystical qualities to it too. In his 1911 *Mars and Its Canals*, Lowell characterises his desert observatory as being less a scientific outpost, more a sacred retreat:

To get conditions proper for his work, the explorer [of the heavens] must forgo the haunts of men ... Astronomy now demands bodily abstraction of its devotee. Its deities [stars and planets] are gods that veil themselves amid man-crowded marts [due to light and smoke-pollution] and impose withdrawal and seclusion for the prosecution of their cult ... To see into the beyond requires purity; in the medium [of unpolluted air] now as formerly in the man. As little air as may be and that only of the best is obligatory to his enterprise, and the securing it makes him perforce a hermit from his kind. He must abandon cities ... Only in places raised above and aloof from men can he profitably pursue his search, places where Nature never meant him to dwell ... Thus it comes about that today ... monasteries in the wilds are being dedicated to astronomy as in the past to faith.[6]

Some of the above words are etched onto Lowell's granite mausoleum located next to one of his domed telescopes on Mars Hill, the hermit sealed away in final retreat from the world forever.[7] Painting himself as a sort of Nature mystic who ventures out in the early hours 'with the snow feet deep upon the ground and the frosty stars for mute companionship' to hear the lonesome coyote howl in the distance as he makes his observations, Lowell portrays Flagstaff as a 'portal to communion with another world'.[8] But, as he admits, 'What the voyager finds himself envisaging shares ... in the expansion of the sense that brought him there.'[9] This almost seems tantamount to admitting that he was seeing things. Even up in the clear, dry air of Flagstaff, Lowell will not have been entirely able to escape the distorting vapour of the Earth's atmosphere; staring for hours through his telescope, waiting for a moment of still air before retreating to his drawing pad to sketch out what he had just seen, there was still room for the imagination to work its magic. Lowell sounds almost as if he was crystal-gazing or scrying, looking for visions and spirits in a mirror – and, eventually, finding them. Near the end of *Mars and Its Canals*, Lowell says this about his quest:

> The less that life [on Mars] proves a counterpart of our Earthly state of things, the more it fires fancy and piques inquiry as to what it be. We all have felt this impulse in our childhood as our ancestors did before us, when they conjured goblins and spirits from the vasty void, and if our energy continues we never cease to feel its force through life. We but exchange, as our years increase, the romance of fiction for the more thrilling romance of fact. As we grow older we demand reality, but ... the stranger the realisation [of this] the better we are pleased.[10]

Something tells me that, when he was hunting Martians out in Arizona, part of Lowell was really still busily occupied in hunting goblins – out in the vasty void of space itself. Curiously, Schiaparelli had also been interested in the topic of psychical research.[11] So were quite a few other astronomers.

## Mental Elf Issues

You might think that such Romantic attitudes would not exist amongst astronomers, physicists and space scientists, who are stereotyped as being inherently logical and rational men, but this is not so. Many astronomers down the years have been a little strange, such as the American Thomas Jefferson Jackson See (1866–1962), who honestly felt that he was immortal – a belief he maintained until the very day he died.[12] I gleaned that valuable nugget from Patrick Moore, upon whose 1972 book *Do You Speak Venusian?* I have drawn in writing this present volume. Moore intended his book as a tribute to what he called 'Independent Thinkers', those persons most would label 'eccentrics'. Moore valued such folk as necessary to the future intellectual development of mankind. Realising that most of the fringe astronomers he interviewed for his book were incorrect in their assertions about Flat Earths and cold suns, Moore nonetheless also knew that the ideas of men like Galileo and Kepler were thought of as being loony by large numbers of people during their own day too. Every new and radical theory had to have been invented by *someone*, and that person would, by definition, have been an Independent Thinker. The continuance of such a tradition was thus of utmost importance to Moore and, whilst his book was basically comic, he treated his subjects with some respect amidst all the ribbing, recognising they were motivated more by the idea of discovery than by profit or fame. True zetetics, they refused to blindly accept the standard worldview of society, and set out to discover new truths about Creation – just like Isaac Newton, but wrong.[13]

It is a pleasing fact that the very word 'eccentric' is derived from astronomy, being used from 1551 onwards to describe the way a heavenly body deviates from a perfect circle in its orbit.[14] Such eccentricities amongst those involved with the cutting edge of science are ably illustrated by a 1954 book called *The Fifty-Minute Hour* by the American psychoanalyst Robert Lindner (1914–56). Lindner was called out to Los Alamos National Laboratory, the centre of US atomic research, to examine a young physicist given the pseudonym 'Kirk Allen'. Allen was a brilliant man, but suffering from a real problem – he thought he could travel into the future. Whenever he felt like it, said Allen, he could mentally advance himself thousands of years through time and inhabit the body of a heroic spaceship pilot, who flew around having wild adventures. This alternative hallucinatory life was so enjoyable to Allen that it was interfering with his work; he spent all his time writing some 12,000 pages about his escapades in outer space, detailing the biology, geology and politics of other planets, christened with stupid names like 'Srom Norba X' and 'Srom Sodrat II'. Lindner indulged Allen in his fantasies, until he almost seemed to begin believing in their reality himself. At this point Allen, worried about the sanity of his own psychiatrist, confessed that he had made the whole thing up to compensate for his embarrassing inability to talk to women.[15]

At some level Allen knew his experiences were not real. I am not entirely sure the same can be said for the American astronomer George Ellery Hale (1868–1938). Besides being the first person to realise that sunspots were really solar magnetic storms, Hale was one of the founders of the highly respected California Institute of Technology (CalTech), and the man who set in motion construction of the world-famous giant telescope which sits atop that State's Mount Palomar. He also claimed to be in direct contact with an elf. During a visit to Egypt, with whose ancient sun-god, Ra, he was obsessed, Hale claimed that, whilst sitting alone in his room one night, a 'little man' suddenly appeared from nowhere and began giving him advice about the conduct of his life. From hereon in, this elf kept on showing up, his presence announced by a strange ringing noise in Hale's ears. Hale sought medical advice, looking for an answer, but was unable to find one.[16]

## Rocket-Men

The early rocket engineer and aeronautics professor Theodore von Kármán (1881–1963) was just as odd, claiming to be the descendent of a man called Rabbi Judah Loew ben Bazalel (*c.* 1512–1609). Loew was a real figure, a Jew living in Prague's ghetto during the sixteenth century, who is supposed to have constructed a magical creature called a *golem*, a figure made from lifeless clay and imbued with animation through the use of magical charms. According to fable, the creature was said to have acted as a guardian for Prague's Jews down the centuries, even into the Second World War, when it supposedly prevented some German soldiers from destroying a synagogue.[17] Rather than treating this as a myth, von Kármán often boasted of his ancestor's legendary achievement to his colleagues! Whilst Hungarian-born, von Kármán was director of an offshoot of George Ellery Hale's CalTech called GALCIT, the precursor of NASA's famous Jet Propulsion Laboratory (JPL) in California, which is still going today.[18] Here, von Kármán was happy to tolerate eccentricity amongst his colleagues in the name of creativity, which was why visitors to CalTech's three-acre rocketry research test site hidden away on the remote Arroyo Seco outside Pasadena might have been greeted by the sight of a student named Apollo M. O. Smith (1911–97) walking around everywhere wearing a large pith helmet with a small electric fan stuck on top to keep him cool in the desert sun.[19] Just so long as Smith did his work well, then why shouldn't he dress like a nut if he wanted to?

Another of von Kármán's employees was a wild young occultist named Jack Parsons (1914–52), called 'a delightful screwball' by his boss, who 'loved to recite pagan poetry to the sky while stamping his feet', and used to say a prayer to the Greek fertility god Pan before each rocket test. Both a disciple of the British practitioner of 'sex-magick' Aleister Crowley (1875–1947) and a chemist of brilliance, Parsons was responsible for conjuring up the Devil at the tender age of thirteen, and for pioneering research into the qualities of liquid and solid fuels, his experiments sometimes being compared to a form of alchemy. Parsons was also interested in the supernatural creation of life, entering into a magical ritual he termed the 'Babalon Working' in January 1946.[20] This was an attempt to create an artificial child through acts of magical masturbation and ritual sex, which supposedly resulted in an outbreak of disturbing poltergeist phenomena.[21] Following the Babalon Working, a new ritual lover named Marjorie Cameron (1922–95) came on the scene, whom Parsons twice impregnated – although, since he allowed her to abort the infant on each occasion, perhaps he didn't want to conjure up a *literal* child. Instead, maybe he simply wanted some new adult sex partner like Cameron to appear, imbued with certain occult properties he hoped to draw down from the macrocosm – he viewed the untamed, red-headed Cameron more as an elemental spirit than an ordinary woman.[22] However, one of Parsons' other occult activities was that of 'impregnating statuettes with a vital force by invocation' before selling them,[23] and when he was killed in a huge explosion at his California home on 17 June 1952, there was genuine speculation that the blast was the result of Parsons' attempts to create a homunculus (a kind of miniature version of von Kármán's *golem*), so maybe he did want to make artificial life of some sort after all.[24] Realistically, his death was probably due to Parsons' trials in rocketry, not alchemy; he had been performing experiments with fulminate of mercury, a powerful but sensitive detonator, trials which he had foolishly continued within the privacy of his own home.[25] Pleasingly, after turning his house into a crater, in 1972 a 'Parsons

Crater' was named after him on the moon – a place to which we may never have been able to fly without him.[26]

California during Parsons' working lifetime was not just the centre of American rocketry research, but also of the dawning New Age philosophy which was later to reach full bloom during the 1960s. Spiritual gurus, drug-taking psychonauts, occultists, outsiders, rebels, freethinkers, sci-fi writers, proponents of free love and unusual ways of life – all those kooks who would eventually come to be labelled hippies and beatniks – gathered around Los Angeles, Hollywood and other nearby bohemian outposts at precisely the same time that GALCIT was starting to take off, meaning there were two separate groups of people in California at the time seeking to make contact with new worlds. The influence of each rubbed off on the other. Certainly, Parsons himself had many unusual acquaintances, and embraced the newly available alternative lifestyle to its full. Leasing a large old mansion in Pasadena in 1942, Parsons placed adverts in local newspapers renting out his spare rooms, saying firmly that only nutters need apply; specifically, he asked for atheists, anarchists and artists, with anyone square enough to have a proper job not being welcome. He ended up providing sanctuary to a silent-movie organist, an opera singer, several astrologers and a variety of young women who liked to dance around coffins in see-through dresses when it was time to party. One evening in 1942 the police were called out by shocked neighbours who claimed to have seen a naked pregnant woman leaping nine times through a bonfire in Parsons' back garden in some weird ritual. Parsons easily managed to convince the officers that, as a renowned rocket scientist, he would never have stooped so low as to have had dealings with any such filth. In fact, the truth was quite the reverse![27] Those who wish to know more about Parsons should try John Carter's excellent biography, *Sex and Rockets*.

## Sun, Sand and Saucers

The presence of the desert was key to the dawning of this alternative Californian society. Practically speaking, it was advantageous to conduct both rocket research and astronomy within desert areas due to the clear, dry air and the always-present possibility of large explosions. And yet, deserts also have a long tradition of mystics and hermits who wander out into them seeking visions, from St Anthony of Egypt to St Lowell of Flagstaff. To that list of sand-dwelling eremites, we can also surely add Jack Parsons. Whilst out in California's own Mojave Desert, Parsons had committed yet more acts of magical masturbation, seeking an encounter with a supernatural entity he termed 'Babalon'. 'The Goddess came upon me', Parsons wrote, possibly at the same time as he came upon her, instructing him via automatic writing to create a magical 'moon child' with his red-haired lover Marjorie Cameron (several of whose relatives, incidentally, worked at JPL – the connections between scientists and New Agers in California being seemingly never-ending).[28] Parsons was commanded by Babalon not to tell Cameron she was the one he sought at first. Instead, he was to wait until she was given a sign from heaven, confirming that she really was the right scarlet woman he needed. This sign quickly came; Cameron saw a large, silver, cigar-shaped UFO hovering over the Californian landscape, marking her out as The One.[29]

Parsons also claimed to have had his own UFO-related encounter out in the Mojave Desert in 1946, during which he had supposedly met a Venusian.[30] A more famous meeting with a man from Venus out in the Mojave is meant to have occurred on 20 November 1952 to a sixty-two-year-old local Polish-American mystic named George Adamski (1891–1965). Adamski was the first prominent example of what came to be known as a 'Contactee' – a human being who had made friendly contact with benign persons claiming to be from other planets. The Contactee movement was primarily an American phenomenon, at least initially, and the aliens themselves much different from the nefarious dwarfish grey abductors we so often read about today. Instead, most, known by the generic soubriquet of 'Space-Brothers', appeared to be modelled upon lost members of ABBA, having as they did long blonde hair, blue eyes, and a generally Scandinavian appearance about them. These 'Nordics', as they are sometimes called, also tended to be telepathic, somewhat androgynous in physique, vegetarian, and, if you read between the lines, borderline Communists, or at least peaceful and tolerant social democrats taking after the post-war Swedish liberal model of society. Such beings are now rarely reported, though throughout the 1950s and '60s it sometimes seemed as if you couldn't move for Space-Brothers, particularly in California.[31] In recent years, speculation has grown online that these Nordics may have been space Nazis from Aldebaran visiting Earth to see how Hitler & Co were doing down in Antarctica, but if so then they don't seem to have had much love for far-right politics in the flesh.[32] Adamski's own Swedish Space-Brother was called Orthon (the resemblance of such names to synthetic fabrics has been noted[33]), and the tale of their meeting in the Mojave was first properly revealed in an international best-seller of 1953 called *Flying Saucers Have Landed*. The book was mainly written by an eccentric Anglo-Irish aristocrat named Desmond Leslie (1921–2001), but its real selling points were the final section telling Adamski's tale of meeting an alien, and the reproductions of several photographs supposedly taken by Adamski of Orthon's flying saucer, or 'Scoutship'.[34] The photos appear no more than charming fakes, but, aesthetically speaking, Orthon's Scoutship looked like something which really *ought* to exist even if it didn't. Somehow it looked right at home in the brave new 1950s world of post-war California, resembling as it did a kind of accidental consumer product of obscure function. Adamski's invention had real power, and was here to stay. As one of Adamski's biographers has put it:

> As an early consumer-object, the flying saucer was perfect; the Adamski version in particular looked like something which had to be bought. It was round, shiny, and looked like a piece off a dream-Cadillac, simply dripping with chrome. It had a wonderful message for all time: that, to the horror of the dull soul, the world refuses to be disenchanted.[35]

Orthon himself also proved to be a highly appealing figure, with lookalikes tricking credulous female devotees of Adamski into letting them inside their houses where they could then 'cross the energies' between Earth and Venus, a process which unsurprisingly involved them having sex. Adamski, too, was most enamoured of Orthon and his kind, having as he did a taste for long-haired adolescents of an androgynous bent, and allegedly developed the occasional habit of sharing out-of-town hotel rooms with such 'aliens' at an hourly rate. The best place to hook up with free and friendly Scandinavian Space-Brothers locally turned out to be the sandy wilds around a place called Desert Centre, although these particular ETs probably catered for those who preferred their companions to be a little more experienced, seeing as Orthon may well have been up to a thousand years old; though, as with so many people in the world capital for plastic surgery, he didn't really look it.[36]

## The Old Man of the Mountain

Adamski was a classic California character. A former soldier, he became a wandering preacher, lecturing on astronomy and metaphysics to bored farmers in a pre-TV age, before settling at California's Laguna Beach in the late 1920s. By 1934 he had established his own small occult group, The Royal Order of Tibet, and secured a spot preaching on local radio, styling himself 'The Professor'. He also developed a self-help course, *Telepathy: The Cosmic or Universal Language*, which he sold as a series of twelve mail-order lessons, guaranteed to turn his students psychic.[37] The reputed true purpose of his Royal Order, however, was allegedly revealed by Adamski some years later:

> It was a front. You know, we were supposed to have the religious ceremonies; we make the wine for them and the authorities can't interfere with our religion. Hell, I made enough wine for half of southern California. In fact, boys, I was the biggest bootlegger around. If it hadn't been for … [the end of Prohibition] I wouldn't have [had] to get into all this saucer crap.[38]

In 1940 Adamski set up a communal ranch where he could continue preaching mystical nothings, before buying land on the slopes of Mount Palomar in 1944 where, under the shadow of George Ellery Hale's giant observatory, his friend Alice K. Wells set up a burger-stand-cum-restaurant called the Palomar Gardens Café, at which he could frequently be seen flipping burgers and helping out (accounts differ as to whether or not this was because he was an actual employee). The road branched off just outside the café, leading up to Mount Palomar Observatory itself, and Adamski quickly latched onto the opportunity by building his own comparatively crude telescope nearby, through which he was claiming to see saucers as soon as they became popular, telling a meeting in 1949 that he had once seen as many as 184 spaceships in a single sighting; it was a wonder they didn't all crash into one another. Many visitors to Mount Palomar presumed Adamski was officially associated with the Observatory, and came away confused that the 'Chief Astronomer' appeared to be a burger-flipping Polish mystic who said he talked to aliens and tried to sell them discounted sausages. Handily, the teachings of the Space-Brothers turned out to be highly compatible with Adamski's own earlier creed – some may say they were exactly the same thing.[39] Adamski claimed numerous further meetings with Orthon, being treated to trips to the moon and other planets, yarns later found to bear uncanny similarities with a sci-fi novel Adamski had previously written, *Pioneers of Space*.[40] One of Adamski's most risible pieces of 'proof' that all this was really happening to him was an unused train ticket; he said that he *would* have used the ticket, but he was kindly given a lift to his intended destination in a UFO instead.[41]

Let us not be ungenerous, however. It has been suggested that at least some of Adamski's experiences may have seemed real to him at the time. Maybe he really did 'talk' to Orthon somehow, in the same sense that George Ellery Hale had once apparently 'talked' with his imaginary elf? It is certainly true that other people believed his claims and got something out of their association with him. One Adamski disciple watching him on TV said she saw him shape-shift into Jesus, for example.[42] There is also the apparent fact that, during his initial meeting with Orthon, Adamski was watched from afar by a whole carful of followers, all of whom said they saw Orthon's spaceship, and one of whom says they saw Orthon himself, through binoculars.[43] Whilst some dispute this version,[44] there can be no doubt that, through the sheer force of his personality and subsequent global fame, Adamski really *did* make people see flying saucers. They may not actually have been there, but the feat still stands. In my view, considering the size of the craze he started, George Adamski deserves to be remembered as one of the most influential writers of fiction of all time. His titles may never have won a Booker, but neither has much else worth reading.

## Bad Day at Giant Rock

Once Adamski had provided people with the basic imaginative template to follow, various other local mystics and fraudsters quickly began to improvise desert dramas all of their own. 'There is a cosmic spell over the desert most of the time' one of the California Contactees once said, and he was right – it had been cast back in 1952, not by the wizard Jack Parsons, but by George Adamski.[45] Orthon need never get lonely when wandering out in the Mojave, not when other such highly developed space beings as Ashtar, Jon-Al, Acta, Baruch, Ermon and A-lan were known to be whizzing around the place too, with their Contactee chums following along merrily behind after them. With so many nutcases on the run nearby, what was needed was a cunning scheme to gather them all together in one place in the name of public safety. This was provided by the other most significant Contactee of the day, George Van Tassel (1910–78).

Van Tassel was a qualified pilot, aeroplane engineer and inspector with Lockheed, who had become enamoured with the life of a local eccentric named Frank Critzer (d.1942), a naturalised German who had been killed in 1942 by trigger-happy policemen who had driven out to his remote home in the Mojave Desert to investigate rumours he might have been a Nazi spy. He wasn't, of course, he was simply a prospector who had retired to a quiet desert area near the Yucca Valley which was known as Giant Rock because … well, guess. Beneath this giant boulder, Critzer had hollowed out a weird underground bunker of the kind some unholy cross between Fred Flintstone and Josef Fritzl might have liked to occupy, and into which the foolish cops hurled some tear gas canisters to smoke out the Nazi. Seeing as Critzer had stored lots of dynamite in his cave, this proved to be a gross tactical error, with the explosion from the gas canister triggering a further blast from the dynamite, and leaving poor Frank Critzer in a state of becoming his own wallpaper. Sick of his job, Van Tassel leased the Giant Rock property in 1947, wiped all the bloodstains away, and set up home, with a dude ranch, airstrip and café attached to allow him a small livelihood. Something enabling Van Tassel to make a rather larger living then occurred one night in 1952 when George, sprawled out on the desert sands meditating, suddenly found his astral body flying towards a giant spaceship orbiting above, where he encountered a band of Space-Brothers called The Council of Seven Lights, who gave him the usual New Age patter about the need for peace, love and understanding. In August 1953, these very same beings returned, giving Van Tassel special instructions for constructing a magical building with the power to vastly extend the human lifespan called The Integratron. This turned out to be a big, dome-shaped structure, painted white, which looked very like Mount Palomar Observatory but without the telescope. To help fund its creation, from 1953 to 1978 Van Tassel held his famous open-air Giant Rock Space-Craft Conventions, which sometimes attracted as many as

10,000 visitors, not to mention the country's most famous Contactees. Here, Van Tassel and others would stand up on stage and tell their tales, sometimes going into trances and channelling down the voices of aliens hovering just out of sight overhead, saying things like 'I am Knut. I bring you love!' It was all rather like a bizarre form of early stand-up where the real jokes were the performers themselves.

As for Giant Rock's host and compere, Van Tassel died of a heart attack in February 1978, with his Integratron apparently unfinished, alas. A Space-Brother named Lo kindly sent down an epitaph for his gravestone via telepathic means – 'Birth Through Induction, Death Through Short Circuit'. Without Giant Rock, and the network of contacts between Contactees that it formed, the movement might never have been such a cult success, so Van Tassel deserves credit. However, the place wasn't just a freakshow. There is no doubt that many attendees were true believers rather than just gawkers after a cheap laugh, with various accounts of guests undergoing apparently mystical experiences beneath the night sky, or seeing saucers out there for themselves. The landscape around Giant Rock had been sacred to the Native Indians and a centre of desert shamanism, once upon a time. For a brief spell throughout the 1950s and '60s, it became sacred once more.[46]

# 8

# Desert Warfare: The Contactees Conquer the World

If the whole landscape of the southern Californian desert was in a sense alien, then the local mindscape was often equally so. Where else but California could the peculiar twentieth-century style of architecture known as 'googie' ever have sprung up? In 1949, the architect John Lautner (1911–94), inspired by the idea of the forthcoming Space Age, designed a rather kitsch-looking café called 'Googie's Coffee Shop' which looked as if it had landed from another planet. This then proved the launchpad for further architectural rockets against good taste. Most famous was the 'Satellite Shopland', a shopping mall with a giant spiky neon glowing model of *Sputnik* perched on top of a large vertical signpost. Another was Lautner's own 1960 'Chemosphere', a strangely shaped house supported by a single concrete column and surrounded by foliage nestled almost impossibly amidst the Hollywood Hills. Seeming to float in mid-air, being somewhat round in shape but with straight edges, like a 50p coin, and having floor-to-ceiling windows in a continuous band all around it, it could very easily be mistaken by the unwary for a flying saucer, especially when lit up at night.[1] Perhaps the Integratron should also go down as an example of this kind of architecture, as yet unacknowledged?

Today, California is also home to Silicon Valley, many of whose richest tech geniuses still dream of exploring and exploiting the wonders of outer space. The most prominent is the South Africa-born PayPal founder Elon Musk (b. 1971), whose love of funding implausible schemes is such that he is often rumoured to be the model for the billionaire inventor Tony Stark in the Hollywood *Iron Man* superhero franchise (though seeing as the initial *Iron Man* comic books were going long before PayPal, this cannot originally have been the case). Musk helps fund an organisation called SpaceX, a privately owned, commercial version of NASA, whose major aim is to set up a one million-strong human colony on Mars by the end of the century. Musk, like many of his peers, is a big fan of sci-fi and fantasy. In 2016, for instance, he announced his considered opinion that the world we live in is actually a gigantic videogame created by some advanced civilisation. Given that pretty soon, in Musk's view, we are going to have V-R games that are 'virtually indistinguishable from reality', it would 'seem to follow that the odds that we're in base-reality is one in millions'.[2] Perhaps this apparent difficulty Musk has in distinguishing reality from fantasy is why, when needing to get some space suits made prior to his planned mission to Mars, the real-life Tony Stark approached not NASA but a Hollywood costume designer. Jose Fernandez had never made a space suit before, but he did design the spandex costumes worn by the assorted mutants, super-villains and freaks seen in the recent *Thor*, *X-Men* and *Batman* franchises, and if they were good enough for the Caped Crusader, then they should also be good enough for astronauts. Fernandez's design brief was to create something 'stylish', 'iconic', 'heroic' and 'bad-ass' yet also 'practical', with the suits becoming something like the interstellar equivalent of a tuxedo.[3] Those who wish to don such stylish garb, says Musk, will have to pay him $200,000 for a ticket on what he calls a '*Battlestar Galactica*-style' fleet of Mars-bound craft, the flagship of which will be called *Heart of Gold*, after a shoe-shaped spaceship from Douglas Adams' (1952–2001) comic creation *The Hitchhiker's Guide to the Galaxy*, whose engine is powered by a nice cup of tea. Musk's own space fleet will probably be powered by something stronger, with plans afoot to launch a new range of BFRs (Big Fucking Rockets) and unmanned robotic landing capsules named after Puff the Magic Dragon up towards Mars as soon as 2018. Those who sign up to be passengers on Musk's planned 200-seater rockets, meanwhile, would first have to answer the following profound question: 'Are you prepared to die?' If so, then welcome aboard! 'If you're going to choose where to die, then Mars is not a bad choice', explains Musk, because employment prospects on arrival would be excellent. 'Mars would have a labour shortage for a long time, so jobs would not be in short supply', he says, particularly if you know how to dig canals.[4]

Meanwhile, Jeff Bezos (b. 1964), the founder of Amazon, has proposed the creation of floating industrial units in outer space. Because in Bezos' view Earth is 'the best planet', he doesn't think we should be polluting it so much. Instead, it would be far more efficient to create gigantic super-factories which orbit the sun, absorbing its rays as a source of clean, sustainable energy. We should still try and settle Mars 'because it's cool', Bezos told a California conference in 2016, but the real future lies in 'building gigantic [micro-]chip factories in space'.[5] Billionaires based or born in the south-west desert regions of the USA don't have to have made their fortune specifically in tech to engage in such fantastic schemes, though, as can be shown by the career of Robert Bigelow (b. 1945), who grew rich in the hotel trade but has since used his billions to fund research into both the paranormal and space exploration. His Bigelow Aerospace organisation is based on the outskirts of Los Angeles – you take a right-hand turning on Skywalker Way and then cruise along Warp Drive until you come to the place filled with inflatable houses destined to grace the moon. To the untrained eye these houses may look like uninflated balloons, but in fact they are Bigelow Expandable Activity Modules (BEAMs) which the hotel magnate hopes to have attached to some of SpaceX's rockets before seeing them blasted off to the moon. Once there, the flat balloons, which do not take up much space, will be filled with compressed air and expand to room size, creating starter homes for astronauts. Bigelow's inspiration was a tale his grandparents used to tell him about having once seen a glowing ball of fire speeding down towards the Nevada Desert, something which convinced him of the existence of alien life. Once he had made his fortune, he then set out to actually find it.[6]

## California Gold Rush

As the success of tech firms like Google and Twitter has demonstrated, what begins in California can very quickly go on to colonise the whole world, and so it was with the Space-Brothers, word of whom soon spread far and wide. In 1967, New York was graced by visits from a six-foot-tall black woman with strange glassy eyes and covered all over in feathers who claimed to be one Princess Moon-Owl, hailing from the planetoid Ceres. She wheezed and gasped constantly, stank of rotten eggs and claimed to be 'seven Oongots' old – 350 years to you and me. Nonetheless, Princess Moon-Owl still managed to give at least one long radio interview, and may or may not have actually been performing some kind of PR stunt on behalf of a UFO convention being held in the city that year.[7] As this implies, the Contactees fast became minor fringe media celebrities, with the New York-based radio talk-show host Long John Nebel (1911–78) finding in such folk a novel angle to structure his late-night programmes around to gain listeners.[8] The Contactees were following the usual path of twentieth-century American entertainment products; invented in or near Los Angeles or Hollywood, California, before being disseminated outwards through the mass media. One of the earliest UFO books was 1950's *Behind the Flying Saucers*, which was penned not, as you might expect, by an investigative journalist, but by a Hollywood gossip columnist named Frank Scully (1892–1964) who, unlike his later *X-Files* namesake, seemed to just *want* to believe. Scully was best known as a humorist, with a weekly comical column in the Hollywood showbiz paper *Variety*. The book itself was the first in a long line to peddle the myth that the US military had retrieved alien corpses from a series of crashed saucers out in the south-west deserts (but not at Roswell, interestingly), claims announced by Scully in the pages of *Variety* in October and November 1949. Naturally, the text turned out to be based upon other people's lies, but it sold well in the first instance, and perhaps that's all that really mattered – the book provided the first solid proof that saucers could be big box office.[9]

If some Contactees thought they were starring in their own film, however, then the vast majority turned out more like cheap B-movies than expensive blockbusters. Consider the tale of Woodrow Derenberger (1916–90), a West Virginia sewing-machine salesman who in 1966 claimed to have met a telepathic spaceman named Indrid Cold. According to Derenberger, Cold hailed from a place called Lanulos, which is not in Wales, but a planet somewhere 'near the Ganymede star-cluster' (Ganymede is in fact a moon of Jupiter). In 1971, Derenberger wrote a book, *Visitors from Lanulos*, in which he told about his trips to the planet, where everyone was a nudist except himself; unlike the well-sculpted aliens, Woody was embarrassed to reveal his flab. Amongst subsequent wonders, Derenberger was rumoured to have become pregnant despite being male, and to have received a vial of space medicine which cured him of stomach ache. Once, said Derenberger, he was kept prisoner at Cape Canaveral whilst the head of NASA grilled him for hours on end about the Space-Brothers. Eventually, the bigwig snapped and told Woody that NASA knew all about Indrid Cold already, but were keeping it from the public for fear there would be an outbreak of mass panic in which 'women would commit suicide [and] throw babies out the window'. Derenberger's main aim appears to have been to try and find fame and fortune, which in a small sense he did. A gullible yet wealthy individual from Massachusetts ended up supporting Woody financially, in the hope of one day being taken up to the nudist wonder-planet himself, for a good look around. Derenberger maintained this dubious relationship right into the 1980s by virtue of forging a series of letters supposedly penned by Cold and his wife Kimi, stringing his dupe along.[10]

An alternative tactic to gain easy cash was to move to California yourself, and con the New Age natives *in situ*. The most egregious example was that of Reinhold Schmidt (1897–1974), a convicted embezzler who relocated to Bakersfield, California, following an alleged 1957 meeting with some German-speaking Space-Brothers from Saturn who flew a propeller-driven cigar ship within which they kept a V-W Beetle for nipping into town *incognito* to top up on food supplies at the local store. After reporting this 'experience' to his local police force, Schmidt was banged up in a mental institution, from which he was sadly later released. As evidence for his claims, Schmidt pointed towards certain oil deposits which had been left in a field by the spaceship. As evidence he was lying, the police pointed towards a half-empty oil can lying in the boot of his car. Clearly, a less questioning audience was needed for Schmidt's claims, and so he moved to Bakersfield. Here, besides selling his book *My Contact with the Space-People*, Schmidt began peddling the lie that, due to his alien contacts, he had discovered deposits of quartz crystals which could cure cancer. If only he had enough money to mine them, he could rid the world of this terrible disease forever – but not before ridding trusting pensioners of $30,000 of their life savings first.[11] Money certainly seems to have been easy to come by for some of these folk. George Van Tassel openly solicited donations from the region's little old ladies to help him with his doings out at Giant Rock – very successfully, so it seems. According to one UFO investigator who travelled there to meet George, some mail had been lining up at Giant Rock for a little while, unopened; once unsealed, these envelopes contained no less than $18,000 in banknotes and cheques.[12] There really was gold in them thar hills!

## The Hair of the Space Dog

My own personal favourite victim of the spreading Californian mindset was Buck Nelson (1895–1982), an apparently retarded farmer from Missouri who claimed to have had a close encounter near the Ozark Mountains in 1954, and who later set up an annual 'Space-Craft Convention' on his farm. You can find pictures of him, standing in his yard clad in dungarees and holding up a hand-stencilled sign advertising this event, in which the 'S' in 'SPACE CRAFT' has been drawn in backwards, giving some indication of Nelson's low mental status.[13] He claimed that his alien pals actually told him to dress and act like this all the time, 'because people can recognise me easier that way'.[14] Nelson made extra money by selling two things. Firstly, there was his plainly titled 1956 pamphlet, *My Trip to Mars, the Moon and Venus*. Secondly, there were envelopes containing what purported to be hairs from an alien dog named Big Bo that Nelson had met upon his travels, but which were clearly just hairs from his farm dog, Ted.[15] By Nelson's own admission 'Analysis has proved it to be hair from a male dog', which he proudly took as being conformation of his story, although a moment's thought must demonstrate that the situation was in fact quite the reverse.[16]

First contact with the Space-Brothers was made on 30 July 1954 when, alerted by the malfunctioning of his radio, Nelson went outside only to see 'a huge big disc-like object' in the sky. Nelson signalled to it with his flashlight, but it shot a ray beam 'brighter and hotter than the sun' down at him, which cured his lumbago.[17] A year later, on 5 March 1955, the saucer returned and out came three men and their space dog, Bo. One of the men was a nineteen-year-old cousin of Nelson's called Bucky who had gone to live in space some years earlier, but the other two were Venusians, Captain Bob Solomon, who was 200 years old but still fresh-skinned, and an old and wrinkly fellow whose name was not divulged. Big Bo, being a space dog, stood up on his hind legs and shook hands with Nelson in greeting. Then, the aliens gave Nelson valuable information about how to keep dust from under his bed because, 'as a bachelor', his house was rather dirty.[18] Nelson's pants must have been somewhat soiled from lack of female attention too, as, upon a subsequent visit on 24 April, the Venusians told him he could come aboard their saucer, just so long as he changed his clothes first. Nelson took Ted the Earth dog with him and, like Moses, was given a list of Holy Commandments. There were Twelve Commandments on Venus, not Ten, but most were close to the usual Earthly ones, albeit with a little more detail. Space-Commandment Number Two, for example read 'Thou shalt not kill (includes accidents and war)'. The brand-new Space-Commandment Number Eleven, meanwhile, read 'Do not eat or drink anything which is not food', which is actually good advice.[19] This information having been absorbed by Nelson, the spacemen let him have a go at flying their ship. Not knowing what to do, Nelson flipped the thing upside-down, but they were all wearing seatbelts so no harm ensued.[20]

Eventually the saucer arrived at Mars, where Nelson took in the sight of Lowell's non-existent canals and noted approvingly that racial segregation still took place there.[21] Then Nelson flew to the moon, where he met some lunar children who rode on the back of the space dog Big Bo 'like a pony'.[22] Then he stopped off at Venus, where he saw flying cars and a machine which turned books instantly into films.[23] Later, Nelson took questions about his incredible experiences. Asked what the aliens looked like, Nelson observed that they all looked just like ordinary country folk from the Ozark Mountains did, except that 'as far as I could see they had nice teeth'. And what did the spacemen eat? Nelson noted that he had been given some meat on Venus, 'or at least [it] looked and tasted like meat', so he supposed they ate that. And how did they dress? Why, in dungarees just like he did.[24] Did the Space-Brothers have any special advice for mankind? They advised us not to worship gold. Instead, we should dig all gold up out of the ground and just leave it lying about in our homes 'because the vibrations of gold are good for the baby', leading to a lifetime of good health.[25] How did their flying saucer work? Nelson was unsure, but drew a helpful diagram of what was inside it. It included such highly technical labels as 'Table', 'Toilet' and 'ENTIRELY EMPTY SPACE'.[26] A diagram of the inside of Buck Nelson's own head may well have shown something similar.

## Crashing and Burning

Or perhaps the interior of Nelson's skull might instead have shown a small industrial kite-mark reading 'Made In California', for so his story surely was. When it comes to UFOs, as in so many other aspects of popular culture, we're all Americans now. One of the most interesting recent commentators on UFOs was an English playwright named Colin Bennett (1946–2014), who had much to say about the Contactee movement. Bennett devised the original notion that the ever-evolving tales of alleged desert saucer-crashes at places like Roswell and Aztec, both in New Mexico, in 1947 and 1948 respectively, constituted a new quasi-literary genre of interactive multi-media myth-making which, in as much as it made numerous people believe in its literal reality and act accordingly, had the power in a certain sense to alter reality itself, or at least our perceptions of it. Go to Aztec these days, for instance, and, whilst no alien spacecraft ever really crashed there, you will enter a landscape, both urban and natural, which has been profoundly altered by the mere *idea* that one did. Stores such as the 'Aztec UFO Information Centre and Gift Shop' do not spring up in towns where spaceships are never said to have landed. Likewise, a mountain bike trail called 'The Alien Run', which circles around the alleged crash site at nearby Hart Canyon, forces an imaginative layer down over the local landscape every bit as much as a network of shrines at a religious holy site like Lourdes may do to the natural landscape there.[27]

In his 1999 essay *The Pixels of Roswell*, Bennett describes photos of badly copied old military telexes with missing words and phrases apparently talking of events at Roswell that have been reinterpreted time and time again by various ufologists, each of whom comes up with a different answer about what they might say. In light of this, says Bennett, 'our idea of what constitutes a "text" must be revised'. As he says, such texts may 'not be much of a read' in and of themselves, 'but the fun comes when we try to fill the gaps'.[28] For example, Bennett cites the US Army Colonel Philip J. Corso (1915–98), author of 1997's *The Day After Roswell*, who speculated that, when all those crashed saucers were supposedly found in the deserts of south-west America, the aliens might have been playing some elaborate trick on us. The alien corpses we are told were hauled from the wreckage might have been crude decoys, with the real extraterrestrials lurking within the crafts' actual circuitry, which was then back-engineered by military scientists. Corso suggests that the modern-day computer, derived from such alien tech, is a kind of 'silicon-based life form in itself', and by manufacturing more and more of them, we are allowing the ETs to colonise us by stealth – an idea which, in these days of smartphones and always-on Internet connections, now seems like an eerily prescient metaphor being presented as fact.[29] In an old-fashioned piece of literary detective fiction, the reader simply gets to try and work out the answer to a pre-defined question – which of the characters performed the murder? In Bennett's new 'pan-dimensional' UFO-related texts, the reader *becomes* the detective and begins asking their own brand-new and inventive questions about the world, just like Corso did. Even better, unlike standard crime novels, the investigation into fictional pseudo-events like Roswell will presumably never end – unless one day someone produces the real-life crashed saucer, which they can't, because there isn't one. These ideas and pseudo-facts, thought of by their proponents as being *actual* facts, then engage in a kind of mental warfare out in the world at large, a sort of 'battle of the memes'. Bennett's opinion is that all such memes, like Corso's silicon-chip-based computer aliens, represent a hitherto unrecognised type of non-organic, purely mental life form, engaged in a process of cultural warfare, with victory achieved via possession of some chunk of the collective mind of humanity. When Buck Nelson, far away in the Ozark Mountains, began babbling on about meeting his own men from Venus, for example, his mind had been partially colonised by the likes of Orthon. This raised an interesting question for Bennett; if memes were psychic life forms, and beings like Orthon acted as memes, then ... might the Space-Brothers in some sense have been *real*? In a 2009 interview with the investigator Nick Redfern (b. 1964) for his useful book *Contactees*, Bennett implied this might indeed be the case:

> Many Orthons have appeared throughout history ... Their sole function is to sow seeds in the head ... These seeds act on the imagination, which replicates and amplifies whatever story-technology [media] is around at the time. People such as Adamski and the rest of the Contactees were, and still are, like psychic lightning-rods for certain brands of information ... The 'space-folk' are sculptured by wars between rival viral memes competing for prime-time [i.e. mainstream] belief. It may be that, as an independent form of non-organic life, memes as active viral information can display an Orthon entity at the drop of a hat ... Over a half-century later, we can no more erase the legendary Contactees from our heads than we can erase Elvis Presley or Marilyn Monroe. Once induced by mere transient suggestion, these powerful images become permanent fast-breeders, turning out scripts and performances in all our heads – even as we sleep.[30]

According to Bennett, aliens might by now have shed their physical bodies and evolved into 'mere part-disembodied tissues of information', with the entire UFO phenomenon being 'a species of live viral information, using all the tricks of metaphor screens in order to stay in business for as long as it can.'[31] 'In this Internet age,' Bennett proposes, 'we may well have to extend our idea of a life form far beyond cells and molecules and conceive of the flow and counter-flow of information itself as a form of life.' Assuming that mankind's experience of first contact with alien beings will necessarily be physical or biological in nature, let alone 'democratic and politically correct', is 'the height of bourgeois intellectual optimism', said Bennett.[32]

## Not All the World's a Stage ...

In order to seem convincing, such character actors as Bennett's living memes need appropriate stages upon which to work their magic. If you claim to have met an extraterrestrial in the bathroom supplies aisle of the Solihull branch of B&Q, then people will call you a nutter. If you claim to have met one in the depths of the Mojave Desert, then people might take you more seriously. A crashed saucer just outside of Bognor Regis sounds stupid; one just outside of Roswell, New Mexico, sounds thrilling. One of Bennett's most interesting ideas was that of so-called 'system-animals', or the way in which certain landscapes, pregnant with myth and history, can seem in some way to be alive, another sort of character in the play of life.[33] Consider the Three Witches who appear upon the Blasted Heath in *Macbeth*; in some sense, they are personifications of that same landscape, and never appear elsewhere, where their presence would not make as much sense. Perhaps Orthon stood to the Mojave Desert much as the Three Witches stood to the Blasted Heath? Just as these Shakespearean projections of the environment bring out the seed of inner ambition which already lurks within Macbeth and allow it to constellate properly within the outside world, so perhaps Adamski's meeting with Orthon allowed some latent fantasy contents from within his own head to be externalised in the form of creative visions of benign spacemen? There are myriad reasons why the Mojave was potentially potent with mythological content, both recent and archetypal. As well as being the setting for various old Indian legends about magical flying machines, numerous real-life futuristic jets, rockets and missiles were being tested by the military out in the Mojave at the time,[34] a nexus between the mythological and the science-fictional which was perfectly embodied through the figure of Jack Parsons and his own quest in search of Babalon amidst the desert sands.

If the American desert really does constitute some kind of gigantic 'system-animal', then it has certainly claimed the minds of many victims. One of Adamski's companions on the trip out to Desert Centre to see Orthon, for example, was a long-time occultist called George Hunt Williamson (1926–86), who in 1959 published a book, *Road in the Sky*, arguing that the saucers' fondness for the dry south-west was no coincidence. Williamson knew his Lowell, and was doubtless aware of the great man's lyrical rhapsodies concerning the supposed similarities between the landscape of Flagstaff, Arizona, and the dying Mars. Maybe, he speculated, as the Martian water supply had first begun to dwindle thousands of years ago, one group of thirsty aliens had decided to fly down and colonise the American desert to escape from the unfolding ecological disaster? If so, then they chose rather a stupid, water-free place to settle in, but Williamson anticipated this argument by saying that the Martians 'only understood the economy of an irrigated area and a desert environment', and so chose not to land in the lush grasslands further north. Therefore, when the Space-Brothers began returning down to Earth in the 1940s and '50s, might they not have been visiting the old-time homelands of their terrestrial cousins?[35]

Another victim of the desert was Clyde Tombaugh, whose decades spent searching for Pluto out at Flagstaff seem to have left his mind open to certain romantic, quasi-Lowellian influences. Tombaugh is on record as having seen several UFOs throughout his career – with the significant caveat that the term 'UFO' does not necessarily equate to 'alien spaceship'. Put simply, Tombaugh witnessed a number of unusual aerial objects he could not definitely identify. One New Mexico evening in 1949, he and his wife both saw an 'apparently solid' and 'yellowish green' object overhead, surrounded by 'about ten rectangular lights arranged in a symmetrical pattern', for example, and whilst working at White Sands Missile Range in the early 1950s Tombaugh witnessed three mysterious green fireballs wending their way across the desert skies. Initially, Tombaugh was dismissive of these visions being spacecraft, but afterwards accused his fellow astronomers of being 'unscientific' in their refusal to at least 'entertain the possibility of [their] extraterrestrial origin and nature'. Seeing as the objects he saw 'appear to be [intelligently] directed', he told a newspaper in 1957, 'their apparent lack of obedience to the ordinary laws of celestial motion gives credence' to the idea they were unknown craft.[36] At the end of F. Scott Fitzgerald's Great American Novel, reference is famously made to the equally great Jay Gatsby seeing a green light glowing across the water, never-approaching, ever-distant, never to be caught or touched. Clyde Tombaugh seems to have seen – and then sought – much the same thing. So do all these mirage victims of the local system-animal.

## The Madness of George King

Another of California's Contactees was an Englishman named George King (1919–97), creator of the world's oldest surviving UFO religion, the Aetherius Society. Whilst the Society was founded by King in London, in 1959 he was directed by space intelligences to move his HQ to California instead, a location which made more mythopoeic sense.[37] A distinctly comic account by Patrick Moore of how it was possible to walk down the Fulham Road until you reached Number 757, which housed the British branch of the movement ('They are quite convinced that the world's problems are much less likely to be solved in Downing Street, the White House or the Kremlin than in the Fulham Road'[38]) showed how it was probably necessary for King to leave England to expand any further. King's description of how his contact with the Space-Brothers began – he was stood at the sink in his Maida Vale flat one afternoon in 1954 washing the dishes when a disembodied voice proclaimed 'Prepare yourself! You are to become the Voice of Interplanetary Parliament!'[39] – sounded more comical than anything, hardly having the weirdly magical quality of Adamski's first contact with Orthon out in the Mojave.

By September 1954 King had established his Aetherius Society, named after a 3,456-year-old entity from Venus whose voice he had learned to channel down through his own body. Soon, King – clad in dark sunglasses – was ventriloquising 'The Master Aetherius' live on-stage at London's Caxton Hall.[40] Whilst hundreds came to be brought grave warnings from space, many were there simply to poke fun. Patrick Moore attended some sessions and, most impressed by Aetherius' claim to be able to speak all known languages in the universe, asked him a question in Norwegian. It turned out Norwegian was the one exception to this rule. Never mind. Moore asked the same question in French. Still Aetherius remained silent. Could Moore possibly try speaking in English instead? He did so, and asked where Aetherius was at the moment. Was he on Venus, inside King's head, or floating above the Earth? Aetherius refused to answer. He said it was a secret.[41] Once some of King's more peculiar teachings began to leak out, mockery intensified. For example, Mars was not only inhabited, said King, it was the workshop of the entire solar system – 'a sort of cosmical Sheffield' – where all the other aliens went to get their saucers built.[42] Worse, it transpired that the Interplanetary Parliament itself, located on Saturn, was staffed entirely by a race of 40-foot-tall white eggs of a perfectly spherical variety. Upon a visit to the Society's Fulham HQ, Moore told an Aetherius spokesman that such descriptions sounded like 'extremely large balls' to him, a statement with which his host innocently agreed.[43] When in 1957 the Society's official newsletter inadvertently printed a series of spoof articles (possibly written by Moore himself) claiming that the use of large quantities of hallucinogenic drugs was an excellent way to sharpen one's sightings of saucers, and featuring testimony from such allegedly noted men of science as Professor N. Ormous, Dr E. Ratic and Mr R. T. Fischall there were red faces all round.[44] After such experiences, it was no wonder that George King decided – sorry, was *told by the Space-Brothers* – to move to Hollywood.

## Hail Mary, Full of Space

It would not be unfair to describe George King as the British George Adamski. Just like the Polish-Californian, King had enjoyed a background in Eastern mysticism long prior to holding any interest in saucers, having studied yoga and allied disciplines for around a decade before being called to take up his seat in Interplanetary Parliament. His mother and grandmother had both been psychics, too, with the former once having run some kind of New Age healing centre.[45] Apparently, King had inherited some of these skills. In his book *You Too Can Heal*, he related how, when aged eleven, his mother had fallen seriously ill, and he retreated to some nearby woods to pray for her. This led to an angel appearing before George and telling him to return home, where his mother would now be in a state of full recovery – which she was.[46] Even the majority of King's initial information about Interplanetary Parliament came not from an alien, it later transpired, but from an advanced Indian yogic adept who had materialised in his living room about a week after he had heard the disembodied voice whilst washing his dishes. Once the yogi had finished imparting his message, he bowed politely, about-turned and walked straight through the locked door of King's flat as if he were a ghost.[47] In 1958, meanwhile, King's mother, awoken at night in her Devon cottage, had her own encounter with a ghost-like being. When the entity entered her home, the place filled with 'the perfume of a thousand flowers', and, even though it was snowing, the eerie humanoid made no footsteps in the whiteness when it left – in spite of which, she still identified it not as a spirit, but a 'Great One from Outer Space, come to converse with me'.[48]

Let's presume for a moment these were real experiences – not *literally* real, but visions of some kind, rather than deliberately concocted lies. Within them we can see something very like Colin Bennett's hypothesised meme wars at work. If Adamski's book had not been published in 1953, then what kind of message would King's visionary ghost-swami have brought him? The same basic script about peace and love would have been imparted, no doubt, but posing as a message from Heaven, not Venus. Given the apparent psychic bent of his mother and grandmother, and his own childhood angel sighting, by rights George King should have spent his life speaking with cherubs – but no! After Adamski, the angels became spacemen instead. Out in California, one of the few female Contactees, Dana Howard, found a similar process to be at work. Dana spent much of the 1950s enjoying sightings of what appeared to be the Virgin Mary – except the Virgin herself claimed to be a Space-Sister named Diane, who hailed from Venus. Howard had first seen a female apparition in 1939, a woman 'of unsurpassing loveliness' leaning against a tree, whose head was 'radiant with a crown of fire', like a halo. You might well have thought it was Mary, Mother of God – were it not for the fact that, hovering 300 feet up above, was 'a beautiful rocket-shaped ship … constructed of some sort of translucent materials, but trimmed in gold and gem-studded'. Down from this craft extended 'an almost invisible ladder' up which the blessed lady then vanished.[49] Howard didn't see her again until 1955, when attending a séance in Los Angeles. Here, the very same spirit materialised, announced her name was Diana, and said that she came from Venus. Before departing, she told her favourite 'Child of Earth' that she should 'Try to make every breath a breath of love' from now on.[50] Surely, had Dana Howard lived a century beforehand, 'Diane from Venus' would have been 'Mary from Heaven'? Not after George Adamski.

If meme wars there are, then it seems as if the aliens from the Mojave Desert might just be winning. Once ideas like those of Adamski are out there, they can never be erased. I am reminded of the old song by '70s supergroup The Eagles, 'Hotel California', concerning a certain type of deluded Hollywood mindset from which, so the lyrics say, 'You can check out any time/But you can never leave'. It's a disturbing thought that, rolling around somewhere inside all our heads even today, are the intellectual offshoots of the assorted insanities of loony old George Adamski. In spite of its dated and quaint ridiculousness, the basic archetype of the Space-Brother lives on still, somewhere within our shared collective unconscious. Orthon has checked out, but he will never leave.

# Red Planets: Communism throughout the Cosmos?

In 1949, James V. Forrestal (1892–1949), the US Defense Secretary under President Harry S. Truman (1884–1972), was found running through the corridors of the Pentagon, screaming that the nation was being invaded and there was nothing he could do to stop it. Taken away for a period of 'rest' in hospital, Forrestal responded to this enforced holiday by jumping out of a window. His death was put down by most to the new emergent state of Cold War paranoia, but saucer-nuts knew better. Forrestal, they said, recognised that the UFOs then invading American airspace were real. Aware that not even the mighty US military had any hope of defeating this fearsome space armada, Forrestal had taken the easy way out.[1] Possibly this was a wise decision, as the forthcoming alien invasion seemed to be both imminent and worldwide.

A typical intertwining of saucer-panic and Cold War hysteria broke out Down Under during 1954. This was the year of the 'Petrov Affair', a significant moment in post-war Australian history in which a diplomat and KGB man named Vladimir Petrov (1907–91) defected to the West and spilled the beans about alleged Russian infiltration of Australian politics. 1954 also saw the first Australian 'UFO-flap', as periods in which heightened numbers of sightings occur are called. In November and December 1953, the *Australasian Post* had serialised George Adamski's *Flying Saucers Have Landed* within its pages, giving the country its initial taste of saucer-mania, and the country's leading UFO investigator, Edgar Jarrold, exploited the interest to get his own ideas into mainstream print. Jarrold predicted that, seeing as 1954 was one of those periodic years when Mars came closer to Earth than usual, we could expect to see a veritable 'invasion' of spacecraft during the year. A visiting French astronomer and Mars expert named Dr Gerard de Vaucouleurs (1918–95) agreed, calling his programme of observation of the Red Planet not only a scientific exercise but also 'a hunt for possible enemies from space'. The Australian UFO-flap began in earnest in late April, right during the height of the fall-out from the Petrov affair. It seems that many persons connected the two prominent news stories, with the Australian Press floating the idea that the saucers whizzing overhead might have been either secret Soviet super-weapons, or some kind of obscure Russian exercise in psychological warfare following Petrov's defection. In such a climate, even Edgar Jarrold was taken somewhat seriously, and a meeting arranged between him and representatives of the Air Force Intelligence unit in Melbourne, although little meaningful came of it.[2]

## Stranges by Name, Stranger by Nature

Jarrold later disappeared from saucer research for wholly prosaic reasons which were quickly mythologised, with his retirement being attributed to visits he had supposedly received from mysterious Men In Black, who may have been either shady government agents or humanoid aliens in disguise depending on your viewpoint.[3] This disturbing idea raised paranoia within the field to a whole new level. What role could such camouflaged extraterrestrials play within the world's governments? Might they have surreptitiously infiltrated them, and be manipulating key Cold War players for their own nefarious ends? The parallel with more plausible 1950s fears about Soviet infiltrators and Communist fifth-columnists hiding within the halls of power is quite clear in such tales, but there was another possibility available for those of a more positive frame of mind – namely, that ET had indeed infiltrated the White House, but was a force not for godless evil, but for cosmic good! The primary promoter of such an idea was Frank E. Stranges (1927–2008), a very odd Contactee who claimed to be close personal friends with a Christian missionary from Venus named Valiant Thor. Stranges' demented self-published 1966 book *The Stranger at the Pentagon* partly details his meetings with Thor, and partly acts as a free advert for a series of tape recordings containing what was termed 'awesome matter', but which turned out to be simply recordings of Stranges talking about things like his theory that the Earth was really hollow.[4]

Stranges' actual text relating the story of Valiant Thor and his benevolent intervention at the heart of American government is an absolute delight in terms of its sheer, unabashed childishness. Thor's encounters with America's most high-up Cold Warriors begin in 1957, when he is intercepted by a pair of cops in the act of landing his saucer, bundled into the back of a patrol car and taken to meet President Eisenhower in Washington DC. 'My God, why couldn't this have happened on my day off?' asks the Air Force Captain in charge of White House security, taking a good swig of alcohol before escorting Thor into the Oval Office, where a strangely unfazed Eisenhower rises from his desk to greet the alien with a shower of risible dialogue:

'Of course, you know we have suspended all rules of protocol. I have a good feeling towards you. Please, sir, what is your name? And where do you come from?'
'I come from the planet your Bible calls the morning and the evening star.'
'Venus?'
'Yes, sir.'
'Can you prove this?' he asked.
'What do you constitute as proof?'
He quickly retorted, 'I don't know.'

The fascinatingly stagy conversation is then interrupted as Vice-President Richard Nixon (1913–94) rushes in, a fine and upstanding man who immediately impresses Thor with his 'fixed eyes and amazing aptitude towards speed and efficiency'. Nixon shakes Thor's hand 'without hesitation' and defuses all tension with a witty wisecrack: 'You have certainly caused a stir … for an *out-of-towner*!' Thor then hands Eisenhower a special message from his superiors at the Council of Central Control on Venus, and retires to a 'beautifully furnished apartment' hidden somewhere within the Pentagon where he lives for the next three years whilst the President thinks the secret proposal contained within the note over. Eventually, Thor agrees to provide Eisenhower with final proof that he really is an alien by taking off his clothes. He wears a one-piece suit of unknown material, coloured 'soft silver and gold lustrous', which proves to be entirely indestructible, with bullets, acid and diamond drill-bits simply bouncing off it. Even a top-secret atomic laser powered by pure rubies has no effect, whilst 'RXT-2 tests', whatever they are, prove that it cannot be destroyed by any means. The conclusion made by the US Government (though not necessarily the reader) is clear – Valiant Thor is telling the truth! Later, as a prominent UFO expert, Frank Stranges himself is then brought to the Pentagon to meet Thor, where he is told that seventy-seven other Venusians have secretly infiltrated America too, and are working to spread the Good News of Jesus Christ.

The Venusians had seen the atom bomb explosions of 1945 and grown wary of what might come next. Their message to Eisenhower contained a promise that, if America could only live in peace with Russia and transform the world's economic system away from today's wasteful form of capitalism, then the Venusians would gift humanity the means to live 'without sickness, poverty, disease and death'. However, Eisenhower was concerned about the number of job losses which abolishing death might lead to, and his political advisors persuaded him to turn Thor down, a ghastly mistake which has been concealed from the world ever since. The politicians' stupidity made Thor sad, and led him to return to his waiting spaceship in 1960. Atomic warfare still remained a threat to both Earth and the universe, however, with dangerous cosmic ripple effects from Hiroshima and Nagasaki having been held at bay in 1945 only by the combined efforts of 100 spaceships hovering above the planet's atmosphere and doing their best to prevent disaster. Therefore, Thor and 287 other space captains had remained behind on Earth to monitor our world discreetly, with Thor's own craft *VICTOR-ONE* being hidden away somewhere outside Las Vegas, 'just south of the Gypsum Plant, about a mile northeast of the junctions of Highways 147 and 166'. Frank Stranges had been on-board this craft himself and so knew many advanced technical details about it, such as the fact that 'it can move about', and that it possessed special technology which meant that short-sighted people like himself didn't need to wear their glasses when they went inside.[5]

Frank was happy to sell you an accurate plan of Valiant Thor's spaceship if you liked, or would gladly attend UFO conventions in California and tell you his life story instead. This grew with the telling, with Dark Forces making several attempts on Stranges' life due to his friendship with the God-bothering Venusian. Once, Stranges was kidnapped by three Men In Black, and driven into the Nevada Desert to be beaten up. Suddenly, however, Valiant Thor and his fellow friend in Jesus Vice-Commander Donn teleported onto the scene and threw two of the MIBs ten feet through the air without so much as breaking sweat, before reaching into their car and magically pulling the third miscreant out through a closed door. Then, Thor raised his hands and dematerialised the sinister trio instantly, sending them away to the salt-mines of Pluto, where the large amounts of salt would drain away all the evil from their souls, like when you pour some onto a wine stain on a carpet. Another time, Frank was involved in a deliberately staged car crash, leaving him paralysed from the neck down. Then, just as all seemed lost, Valiant Thor materialised beside the stricken man's hospital bed, laid hands upon him and said a little prayer. Immediately, Frank 'Lazarus' Stranges got up and went home, with nary a scratch on his body.[6] In a time of Cold War tensions, it was good to know that such supremely powerful beings were keeping a watchful eye over humanity.

## Marxist Martians

Reading between the lines, however, Valiant Thor might not have been such an ideal citizen of 1950s 'apple-pie' America after all. What was it about his scheme for altering society which led Eisenhower and Nixon to reject it? Why would adopting it have meant destroying the underpinnings of the entire US economy? Might it possibly have been somewhat ... *socialist* in nature? Maybe so. There seems to have been a general presumption, long preceding the Space Age, that any intelligent aliens would probably be Communists. It all has to do with Karl Marx (1818–83) and Friedrich Engels (1820–95), history's premier left-wing theorisers, and their idea, expressed under the rubric of 'dialectic', that there was a kind of inevitable natural law of evolution at work within every human society, during which it would progress upwards from serfdom to capitalism to socialism and then, some glorious future day, on to full-blown Communism. If this supposedly universal rule held true for our own world, then it should really hold true for others, too. Marx liked to bill his theories as being 'science', and scientific laws, like those of gravity, are as true on Mars or Venus as they are on Earth. So, by this logic, all you had to do was find a planet which was both inhabited and older than our own and, all things being equal, its inhabitants would be Communists. During Marx's day, the influence of a now-discredited theory known as the 'nebular hypothesis', which held that the further out from the sun a planet was, the older it was, was very strong.[7] Seeing as Mars was the fourth planet from the sun, and Earth the third, it seemed to follow that, should it harbour life, Mars would turn out Red in more ways than one.[8] Furthermore, the fact that Mars is a smaller globe than Earth was taken as making it more suitable for the quick development of Marxism. If resources on a smaller world were scarce, then surely the Martians would have realised early on in their history that collectivising land and agriculture was the best way to share everything out fairly and end all conflict?[9] When Mars' tiny twin moons Phobos and Deimos were discovered by the American astronomer Asaph Hall (1829–1907) in 1877, such thinking got a second wind, it being speculated that Marxism would have run riot within such confined spheres, leading to early instances of alien utopias being built.[10] Maybe it was through Communism that all those wonderful Martian canals had been constructed, too? Percival Lowell himself implied so:

> The [canal] system betrays a wonderful unity of purpose, for it girdles the planet from pole to pole. Race prejudice, national jealousy, individual endeavour are there all subordinated to the common good. The network is harmoniously one, from one end of the planet to the other, each line running into the next in the most effective manner all round the globe.[11]

There is actually a region on Mars called 'Utopia Planitia', or 'Utopian Plains'; it sits at about the 65th parallel and was named by Schiaparelli.[12] Down here on Earth there is not, and never will be, any genuine utopia to be found (the word actually means 'no-place'), but away up on other planets it was easy to imagine the existence of such unlikely paradises. In the wake both of Marx and Engels, and of Lowell and Schiaparelli, a whole new genre of Martian utopias appeared on the market, a once-flourishing literary craze killed off by the rather less optimistic view of what Martian life might have evolved into given in H. G. Wells' (1866–1946) hugely successful 1898 sci-fi classic *The War of the Worlds*. In Wells' own book, mention is made of the nebular hypothesis, and the smaller size of Mars leading to an accelerated rate of evolution, but to him all this means is that the Martians rapidly develop 'intellects vast and cool and unsympathetic' enough to wipe us Earthlings out in an act of high-tech interplanetary genocide without a moment's hesitation.[13]

## Pious Fiction, Not Science Fiction

Unlike Wells, the authors of these more optimistic Martian books are now long forgotten. Typical was Henry Olerich (1851–1927) a Nebraskan lawyer, farmer and author who also once invented a new kind of tractor; had he lived long enough to have visited the USSR, he could have seen an entire factory full of such devices.[14] Olerich's ideal fictional utopia, however, was found not in Russia but on Mars, as described in his 1893 book *A Cityless and Countryless World: An Outline of Practical Co-operative Individualism*, which used the device of a Martian named Mr Midith coming down to Earth as a means of getting Olerich's Lefty message across. Mr Midith was a seller of improving literature door-to-door, and the book has him call at the home of the American Unwin family, who invite him in to share a meal, during which he accidentally lets slip he is a 'Marsian'. Mars, he tells the family, is a haven of 'co-operative individualism' (re: forced collectivism), whose high level of social development was caused by inevitable factors already discussed. In Mr Midith's words:

> According to your "nebular hypothesis", which is true ... Mars was detached from the sun ages before the Earth was born; for Mars is further from the sun ... Mars is also much smaller and less dense than the Earth, in consequence of which it cooled ... more rapidly. Mars, then, is older astronomically and geologically ... We can readily see, then, that according to these data, other things being equal, Mars must have an older and more advanced animal and vegetable life. The Marsian social and industrial organisations must be much more perfect than yours.[15]

In what way was life on Mars more perfect? According to Mr Midith, the Martians' great idea was to build giant houses holding 1,000 people, thereby creating a series of collectivised farms and self-contained little mini-societies. Arranged in a rather stilted conversational Q&A format, and coming complete with architectural plans and diagrams, Olerich's 'novel' virtually eschews narrative and is essentially an instruction manual for building a new society. The book's message is one of pure Communism, and Mr Midith's Martian background is in effect irrelevant;

he may just as well have been called Mr Marx and come to America from Germany, via Manchester.[16] Readers of the book had no choice but to stomach such exceedingly unliterary speeches from Midith as the following:

> Our "big houses" are built about a half a mile apart all around rectangular fields 24 miles long and six miles wide, containing according to your measurement four geographical townships, or 92,160 acres each ... There are double-tracked, electric-motor lines running all around these large divisions of land, so that every "big-house" is situated on a motor-line. These large divisions of land, together with the houses and people that live on them, we call communities. A community, then, has an area of four townships, more or less, and a perimeter of 60 miles, on which a big-house, containing about a thousand inmates [a telling word?] is situated at intervals of about half a mile. This gives a community a population of about 120,000 persons. These motor-lines connect with railroads at intervals of about a hundred miles or more, as the case may be. Our railroads are nearly all straight and almost level, with heavy steel-composition rails, laid on a solid roadbed, and the time of many trains exceeds a hundred miles an hour.[17]

Other books within the same stodgy genre often proved just as reluctant to feature any real narrative content. The Anglican priest Wladislaw Lach-Szyrma's (1841–1915) 1883 *Aleriel, or a Voyage to Other Worlds*, for example, was notable not for its plot, but its description of a Mars without money, war, meat-eating or discrimination between the races and genders, in which communal living is universal, and where the day, as advocated by the Welsh socialist Robert Owen (1771–1858), is divided into one-third work, one-third play, and one-third sleep. Like Stalin's Russia, Lach-Szyrma's Mars still had the death penalty for those ungrateful dissidents who refused to be happy with life inside their enforced paradise, however![18] Every bit as dreary-sounding was 1883's anonymous *Politics and Life in Mars*, whose preface used the canals as evidence the planet was inhabited, but which otherwise contained a lot of pious cant about the abolition of the Martian equivalents of the monarchy, House of Lords and British Empire, together with an explanation of how workers on Mars had managed, as Marx urged, to seize the means of production.[19] A later book in the genre, 1911's *To Mars via the Moon*, was specifically dedicated to Percival Lowell by its English author, Mark Wicks (1852–1935), but goes beyond mere talk of canals. As usual, Wicks' Mars is a socialist paradise, with telepathic Martians whose psychic affinities render them all in perfect sympathy with one another. Thus, when election time arrives, they all vote instantaneously, via psychic means, for the same candidate – sounds not a million miles away from North

Korea.[20] The academic Robert Crossley (b. 1945), in his own survey of such literature, states of Wicks' book that:

> We are offered an inventory of all the various social, medical, ethical and technical advances that the Martians have achieved, including excellent housing for all, full employment, political unity, rational urban-planning, chaste and loving courtships, total care for elder citizens, radiation treatments for diseases, and nominal equality of the sexes. The standard list of utopian absences also appears: of poverty, social classes, war, hereditary privileges, prisons, divorce, tariffs and alcoholism.[21]

If only the real-life Communist 'utopias' down here on Earth had proved to be so benevolent – or so boring.

## Red or Dead

Following the Russian Revolution of 1917, the coincidence of Mars being the Red Planet was eagerly exploited by writers of Soviet sci-fi. The most famous Russian novel about Mars was *Red Star* by the early Bolshevik Alexander Bogdanov (1873–1928), a 1908 work telling the story of a Russian scientist named Leonid who is taken on a tour of a post-Revolutionary Communist Mars in a manner not unlike that of Westerners such as G. B. Shaw (1856–1950) being shown around the new USSR in the years following 1917. There was a theory at the time that Mars was red not because of its soil, sand and rocks, but due to its surface being covered in masses of thick red vegetation. So it is on Bogdanov's Mars, as Leonid points out:

> What surprised me about Nature on Mars, and the thing I found most difficult to get used to, was the red vegetation. The substance which gives it this colour is similar in chemical composition to the chlorophyll of plants on Earth and performs a parallel function in their life-processes, building tissues from the carbon-dioxide in the air and the energy of the sun.[A Martian] suggested that I wear protective glasses to prevent irritation of the eyes, but I refused. 'Red is the colour of our socialist banner,' I said, 'so I shall simply have to get used to your socialist vegetation.'
>
> 'In that case you must also recognise the presence of socialism in the plants on Earth,'[the Martian] remarked. 'Their leaves also possess a red hue, but it is concealed by the stronger green colour. If you were to don a pair of glasses which completely absorb the green waves of light but admit the red ones, you would see that your forests and fields are as red as ours.'[22]

In other words, as we are on Mars today, so shall you all be down there on Earth, tomorrow. All it requires is a shift of perspective, and you humans too shall occupy a second Red Planet. This is presented as more-or-less inevitable. At one point during his tour, for example, Leonid is fed yet another variant of the nebular hypothesis as an explanation for Mars' advanced state of socialist evolution.[23] In any case, the Martians, whose resources really were running out, just as Lowell had surmised, had a plan to colonise Earth which would have involved either converting humanity to Marxism or, should this prove impossible, simply killing us all outright instead. Upon his return to Earth, Leonid vows to rally the forces of revolution so as not to disappoint the Martians. The problem is that humankind is too attached to outmoded capitalist notions of private property and land-ownership to do the sensible thing and share their planet with the dying race. According to a transcript of a Martian political meeting about the issue which Leonid uncovers:

> The peoples of the Earth ... will under no circumstances relinquish any significant portion of its surface. Such reluctance derives from the very nature of their culture, which is based on ownership protected by organised violence ... [Due to the incompatibility of our economic and social systems] we would be forced to evict the Earthlings from all the territories [to be] occupied by us ... Given their political system, which does not recognise any principle of fraternal mutual assistance, given their social order, in which services and aid are paid for in money, and finally, given their clumsy and rigid system of production ... the overwhelming majority of these [refugees] would be doomed to slow starvation. The survivors would organise themselves in cadres of embittered, fanatical agitators working to incite the rest of the population ... Deep racial hatred and fear that we would seize more territory would unite all the people of Earth in wars against us.[24]

One alternative, says the speaker, would be for the Martians to join ranks with enlightened human revolutionaries around the globe, but this would also prove impossible, due to their small number. Sadly, 'the dying remnants of the petty bourgeoisie', together with 'large property-holders' and the 'reactionary' elements of the working class would vastly outnumber 'the vanguard of the proletariat' and make their defeat inevitable, even if they were armed with advanced Martian weaponry.[25] Thus, seeing as 'a higher form of life cannot be sacrificed for a lower one', the speaker concludes that the Martians have no option but to wipe out humanity, in order to save '*our* socialism' and not the 'embryonic' human variant just about to be born on Earth.[26] Other speakers then argue against this position, with the decision being made that Mars will try to colonise Venus first of all, but that if this does not work out, then they will have no option but to land on Earth.

The ultimate conclusion is inescapable. If Leonid – and by implication the real-life global proletariat – do not manage to successfully instigate world revolution, then one day the planet will be invaded by angry aliens, and humanity destroyed! Just as H. G. Wells' alien apocalypse was intended as a kind of otherworldly punishment for Western man's sins of colonialism, with his technologically advanced Martians treating white Europeans how they sometimes treated their brown-skinned subjects,[27] so Bogdanov's threatened Martian genocide stood as a warning to that very same Western world about the need not to stand in the way of the 'historically inevitable' advance of Soviet Marxism. To do otherwise would be to transform *ourselves* into the dying race, not the Martians.

# Engineering the Future: From Soviet Science Fiction to Science Fact?

How seriously were readers meant to take stories like *Red Star*? Disturbingly seriously, it turned out. The official intended function of Soviet literature was not simply to entertain – that would be *far* too bourgeois! – but also to reflect and somehow help to create reality itself. Strange occult ideas began to appear during the years leading up to 1917's Revolution, concerning authors' alleged ability to transfer their thought-energy into the pieces they wrote, or claiming that words themselves had inherent semi-magical powers, linked to the motion of the stars.[1] The neurologist Vladimir Bekhterev (1857–1927) speculated that the 'thought-energy' expended by any given person in a speech, newspaper article or novel was a genuine physical thing which lived on even after that person's death, as other people read or heard about their ideas and so absorbed their 'particles of universal human creativity', leading them to alter their own opinions and way of life.[2] The writer Maxim Gorky (1868–1936), one of the leading figures in the 'Socialist Realism' movement within Russian letters, also believed in the capacity of literature to act as a medium for 'thought-transference', and thus a powerful means of swaying the masses by offering them 'a glimpse of tomorrow' through fiction. If sci-fi stories portrayed a utopian world in which Communists had conquered outer space, for example, then their readers would become exposed to a kind of literary hypnosis; if a whole generation grew up reading about brave cosmonauts prepared to spread the doctrine of Marx to other galaxies, then this could provide the young with enough inspiration to study and work hard, building their own real-life spaceships in the future. When Stalin famously proclaimed that writers were 'engineers of human souls', this is what he meant.[3]

The pace of change in the formerly backwards country – where sefdom had only been abolished as late as 1861 – was extreme, with the sense of a world suddenly transformed being captured in much Russian sci-fi of the day. With the nation industrialising so rapidly, potential salvation from want and disease seemed to come no longer from God but from science, a new deity whose capacity for wonders seemed almost limitless; electricity, railways, motor-cars, medicine … whatever next? Within such a climate, what had always seemed like mere dreams of settling other planets could now come across as being 'rational blueprints to shape the future', as one recent historian has put it, there being only 'a short step from science fiction to *Sputnik*'.[4] Due to its new-found technological ambition, the Revolutionary leader Leon Trotsky (1879–1940) could declare Russia to be 'at the same time the most backward and the most advanced nation', with the sheer pace of change giving rise to the possibility the USSR might even be able to whizz ahead of the Western countries, in spite of their prior industrial head start. Amazingly, Russia had managed to skip a whole stage within Marx's predicted programme of dialectical change, passing straight from Tsar to Bolshevism without any intervening period of bourgeois capitalist rule in-between. Might they also be able to skip a stage or two in their technological progress, going straight from steam engines to starships?[5]

Two significant figures whose particles of 'thought-energy' echoed down the ages and ended up becoming the intellectual fuel behind the later Soviet space programme were Nikolai Fedorov (1829–1903) and Konstantin Tsiolkovsky, the latter of whom we met earlier, talking about living atoms and space angels. Fedorov was the founder of an influential intellectual cult nowadays known as 'Russian Cosmism', a bizarre strand of pseudo-scientific utopian mysticism which held that such eternal enemies of mankind as time, death, aging and disease were ultimately conquerable through science. The writings of Fedorov are full of implausible plans to resurrect the dead (*all* of them, right back to Adam and Eve!) and transform Earth into a gigantic self-propelled spaceship, with his ultimate goal being to achieve full 'knowledge and control over all atoms and molecules' in Creation. Though he himself had neither the skill nor equipment to achieve any of these aims – his practical efforts at resurrecting corpses were limited to making lists of obscure dead people so that they wouldn't be forgotten about come the Day of Resurrection – Fedorov nonetheless really did expect these wonders to occur, step-by-step, over the following few centuries. He termed his plan to conquer Nature 'supramoralism', and wanted to convert people from being mere observers of Nature to masters over it, also a key tenet of Bolshevik thought. The philosopher dreamed, for example, of 'meteorology' becoming 'meteor*urgy*', with forecasting and study of the weather superseded by human weather control.[6]

Sometimes Fedorov expressed these desires in a strange, rather stilted form of sci-fi, as short stories whose 'plots' were virtually non-existent, consisting as they did of little more than descriptions of super-scientists of the future discussing their theories with others. For example, his 1892 work *Karazin: Meteorologist or Meteorurge?* is a brief account of the ideas of the titular Karazin who, whilst manufacturing saltpetre, has an idea that instead of doing it the old-fashioned way, 'potassium nitrate could be made with electricity captured from the atmosphere's highest strata by specially constructed balloons, which would be anchored to the Earth with metallic cables.' Extrapolating, Karazin realises that atmospheric electricity could potentially be exploited in a number of other highly useful ways, too. Therefore, he begins a letter-writing campaign, posting pleas off to various aristocrats and officials, as well as the Russian Academy of Sciences, asking that they consider funding his idea – but they don't. The End. Fedorov himself calls the two-page piece an 'essay', not a story, and so it is.[7]

## Social Engineering

Such stuff may seem didactic and un-literary in the extreme, but its influence amongst a select few acolytes was great. One of Fedorov's proposed branches of supramoralist science, for example, was called 'astronautics', and represented the next step up from mere astronomy – instead of watching the stars through a telescope, the so-called 'astronauts' of Fedorov's imagination were to fly up into outer space itself. Most people who heard Fedorov outlining such schemes from amongst the bookshelves of Moscow's Rumyantsev Library, where he worked hard for a pittance, must have thought him mad. Aeroplanes did not even exist at the point he was voicing such ideas, let alone spacecraft. One of Fedorov's students, however, happened to be none other than Konstantin Tsiolkovsky, the father of Soviet rocket science, who went to pay his respects to the madman in Moscow during 1873.[8] Amongst other intellectual inheritances from Fedorov, Tsiolkovsky owes a great debt to the older man in terms of literary style. Inspired by the 'scientific romances' of Jules Verne (1828–1905), in particular 1865's *From the Earth to the Moon*, Tsiolkovsky himself liked to write sci-fi for two separate purposes. On the one hand, in line with the thinking of Maxim Gorky, Tsiolkovsky wished to spread his 'thought-energy' out into the proletariat. As he wrote in 1935:

> Science fiction stories on interplanetary travel carry new ideas to the masses ... They excite interest ... and bring into being people who sympathise with, and in the future engage in, work on grand engineering and technical rocketry projects.[9]

This is quite true; Jules Verne's novels were often cited as one of the main inspirations behind their work by the people behind NASA's *Apollo* missions. As the American sci-fi writer Ray Bradbury (1920–2012) once put it, 'Without Verne there is a strong possibility we would never have romanced ourselves to the moon.'[10] Quite aside from using the genre as a vehicle for inspiring the next generation, though, Tsiolkovsky also used sci-fi as a kind of cheap-and-cheerful imaginative laboratory, in which he could work out his ideas more fully within a basic narrative framework. Imagine if Newton had written a story about an apple falling on his head, and you get the basic idea. 'Many times I assayed the scientific concept through the task of writing space novels, but then would wind up becoming involved in exact complications and switching to serious work' he once wrote.[11] The end result were yet more almost plotless stories modelled after those of Fedorov, often with titles so plain as to be comical. *The Exploration of Space by Reactive-Propelled Devices, Changes in Relative Weight, Living Beings in Space, Biology of Dwarfs and Giants*, all were the poorly named work of Tsiolkovsky's own otherwise ingenious pen. Nonetheless, as utilitarian as they were, the books featured the first fictional appearances of such now real-life developments as self-propelled multi-stage rockets, floating orbital space stations (or 'space cottages') and solar batteries. His novel *Beyond the Earth*, meanwhile, published in 1923, contained strong hints that only centralised, state-run command economies along the Soviet model could ever truly hope to put men into space.[12] *Beyond the Earth* gave 2017, the 100th anniversary of the Russian Revolution, as the date when a man-made object would finally break through Earth's atmosphere. As it was, such an event actually happened in 1957 with *Sputnik* – a year which also marked the centenary of Tsiolkovsky's birth. The Kremlin being unwilling to miss out on such a propaganda opportunity, a big thing was made out of the coincidence, with President Khrushchev bestowing the folkish title 'grandfather space' on Tsiolkovsky's corpse and ordering festivities to take place in his honour.[13]

Alexander Bogdanov's writing might have fitted well within such a programme, too. In 1913 he published a prequel to *Red Star*, called *Engineer Menni*. This told of life on Mars during the final dog days of alien capitalism through the eyes of the titular engineer, Menni, whose own endeavours in building the Martian canal system to save the planet from environmental destruction led ultimately to ushering in the final victory of Communism over capitalism on the Red Planet. Never before had the alien economics (Mars' greatest economist is called Xarma – Marx spelled backwards, nearly) underpinning the construction of Lowell's imaginary canals been laid out in such detail:

> Menni demonstrated that [for the government] to continue construction [of the canals] … through loans [from capitalists] repaid by income from the reclaimed deserts would mean prolonging the Great Project over several centuries. The new financial plan which he proposed instead showed that he could also be a revolutionary outside of his own special field [of engineering]. It was a plan for nationalising the land.[14]

This was not just idle talk. Following the beginning of Josef Stalin's (1878–1953) brutal dictatorship in 1924, Bogdanov's books came back into vogue and were republished in mass editions, which were read and openly praised by several of the engineers, planners and architects behind Stalin's notorious 'Five-Year Plans' to revolutionise agriculture, build up whole new industries from scratch, or create perfect new cities within impossibly short time-frames. Engineers like G. M. Krzhizhanovsky and city planners like L. M. Sabsovich were united in comparing Stalin's crazy schemes to the building of the canals on Bogdanov's Mars. If Engineer Menni could manage such a feat whilst still having to fight against the last stand of alien capitalist parasites on Mars, then why couldn't they do something similar under the much more favourable conditions of a Communist government down here on Earth?[15] The use of forced labour and utter indifference towards human life shown during the various Five-Year Plans, of course, ensured that such wonders really *were* possible to perform. Stalin's monumental White Sea Canal project, in which 125,000 political prisoners were forced to dig a (largely useless) 140-mile canal within record time might have made even Engineer Menni red with socialist pride – just so long as he didn't peer too closely into the concrete, which he would have found had been artificially reinforced with the bones of all those who had been worked to death during the canal's construction.[16] Vladimir Bekhterev and his disciples became convinced that, at times of great social upheaval, the masses became literally possessed by the 'thought-energy' of some radical new idea or other, and went temporarily mad, becoming hypnotised into seeing some grand new utopia opening up before their eyes when in actuality all was bloodshed and chaos.[17] Instead of seeing White Sea Canals all around them, you might say, by Stalin's day the Bolshevik dupes had been conditioned into seeing Martian Canals everywhere they went. Science fiction played an unfortunate part in this process; far from being the clean-cut hero portrayed within Bogdanov's book, Engineer Menni actually had blood on his hands.

## The Posadan Adventure

One group of Communists who appeared to take speculation about socialism on other planets to heart were the Posadists, an obscure Trotskyist sect with members scattered across Europe and South America, who are often held up these days as an example of a Marxist saucer-cult. Seeing as the group's founder, J. Posadas (1912–81), is on record as saying things like 'We admit to the existence of extraterrestrials as a conclusion of dialectical thought', this is not altogether surprising.[18] However, is the accusation entirely accurate? Posadas was a former professional footballer from Argentina named Homero Rómulo Cristalli Frasnelli, who had once been considered good enough to play for Estudiantes, arguably the country's biggest club. By 1962, however, he had drifted far away from his sporting youth and established his Posadist faction to carry the flame of Trotsky's doctrine of 'perpetual revolution' throughout South America. Extreme in ideology, the Posadists ended up being banned from many countries in the region, with Posadas himself exiled to Italy in 1968. Here, the group became a shadow of their former selves. 'Meetings' seemed to consist of little more than football kick-abouts between Posadas and his friends in the countryside near Rome, followed by drink and tasty home-made pastries. In Britain, some Posadists, trading under the name of the Revolutionary Workers' Party, tried to infiltrate the Labour Party and succeeded in fostering union militancy amongst workers at the Vauxhall and Austin car factories, but that was about as far as they ever got in terms of toppling the hated capitalist hegemony.

Banned from engaging in political activities as a condition of his asylum, many texts which subsequently appeared under the name of Posadas were henceforth dated '1968', to give the impression they had been written prior to his arrival in Italy. He would further frustrate authorities by making undateable speeches into a tape recorder, which were then transcribed in the various party newspapers or affiliated journals by others. Unfortunately for the Posadists, some of the Dear Master's ideas appeared somewhat comical to those not blessed with a perfect understanding of the dialectic. Posadas' most dearly held doctrine, for example, was to encourage Russia and China to immediately launch all-out nuclear warfare with the West, thus destroying civilisation itself. From the ruins of this dead world Posadas said would scramble 'within a few hours' the last hardy remnants of the global proletariat, who would immediately wipe themselves down and begin building a brave new Trotskyite world out of the radioactive rubble. As well as his Trotsky, Posadas appeared to have been reading his Fedorov and Tsiolkovsky, however, with his speeches containing various echoes of the extreme scientific utopianism characteristic of Russian Cosmism. For example, just like Fedorov, Posadas hoped that one day Marxist science would abolish 'miserable, abominable sexual excitement' in favour of allowing us all to reproduce asexually like amoebas, and was particularly interested in the notion that one day the means could be found to enable us to talk with dolphins or live as long as elephants did (260 years, in his view). In short, mankind had to show 'Audacity in the face of Nature!' and then conquer it, the keynote idea of Cosmism.[19]

Obsessed with the Sino-Soviet space programmes of his day, Posadas' most overtly Cosmist text was called *Childbearing in Space: The Confidence of Humanity and Socialism,* and concerned an alleged Russian plan to order female astronauts to become mothers to special 'space babies' which would have been conceived in outer space as well as born there, in acts of zero-gravity intercourse. 'This', Posadas said, 'could be one of the forms love takes in the future', so ought to be investigated. What will happen to babies generated in space through floating sexual acts? Nobody knows, and only the Russians were brave enough to try and find out. The desire to create space babies 'can only arise in a Workers' State [where] humanity has already reached a high degree of social self-confidence', Posadas explains, because in lesser and more primitive capitalist lands, all people care about is gaining higher living standards. The scientific experiments of Communists have 'a profoundly human character' as they are 'based on human love' in all its forms, whereas capitalist experiments in science revolve only around a desire to 'see how better to sell goods, eliminate human beings and dispatch the proletariat'. Because of capitalism, Western science has become 'oppressed by the straightjacket' of the profit-motive. However, Posadas plans to liberate science from such shackles through revolution ('it will not be long, in only a few years') at which point, apparently, 'we will wipe out all the problems' in the world, like 'floods, hunger and misery'. Shorn of the need for funding to pursue their ideas in a new, moneyless world, every person on Earth will be free to experiment scientifically in any way they desire, so much so that 'everyone will be an architect, an engineer, a doctor, and the like'. Due to the state of perpetual scientific revolution which will be unleashed amongst the masses, invention after invention will come to pass, no matter how implausible. In this paradise, 'we are going to find electronic means, or equivalent, capable of sending and receiving not only human thought, but [finding] the specific weight of each thought', Posadas predicts. The forcible impregnation of female astronauts is the ultimate sign that this Trotskyite technological utopia is now at hand. The creation of space babies demonstrates that mankind is finally becoming an interplanetary species, and will 'tighten our relationship with the universe', leading to 'improved relations with other worlds'. Only true Marxists could ever conceive such a scheme, said Posadas, as it had no financial benefit to it; the plan 'was not reached through calculations and programmes' of a petty mercantile nature, but in a spirit of unbounded idealism. The very idea of space babies, said Posadas, 'shows that humanity is secure. Secure in the notion that every problem can be solved, absolutely! And easily!'[20] If only.

## Power to the Space People!

This all makes Posadas sound mad, but we may be getting the wrong impression. If you look through issues of the cult's official newspaper *Red Flag*, you will find more headlines along the lines of (to give the most famous example) 'LONG LIVE THE CONTINUATION OF THE DUSTMAN'S STRIKE!' than about aliens. Some suggest that other far-left groups, notorious for the bitterness of their infighting, simply picked up on Posadas' saucer-beliefs and gave them undue publicity, hoping to paint his followers as a bunch of nutters. According to the testimony of one of Posadas' rare Italian followers, Luciano Dondero, the man himself wasn't even particularly interested in UFOs. Rather, it was one of Posadas' fellow Argentine exiles, Giuseppe Bonotti/Miguel Arroyo (both are pseudonyms), who was obsessed with the subject, continually sending sighting reports to the Dear Master. It was in response to Bonotti's endless prodding that Posadas produced his only substantial comment on the subject, 1968's *Flying Saucers: The Process of Matter and Energy, Science and Socialism*, which was intended as a way of telling Bonotti to shut up and concentrate on more important issues like fomenting nuclear war.[21] Nonetheless, the piece didn't actually argue with Bonotti's view that flying saucers were real; instead, it confirmed it!

For Posadas, capitalism constrained science, whereas Marxism liberated it. If aliens really were visiting us in spaceships, then by definition their technology must have been more advanced than our own. However, seeing as there is no immediate financial benefit to be derived from space travel, as the moon landings taught us, then how can the aliens have managed to fund the R&D costs for their craft? The obvious answer was that ET was a Red. On other, more Trotskyite worlds, perpetual scientific revolution may already have produced a situation whereby extraterrestrials were able to 'produce light, attract, reject or organise energy' simply by raising their hands. Where Chairman Mao (1893–1976) ordered peasants to move mountains with a 'pickaxe and spade', Marxist aliens might have been able to simply use 'the energy contained in the mountains' to make them move themselves. Also, aliens may no longer have to eat food, thus dealing a blow to the sub-human robber-barons who control agriculture. Under true alien socialism, 'existence will mean progress' automatically, unlike under human capitalism, with universities being abolished due to lack of need. Here on Earth, though, those who proposed wonderful schemes to exploit Nature, like Fedorov's fictional hero Karazin the Meteorurge, were mocked. As Posadas explained:

> It is *social organisation* that determines the scientific capacity of the human being. In [a capitalist system based upon] private property, social organisation is very limited because the élan, the courage and the audacity it allows are governed by interests linked to individual appropriation [of goods or capital] … Those who go further than this feel cast out. They come under social limitations that restrain their social capabilities.

This was certainly true of the neglected scientific geniuses amongst the Posadists. True sons of Nikolai Fedorov whether they knew it or not, in 1961 they had proposed turning natural disasters to mankind's advantage by using earthquakes to generate electricity, something which could be done easily just by 'injecting into the Earth systems somewhat like radars, to measure all the movements and gases below', thereby telling us when the best time to attach the crocodile clips to these free geological batteries might be. People had called the Posadists mad at the time, but on a Marxist alien wonder-world with no money and limitless access to the necessary resources, they could just go ahead and carry out such trials regardless, and have the last laugh. 'Science is not independent', Posadas proclaimed; it lacked independence not only from capital, but from time itself, which was an inherently 'bourgeois' concept. Aliens may have mastered science to allow them to live forever, becoming 'seeds' which grow again, 'feeding into something else [a second life] without having to start from ashes', hoping to 'reappear in another form'. Such immortal aliens could pursue long-term experiments, like spending a million years zooming between star systems in their spaceships, on a whim. Why should aliens have 'deadlines and billings, [or] debts to pay by due dates' asks Posadas?

But why were they here? Perhaps they came to cause humanity to be 'integrated into [a] higher form of progress'. The workers of the world have nothing to lose but their chains, said Marx, a phrase which Posadas thought 'can be applied to everything', even the idea of alien visitation. Showing familiarity only with a certain kind of cosy 1950s type of Contactee encounter, Posadas claims that all aliens encountered by humans during his own lifetime 'arouse no feeling of alarm but a feeling of serenity', there having been 'no manifestations of the wish to attack, rape, steal or possess' upon the ETs' behalf, something indicating their advanced socialist nature. Whilst they appear peaceful, though, our otherworldly visitors must nonetheless have access to advanced weaponry, and this should be exploited by any true revolutionaries who happen to meet them. Anyone who went in a flying saucer just for a free tour of the solar system was nothing more than a class traitor. Instead, the aliens 'could be immediately useful on Earth' if they could only be called upon by their more class-conscious abductees to 'collaborate with us for the suppression of misery' by immediately killing all capitalists. Ironically, the fact that the aliens hadn't done so already was another sure sign they must be Communists, as throughout Earth's history all technologically advanced nations had traditionally conquered less advanced ones in the name of imperialism. The 'Yankee murderer' General Douglas MacArthur (1880–1964) may once have made the strange proclamation that 'The next war will be fought against evil beings from outer space', but he was deluded.[22] The tactics of divide-and-rule would *never* act to successfully separate the Space-Brothers from their Red brethren down here on Earth! Posadas' message to the socialist saucer-men was as follows:

> The fundamental obstacle is the capitalist system. We must destroy the military force it has today. Destroy all their atomic weapons. Destroy all the military power of the capitalist system, of Yankee, British, French imperialism. Appeal to the masses! Give them the means to destroy capitalism immediately, overcome the Workers' States' bureaucracy and build the new society of socialism!

The incompatibility of flying saucers with capitalism, said Posadas, could be seen in capitalism's marginalisation of the topic. In an observation which rather dates his argument, Posadas dismisses the subject of UFOs as having 'no commercial … benefit' to it. Had he lived to see the *X-Files* and Roswell-mania of the mid-1990s, he might have begged to differ! More cynically, Posadas proposes that Western scientists and authority figures ridicule the saucers simply in order 'to stop people thinking that there are superior forms of [social] relations beyond the reach of capitalism'. What makes Western societies 'turn away from the study of UFOs … is the reading they get there of their pending elimination'. By this logic, fostering belief in saucers might simultaneously foster revolution, but as yet no plausible Far Left group appears ever to have done so. A real missed opportunity for the historically inevitable advance of the global workers' collective![23]

## Going Dutch

The Posadists do still technically exist, albeit in tiny numbers, but don't seem to talk about UFOs much anymore. The final keeper of this particular flame was a Communist German metalworker named Paul Schulz, who spent some years in Argentina where he encountered the Posadists. Returning to Germany in 1991, he attempted to set up a local branch of the sect but failed, making do instead with creating a web-based organisation called *Social Reform Now!*, analysing UFO reports and alien encounters from a left-wing perspective, before eventually coming to claim that he was the official Earth ambassador for an advanced Marxist civilisation from another world. Unfortunately, Schulz's 2001 book *Official Contact by an Extraterrestrial Civilisation with Us Earthlings is Nigh; Let's Show Ourselves Worthy of this Exceptionally Joyful Event of Epochal Significance* has not yet been translated into English.[24]

One man who may have wished to join the Posadists had he known about them was the Dutch businessman Adrian Beers, also known as 'Stefan Denaerde', or 'Steve of Earth' as his alien friends liked to call him, the author of the strange 1969 book *Extraterrestrial Civilisation*. A smash hit in Holland, selling 40,000 copies, it was first published in English in 1977 as *Operation Survival Earth*, then in expanded form in 1982 under the new title *UFO Contact from Planet Iarga*. Curiously, initial editions of the book promoted it as a work of fiction, whilst later ones said that every last word of it was true. Whilst the Iargans, as the alien race were called, claimed to be neither Communist nor capitalist, but some kind of superior synthesis of both, to the average reader Iargan civilisation comes across simply as a more successful, humane and non-belligerent version of the USSR. Denaerde himself was definitely familiar with the workings of the free market, seeing as he was the director of a company responsible for importing lorries made by the Swedish firm Scania into Holland, a position which had left him very well-off.[25] The book reads not unlike the mildly exasperated rant of a man who has devoted his whole life to commerce and, frustrated by its inefficiencies but unable to do anything much about them, has decided to let off steam by imagining how perfect society would be if he was in charge of things.

Beers' book shares the exact same structure of Lowellian-era tales of Martian utopia like Henry Olerich's *A Cityless and Countryless World*; it is just that, instead of using the dated framing device of having a Martian like Mr Midith explain what life on other worlds was like over a nice dinner with a human family one evening, Beers chose to exploit the more contemporary notion of claiming to have enjoyed a close encounter with a flying saucer. The tale begins with Beers and his family enjoying a boating holiday on the Oosterscheldt, an estuary in south-west Holland, when their pleasure-craft collides with what later transpires to be a submerged spaceship. Seeing a body floating in the water, Beers jumps out and saves the person, who is strangely dressed in a 'metallic suit' and with 'a kind of rubbery ball' around his head as a helmet, obscuring his face. Then, another figure appears on the scene. The only problem is that this figure is not human, but some kind of monster with 'an animal-like face' and 'large square pupils in the eyes', leaving Beers terrified. Fortunately, the square-eyed alien speaks English, thanking Beers for saving the other ET's life. The first sign that the aliens might be left-wing in nature comes when they surprisingly then start lecturing Beers about race relations, calling Westerners like him undeveloped children and saying that 'any English-speaking negro, Chinese or American Indian' could tell white European males like him a thing or two about the dangers of encountering hostile races with access to advanced technology. Fortunately, the Iargans, just like the Martians of earlier Marxist fiction, are not like that, and offer Beers the chance to come aboard their ship where they will show him how they live on their home-world. Beers agrees, and the scene is set for the rest of the book to abandon narrative altogether in favour of a long, novelised description of an illustrated alien video lecture.[26]

## A Design for Living

Once aboard, Beers is quickly identified by the Iargans as 'a representative of the directing class of the Westbloc' and 'a supporter of a free economy'. This is made clear by the distaste Beers displays for certain aspects of their civilisation as it is explained to him:

> Their weak point is the development of their individuality. They do almost everything in groups. They think collectively, and they obey the laws of their society to the letter. They live for and through the friendship and love within the group.

On the other hand, Beers is now able to get a closer look at the aliens and, shocking though their appearance may be – illustrations show squat brown-skinned humanoid mammals with truncated horse snouts and sad yellow eyes – he is rather fascinated. Beers is impressed by the stocky and muscular build of the Iargans, which derives from their past amphibian nature. Whilst they have now evolved into semi-aquatic mammals like otters, they have retained their ability to kill small whales by 'ramming [them] like a torpedo' with their heads when swimming, making them tough indeed. Furthermore, their need to have streamlined bodies for swimming means that they lack protruding breasts or penises, and mate based purely upon desire for love, not lust for sex. Intrigued, Beers agrees to be given a long and semi-interactive holographic film lecture about Iargan history, which proves most illuminating. It turns out that, just as had once been predicted about Mars, the smaller size of Planet Iarga has led to an accelerated rate of social evolution amongst its inhabitants. Most of the planet was covered by water, leaving little land for living on, meaning that peaceful methods of collective living had to be adopted early on in Iargan history, with all differences of race, religion, nation and class now being long-forgotten. 'Efficiency' is the main mantra on Iarga, with massive blocks of flats known as 'house-rings' being devised to hold 10,000 aliens per unit, like Mr Midith's 'big houses' on Mars, which held 1,000.

'The uniformity appals me!' says Beers, and asks why they don't build 'simpler, smaller houses with more privacy?' Because such thinking is inefficient, the interactive film informs him, as is everything else on Earth. Letting people have different types of housing would be unfair and lead to jealousy arising amongst the populace. By living communally, all class differences are eroded as people are forced to live 'a different, more social way of life'. Similar attitudes could be seen in relation to public transport, where the railways had been nationalised and a series of 'slim torpedo trains that move without creating friction' criss-crossed the planet in such a regular and reliable way that timetables were not even necessary. However, seeing as there was no such thing as money on Iarga, such marvellous railways were not expensive to create or maintain, as the very term 'expensive' had no meaning there. Instead of money, things cost only 'production capacity', an arrangement which the Iargans termed the 'cosmic universal economic system'. This blended the best elements of both capitalism and Communism, and was the gateway to civilisational utopia. In a world without expense, the welfare state could be all-encompassing, with 'the least fortunate man' being as comfortable as the most, seeing as such a state of 'collective unselfishness' would cost nothing at all to administer. At this point Beers, his old capitalist beliefs not yet vanished, still holds onto 'a horror at the thought of such monotony', but his mind is soon changed.[27]

When the hologram film informs him that 'personal property is an indication of a very primitive level of culture', Beers is shocked, sitting 'bolt upright in my chair' and exclaiming that 'These beings are pure

Communists!' However, as the lecture continues, Beers gradually begins to realise that, when you think about it, maybe Communism is a good idea in theory, it is just the way it has been implemented by the brutal leaders of the Soviet Union that is the problem. If only, like the Iargans, we could learn to be more *efficient* in our socialism, then maybe one day utopia could be created here on Earth, too? The Iargan economy is managed for a maximum rate of efficiency thus:

> The total production of goods and services is ... in the hands of a very small number of huge companies, the "trusts". Nothing is paid for on Iarga, only registered. What a consumer uses is registered in the computer-centre ... and this may not exceed that to which he has a right ... You cannot *buy* anything ... [although] the right of use remains for life. This is almost the same as personal ownership, except that in the event of death, the goods are returned to the trusts ... You may not have more "in stock" than is reasonable for your own use, otherwise the surplus can be confiscated ... Legally, all goods remain the property of the trusts that supplied them ... Not only are they responsible for their upkeep, they also take the total risk of loss or destruction. Thus, all the articles are made to such a high standard that repair is never necessary ... The trusts work on a cost-price basis, whereby the term "profit" is replaced by "the cost of continuation [of manufacture]" ... Their economy was as stable as a rock .... The trusts competed with each other, and the prices were determined by the law of supply and demand, the principle of the free market. But their cost-price was computed on the standard work-hour, the ura ... What is conveniently called "price" is in fact purely a method of expressing the production-time demanded by a certain article.

The end result of this? 'On Iarga we are all rich!' the video lecture declares. By now, Beers is ready to denounce his former way of life, condemning capitalism, or 'our stupid way of sharing prosperity', in the following terms: 'Perhaps we can't help it because our mentality is wrong, but no matter how you try to excuse it, it remains stupid.'[28] But isn't it also a bit stupid to present a supposedly true experience in the form of fantasy? Towards the end of his book, Beers is specifically instructed by the Iargans to present his novel as science fiction, upon the following logic:

> You must leave people free to believe it or not, as they choose. If anyone should ask you if it really happened, you must deny it and say that it is pure imagination. The people for whom this book is destined will say: "I am not interested whether it really happened or not; for me, it is true. I have changed my insight and now I live consciously. I know the meaning behind life."[29]

Or, in other words, Beers made it all up. I suppose I am probably not one of 'the people for whom this book is destined', but such optimistic, well-meaning folk do exist. One such man is Gerard Aartsen, a Dutchman and 'student of the Ageless Wisdom teaching for over thirty years', who is apparently big in the world of exopolitics, and has examined the accounts of men like Beers, Adamski and other such Contactees with a more trusting mind than my own. Rather than having to do with government-related conspiracy theories around UFOs and suchlike, Aartsen now thinks that exopolitics should be defined as follows: 'People from other planets showing humanity alternative, saner ways of organising society, without imposing their views.'[30] Traditionally, 'aliens' had performed this task through fiction like that of Henry Olerich and Alexander Bogdanov. Post-Adamski, they seem instead to have begun doing it for real – in the eyes of the gullible, at least.

# II

# Atomic Aliens: The Interplanetary Campaign for Nuclear Disarmament

Another standard trope of the Left commonly to be found within early alien-encounter cases is the idea that the West should unilaterally surrender its nuclear weapons. The peak of this kind of alien encounter came during the formative years of the Cold War, with several cultural products at the time hammering home the thought that ET is probably watching, and he might just be a peacenik. Hollywood was at the forefront of spreading such a message. Most famously, there was the 1951 movie *The Day the Earth Stood Still*, in which human-looking Space-Brother Klaatu and his invincible robot Gort threaten the planet with destruction if it doesn't renounce its atom-splitting ways immediately, a storyline many Contactees of the day ripped off shamelessly.[1] The other side of this coin, meanwhile, could be seen in paranoid classics like 1953's *Invaders from Mars* and 1955's *Invasion of the Body-Snatchers*, in which the aliens become metaphors for potential Commie infiltrators during the age of McCarthyism. With such ideas in mind, might the Contactees themselves have represented a kind of Red Menace within, some wanted to know? Men like Gabriel Green (1924–2001), Head of the Amalgamated Flying Saucer Clubs of America, who ran for President as the 'space candidate' in 1960, and then to be California's State Senator in 1962 on an anti-nuclear platform, hardly sounded like true, red-blooded Americans. In telepathic contact with men from the planet Korendor, Green passed on their controversial message to the US electorate that the 'true way to the stars' lay in love and pacifism, not 'missile-fizzles and launching-pad blues'.[2] Was Green serious, or did his true contacts sit in the Kremlin, not on Korendor? Many Contactees were mocked, ridiculed and even feared by mainstream 1950s society not only as kooks and loons, but as possible

Communists and dangers to society, too. Such fears were not wholly unfounded. This, for example, is George Adamski writing about life on alien planets in 1957:

> Their means of exchange is a commodity and service-exchange system, without the use of money. All production is for the benefit of everyone, with each receiving according to their needs. And since no money is involved, there are no rich, there are no poor. All share equally, working for the common good.[3]

Even when Adamski's Space-Brothers took their Space-Sisters on a night out to the local hop, it appears that they danced in an inherently collective, socialist manner, disgracefully disowning the great individualistic 1950s American invention of rock and roll. According to one blonde alien doll-woman of Adamski's acquaintance, named Kalna:

> Our social dancing is usually of a group-pattern ... as the poem-form in words can suggest deep feeling not possible to the prose-form, so it is with the perfect rhythm expressed in the movement of a body dedicated to a dance of [collective] worship ... We could derive no joy from the kick, wiggle and hop we have observed on your Earth, during which a man and a woman clutch each other ferociously one moment and fling each other off the next.[4]

Hanging around with such uncool, un-American cats like that, it was no wonder that Adamski ended up being investigated by the FBI; most of the early 1950s Space-Brothers sounded as if they would have been much happier landing in Red Square than at Roswell. Indeed, one ungenerous view of the Contactee movement was that it was simply a front being run from Moscow, intended to spread the Bolshevik message to the gullible. For example, in 1960 the FBI received a phone call from an attendee of Adamski's lectures who was growing increasingly concerned at the nature of the message he was preaching. According to an official FBI report on Adamski, this woman had recently begun to wonder if her former idol was 'subtly spreading Russian propaganda':

> She said that, according to Adamski, the 'Space-People' ... told him the Earth is in extreme danger from nuclear tests and that they must be stopped, [and] that they have found peace under a system in which churches, schools, individual governments, money, and private property were abolished in favour of a central governing council, and nationalism and patriotism have been done away with ... It was hinted that the Russians receive help in their outer space programmes from the 'Space-People' ... It occurred to [our informant] that the desires

and recommendations of the 'Space People' whom Adamski quotes are quite similar to Russia's approach, particularly as to the ending of nuclear testing, and it was for this reason she decided to call the FBI.[5]

As this implies, the most common concern about the Space-Brothers was that they all appeared to be a bunch of long-haired, sissy peaceniks – a derogatory term derived, of course, from *Sputnik*. There was real fear that Russia might be somehow exploiting the Contactees to try and gain public support for banning further US atom bomb tests, thus allowing the Soviets to gain an upper hand.[6] According to another FBI informant on Adamski, not only had he told his followers in 1950 that the US Government had secretly met the Space-Brothers themselves and knew they were all Communists, he had also predicted that, pretty soon, Russia would grow to 'dominate the world'. Happily, this would then lead to a period of global peace lasting for 1,000 years. Before this could happen, however, America itself would have to be destroyed. Adamski claimed in conversation with the FBI's mole to be party to secret knowledge that the Soviets already had the hydrogen bomb, and were planning to drop ten of them on the US' most important population centres, in a surprise Pearl Harbour-style attack intended to cripple the nation before it could respond in kind. America deserved its fate, said Adamski, as capitalists were simply 'enslaving the poor', and the USA had grown corrupt and ready for a fall, just like the decadent late Roman Empire.[7]

## Agents of Absurdity

George Van Tassel was also investigated by the FBI, although such was his naivety during an official interview with two of its agents held in November 1954 that he mistook the nature of their interest entirely. Presuming they must have been fellow saucerians, he eagerly arranged to send them out some free subscription copies of his regular newsletter detailing his doings with the Space-Brothers, considering this a certified sign of belief in his claims from the government.[8] In fact the FBI were scouring through his writings in search of evidence Van Tassel was really a Communist, which is also why they sent undercover agents to listen to his lectures, posing as True Believers. At one 1960 speech, after sitting through a load of mad revisionist rubbish about Jesus and Mary being aliens and potatoes being brought down to Earth from other planets in a 'flaming canoe', one FBI man suddenly sat up in his seat and began to pay more attention. The Space-Brothers, said Van Tassel, were becoming very concerned about mankind's current atom bomb tests. All these explosions, he said, were slowly but surely shifting the Earth upon its axis, which would ultimately lead to a kind of polar shift, whereby 'the tropical climates become covered with ice and vice-versa', killing millions. The aliens had tried to warn the US Government, but they were too stubborn to listen. Whilst careful not to encourage his audience to take any specific action against the authorities, he nonetheless implied that, as 'intelligent people', they would all know what the right thing to do was 'when the critical time came'.[9] Was this an attempt at fostering revolution? Ultimately, the FBI concluded not. Their final assessment of Van Tassel was that he was simply 'a mental case', not the new Lenin.[10]

Over the pond in Britain, George King and his Aetherius Society were also attracting attention from the authorities. King had long been a pacifist and was registered as a conscientious objector during the Second World War, when he acted as a section leader with the National Fire Service during the Blitz.[11] Furthermore, he had already seen the Earth come perilously close to oblivion once before, when at some point during the 1950s a race of friendly Martians had kindly diverted a gigantic life-destroying missile from our path. It transpired that this interstellar ICBM had been aimed at our planet by a race of alien fish-men whose own dying planet was rapidly succumbing to a state of Lowellian super-drought. Greedy for *lebensraum* in our own capacious oceans, the fish-people had nearly succeeded in their scheme because Aetherius had happened to be away on holiday at the time, but thankfully the Martians had stepped in at the last moment to fire a giant thunderbolt at it, or so King said.[12] The Martians were already well disposed towards Earth because some years earlier, during a trip to the planet using his astral body, King had personally destroyed an enormous sentient super-asteroid, 'the size of the British Isles', which was heading towards the Red Planet with evil intent. To achieve this feat, which had proved beyond all advanced Martian war machines, King had simply used his own personal 'weapon of love',[13] a version of which he was now shamelessly whipping out in public, live on stage at London's Caxton Hall, in the shape of a long string of anti-war messages being channelled through from The Master Aetherius. As shown by the researchers David Clarke and Andy Roberts, Scotland Yard's Special Branch took a close interest in King's activities throughout the late 1950s, albeit only after being prodded into action by the ultra-right-wing newspaper *The Empire News*. One of their journalists had looked through an issue of King's regular newsletter *Cosmic Voice*, and found the following channelled statement from a rather Jeremy Corbyn-like being called 'Space Sector Six':

Have not the latest peace moves come from Russia? You in the West blame Russia and say it is necessary to make these [nuclear] weapons to protect yourselves from them. YOU IN BRITAIN ARE IN A FAVOURABLE POSITION TO SHOW THE LARGER COUNTRIES THE WAY.[14]

An employee of *The Empire News* then took a copy of this newsletter to Scotland Yard, said it showed 'a bias towards Communism' and asked officers if they were looking into King and his Society. The police said they were not, but the journalist left his copy of *Cosmic Voice* with them anyway, for further examination. Then, two days later, the newspaper used the fact of their own denunciation to claim that the police were investigating the society upon suspicion of being Soviet dupes! Scotland Yard then sent a man out to interview King, but the detective reassured him he was not in any trouble, reporting back to his superiors that King was simply 'a crank, albeit a harmless one'. Nonetheless, the Yard kept a discreet eye on King until he finally left for Hollywood in 1960, where his pacifist but hardly illegal message – which the police concluded was indeed 'closely allied with that of the Communist Party' – would find a more receptive audience.[15]

## A-lan the Alien

Pacifist aliens had a longer history in the south-west deserts of the USA than anywhere else, with the most significant CND-style saucer-story of the early days being told by a man named Daniel Fry (1908–92), the back cover of whose 1954 book *The White Sands Incident* describes him as being 'recognised by many as the best-informed scientist in the world on the subject of space travel', which makes it all the more surprising that he had to resort to buying his PhD from a fake university in London.[16] Fry (or so he said) had once been an engineer working for the prestigious Aerojet Corporation on the US military proving grounds at White Sands, New Mexico, where various rockets and other new-fangled propulsion systems were tested out away from the prying eyes of the public, amidst the unforgiving glare of the desert sun. After missing a bus out of town on 4 July 1949, Fry had gone out for a walk in the lonely Organ Mountains, when he unexpectedly encountered 'a spheroid ... like a soup-bowl inverted over a sauce dish' floating in the sky above. Then, a disembodied voice calling itself A-lan asked him if he would like to climb inside the sphere – which was really a kind of large, hollow *Sputnik* in the days before *Sputnik* had even been launched – and go for a ride. Fry said he would, and was taken on a flight all the way out to New York and back within the space of half an hour. During his journey, A-lan filled Fry in on the secret history of the Earth. It turned out that the sunken continent of Atlantis had been a real place, as had its twin counterpart Lemuria, upon which A-lan's ancestors had once lived. However, rivalry between Atlantis and Lemuria had grown, just as it was then growing between America and Russia, leading to the two lands destroying one another with nukes, and Earth's subsequent abandonment. With this terrifying history in mind, A-lan left Fry with a parting message: 'it should always be remembered that an ounce of understanding is worth a megatonne of [nuclear] deterrent'.[17]

Fry later claimed to have heard A-lan's voice upon subsequent occasions down here on Earth, their conversations proving the basis for a cult he started in 1954, called Understanding Inc, which at its height in the early 1960s possessed 1,500 members and 55,000 acres of land in Arizona, filled with buildings resembling classic disc-based flying saucers.[18] Why did his story touch such a nerve? I am led to wonder if, once again, the secret lies not merely within Fry's – sorry, A-lan's – message, which was after all rather banal, but in its interplay with a wider mythic landscape, or 'system-animal', as Colin Bennett might prefer. As mentioned, White Sands was the home of AeroJet, a company which had been co-founded by none other than Jack Parsons, the wildman occultist and father of rocket-fuel technology whom we met earlier. Aerojet was originally based in California, where in 1942 Parsons had devised special types of fuel and fuel preparation which, following subsequent modification by others, eventually ended up being used both to help launch the US space shuttle, and to fire off America's *Minuteman* and *Polaris* missiles – the very same nuclear-tipped ICBMs which would have been used to destroy Russia and China, should the Third World War ever have actually broken out.[19] In other words, White Sands was one of the epicentres of the new, emerging post-war military-industrial complex, as were many of the desert areas of the American south-west; Clyde Tombaugh, who spent the war years teaching navigation to naval personnel, also worked at White Sands during the early 1950s, having several UFO sightings there.[20] Nuclear weapons themselves had been developed out at Los Alamos in New Mexico, but the means for actually launching them directly at foreign powers was created in areas like White Sands. Perhaps the real elemental force Parsons had helped conjure up in the Mojave Desert during his sojourns there was not Babalon, but Armageddon?

## A SIGN of the Times

Is it any wonder that, apparently as some kind of compensatory mental factor, so many of these new super-advanced, yet entirely peaceful, flying machines from Mars and Venus emerged initially from within the very same desert landscape that the new hyper-destructive military technology of the 1940s and '50s had come? The schizoid madnesses of men like Adamski, Van Tassel and Fry now read like a kind of imaginative revenge by the nuclear-scarred landscape upon the society which had so disfigured it. Connections between early ufology and the US military were manifold. Alfred Loedding (1906–63), of the Army Air Corps, was both the military's first ever UFO investigator, helping head up a scheme called Project SIGN in 1948, and the man who had been sent out to inspect some of Parsons' work at Aerojet to see if it was really of any value to the war effort.[21] But why were the military during this period interested in investigating UFOs? Because, during the initial wave of American saucer sightings in the late 1940s, very few people thought that they were spaceships at all. In 1947, the polling organisation Gallup released results of the first public survey into what people thought the saucers were. Most Americans who didn't simply respond 'don't know, go away' dismissed them as optical illusions or hoaxes, but the next most recorded response was that they were new secret weapons of some kind, probably created by their own government. The notion they were from outer space didn't even register.[22] However, Washington itself knew full well it hadn't created the saucers, and a number of investigators involved in Project SIGN concluded that the Earth may well be being visited by extraterrestrial beings. But if so, why now? A formal report on Project SIGN's findings, released on 27 April 1949, whilst rejecting the idea of alien invasion, outlined a hypothetical scenario in which:

> Martians have kept a long-term routine watch on Earth and have been alarmed by the sight of our A-bomb shots as evidence that we are warlike and on the threshold of space travel ... The first flying objects were spotted in the spring of 1947, after a total of 5 atomic explosions [on Earth] ... Of these, the first two were in positions to be seen from Mars ... It is likely that Martian astronomers, with their thin atmosphere, could build telescopes big enough to see A-bomb explosions on Earth.[23]

In the years since this memo was published, it has become a staple idea that the dawning of the nuclear and Space Ages together during the 1940s and '50s might have spurred aliens into taking a closer look at our troubled planet. Roswell, for instance, was home of the 509th bomber group, the same unit which had dropped the atom bombs on Japan in 1945,[24] and 1947, the year of Kenneth Arnold's first saucer sighting, also saw the invention of the transistor and the breaking of the sound barrier by US pilot Chuck Yeager (b. 1923).[25] Technology was advancing at an ever faster rate, indicating a world gone mad, in which the myth that captured alien tech had been back-engineered by the US military became one of the few ways of accounting for it all. Another myth was that alien eyes saw all this going on from afar, and were growing increasingly concerned. By October 1950, Britain's *Sunday Dispatch* newspaper was announcing as its official editorial line that aliens had been watching us from space for at least 200 years, with observations being upped following the 1945 nuking of Japan.[26] On 1 November 1952, meanwhile, America tested its first hydrogen bomb, whose force was greater than that of all the bombs dropped in the two World Wars combined.[27] Less than three weeks later, Orthon had made first contact with George Adamski. By the time of the Cuban Missile Crisis of 1962, when the world teetered on the edge of total oblivion, there were even dubious reports of aliens being seen watching TV coverage of the affair on their saucers, making mournful comments to the effect that 'That's how our planet ended, too'.[28]

However, it would be simplistic to say that the tensions of the early Cold War 'caused' the whole UFO phenomenon. Strange things have always been seen in the skies down the ages and interpreted accordingly, whether as dragons or as angels and, as the 1947 Gallup poll showed, during the phenomenon's early days mainstream public opinion remained firmly set against the possibility of alien intervention. In 1948, even US Naval Intelligence preferred to speculate that the saucers were some strange form of Soviet psychological warfare, aimed at fooling America into thinking that their precious atom bombs were not the ultimate weapons they had presumed them to be.[29] Indeed, the UFOs seen during the first major wave of post-war sightings of strange things in the sky were not alien saucers at all, but phantom Soviet rockets. Throughout 1946 in Sweden, the coming global UFO craze was prefigured by a wave of mass hysteria in which tens of thousands of citizens reported seeing Russian missiles infiltrating their airspace. Towards the end of the Second World War, Soviet troops had captured Peenemünde, the Nazi centre of ballistic missile research, from whence Wernher von Braun's V-2 rockets had been launched against London. As Swedes all across the country began reporting strange objects in the sky, speculation gathered that Stalin was ordering the captured rockets to be fired into Swedish territory as a precursor to invasion. The fact that such a tactic would make no military sense was ignored amid the panic. Later analysis showed that the peak time for sightings of the Soviet 'ghost-rockets' – so called because no trace of them having landed could ever be found – coincided with a series of unusual astronomical events like large meteor showers and spectacular auroras, suggesting that many witnesses really were seeing strange things in the sky, but then misinterpreting them in light of then-prevailing social fears.

That people were seeing what they most expected (or dreaded) to see during the panic can be illustrated by the fact that the first specific rocket sighting, on 24 May, was described by one set of witnesses simply as 'a fireball with a tail', whilst another swore blind the object had been 'a wingless cigar-shaped body' like a missile, spraying out exhaust sparks. Presuming the object really was something like a meteor, the first account would have been the more accurate one, with the second a more imaginatively charged distortion of reality. The Swedish military began actively soliciting sighting reports from the public to check them out, giving the scare apparent credibility, something aided further by uncritical Press reports. The ghost-rocket scare even had its own equivalents of American pseudo-events like Roswell, with investigators called out to examine supposed missile crash sites. Seeing as a few stray German V-2s had actually landed in Sweden during the war, leaving behind large craters, people should have known what to look for, but Swedes began calling the authorities out to examine all kinds of innocuous marks on the ground and items of random scrap. Trivial events like collapsed barns and fires in henhouses were attributed directly to the V-2s, as if the USSR had declared war on farmyard animals, and when three cows dropped dead in Jamtland, their owner blamed a Russian missile for spraying out poison gas.

A caterpillar infestation that struck southern Sweden during July, meanwhile, was blamed by one paranoid correspondent of the Swedish Defence Staff upon their eggs being scattered from the air by the mystery missiles in some new Soviet form of biological warfare. In the end, the affair fizzled out as it became increasingly clear that the whole thing was simply an early example of Cold War hysteria.[30]

## Jung Love

The most perceptive early commentator upon UFOs was the great Swiss psychoanalyst C. G. Jung (1875–1961), whose 1958 book *Flying Saucers: A Modern Myth of Things Seen in the Skies*, is still an absolutely central text of ufology. Jung saw UFOs as being the consequence of a Cold War world torn asunder between the two competing factions of East and West. Under the novel and constant threat of nuclear annihilation following the USSR's successful testing of an atom bomb in 1949, Westerners in the 1950s still retained the traditional religious desire to look up towards the Heavens in search of salvation, Jung argued. As he had it, 'The present world situation is calculated as never before to arouse expectations of a redeeming supernatural event.' However, religious belief was on the wane in the developed world, said Jung, meaning it was hardly likely people were going to have visions of angels descending to save them from destruction. Instead, a new image was needed, one more apt for our modern-day, materialistic society – and the saucers filled this role perfectly. 'Anything that looks technological goes down without difficulty with modern man,' he observed. In an era of unparalleled scientific advance, the UFOs were gods for a new, godless age which would not accept them – unless, that is, they came disguised as something other than gods.[31]

When things split asunder, says Jung, a compensatory image emerges from within the collective human psyche, mediating between the two split halves and pointing the way back towards wholeness.[32] A student of alchemy, Jung could not help but notice that the insignia of the American and Russian air forces – whose planes, during the early nuclear age, would have been used to drop atomic payloads, not ICBMs – were two five-pointed stars, one white, one red. In alchemy, substances of these contrasting colours were considered to be feminine and masculine respectively. To mix such materials within the alchemist's glass vessel would be to unite them symbolically in a so-called *coniuncto oppositorum*, or 'marriage of opposites', producing some new substance superior to either ingredient individually, with the whole naturally being greater than the sum of the parts. To unite Red Square in peaceful marriage with the White House was therefore what was needed if the world was to survive.[33] The ubiquitous UFOs, proposed Jung, were like harbingers of this desired marriage in some way, circles being traditional symbols of both wholeness and of God Himself – think of the circular marriage ring, or the old medieval model of the universe we examined earlier. God, said Jung, was 'a totality symbol *par excellence*, something round, complete and perfect', capable of uniting even the 'irreconcilable opposites' of Marx and market.[34] Specifically, Jung calls the UFOs *mandalas*, ancient circular Indian Buddhist symbols of God and spiritual wholeness whose name is derived from the Sanskrit for 'circle'. The UFO *mandala*, says Jung, has been 'superimposed on the psychic chaos' of modernity by mankind, in an unconscious plea for peaceful unity.[35]

However, rather than wedding bells, warning sirens were now in the air. The splitting of the atom and the splitting of the world in two were aspects of the precise same problem. As such, the atom too had to be reclaimed as a symbol of wholeness and peace, not explosive division. Jung knew the tale of a California Contactee called Orfeo Angelucci (1912–93), author of the 1955 book *The Secret of the Saucers*, and noted with interest the fact that, following his own claimed experiences with the Space-Brothers, Angelucci had developed a strange burn mark on the left side of his chest, shaped like 'an inflamed circle with a dot in the middle'. Jung viewed this as embodying the old saying that God is 'a circle whose centre is everywhere and circumference nowhere',[36] but Angelucci's own analysis of the round burn was that it was 'the symbol of the hydrogen atom'.[37] What was this stigmatic sign trying to say? Well, the destructive potential of the atom split asunder was huge indeed; but its proximity towards Angelucci's heart spoke of the healing potential of love to draw what has been torn apart back together again. Angelucci worked as a mechanic at the Lockheed aircraft plant in Burbank – the very same place where George Van Tassel had himself once toiled away prior to quitting to go and seek saucers out at Giant Rock. Lockheed, of course, makes fighter jets, and so was one of the major players in the newly emergent military-industrial complex whose growth was then being encouraged to keep America safe. In short, Angelucci was engaged in the upkeep of Washington's war machine, whether he liked it or not – which apparently he didn't. Angelucci secured employment at the Lockheed plant on 2 April 1952. He first met the Space-Brothers on 23 May 1952, less than two months later.[38] Most people who find their jobs unbearable quit and leave town. Angelucci went one further, and left the planet.

Angelucci's absence turned out to be only temporary, because one day in 1953 he suddenly 'awoke' in the Lockheed factory, tools in hand, after spending a week's mental vacation on a small planetoid known as Lucifer. It was only by examining the date on that day's newspaper he was able to realise how long he had been away. Apparently, during his absence Angelucci had still been turning up to work, interacting with colleagues and fulfilling his duties, but this was only because one of the Space-Brothers had taken remote control of his physical body whilst his soul was away amongst the stars. Lucifer, which was hidden craftily somewhere between Jupiter and Mars, was a veritable Heaven, with glowing flowers, incessant angelic music, no pollution, crystalline colour-shifting buildings and, naturally, no sign of atomic warfare. Moreover, all of Lucifer's inhabitants were 'statuesque and majestically beautiful', particularly a certain Space-Sister called Lyra, with long golden locks and beautiful big blue eyes. Angelucci liked Lyra and began to have lustful thoughts about what might be lurking beneath her Ancient Grecian gown, but unfortunately it turned out that both she and everyone else on the planet was highly telepathic, leading to much embarrassment when they all realised what he was thinking about doing to her.[39] Fortunately, as an inferior animal-like being incapable of not thinking such vile thoughts, Lyra forgave him much as a pet owner might forgive a dog for trying to mate with her leg, and tried to teach him that such a thing as a spiritual union might be possible instead of any sexual one. At this point, said Angelucci:

> I turned and looked into Lyra's wonderful eyes shining with sympathy, compassion and purest love. My own heart swiftly responded. Then suddenly, miraculously, we were as one being, enfolded in an embrace of the spirit untouched by sensuality or carnality.[40]

Jung interprets this shared rhapsody as being another *coniuncto oppositorum*, like that which he felt had to happen between the US and USSR – or, indeed, between Truman and Khrushchev, if you can bear to picture such an image.[41] Instead of splitting the atom marked out above Angelucci's heart, could its bonds not be made stronger through such a form of geopolitical marriage? During their various other encounters down the years, the Space-Brothers told Angelucci that mankind's level of moral development had not kept pace with his technological advance, but held out hope for the future, saying that they too had once had to endure such a period of 'growing pains' and that as such they looked down on Earth from above with 'deep compassion and understanding', being 'simply older brothers' who wished us well.[42] The implication is that one day humanity will shed its warlike ways and evolve up towards the Luciferians' own high level of social and ethical development, with all threat of nuclear oblivion being banished from the globe forever. Such benign Space-Brothers are like the subsequent personifications of Jung's *mandala*-like god-discs, growing from out of them amidst the madness of the Cold War to spread a message of unity and love. In short, it seemed that the *Sputniks* had begun spewing out peaceniks!

# 12

# From Sputniks to Beatniks: Getting out of This World by Whatever Means

Another Contactee peacenik, George King, had correctly anticipated the spread of the Cold War battleground into outer space and taken steps to neutralise the negative emotions which were later to be released by *Sputnik* several years before the terrifying device was even launched. King, you see, was already aware of another such object which had been orbiting away up there in Earth's atmosphere for some time now, called 'Satellite Number Three'. This was nothing less than a 'satellite of love' placed above Earth by the ever-watchful Space-Brothers, and King, during a trip he made into outer space using his astral body on 23 March 1956, had actually been there! Overseen by a mysterious Martian scientist, the purpose of the satellite was to detect the positive waves of energy beamed out by any Earthlings who should happen to perform selfless acts during its orbit overhead before using crystals to amplify them by a magnitude of 3,000, bathing our planet in love. According to King's own eyewitness account, Satellite Number Three is 'colossal' in size, oval in shape and filled with shifting coloured lights, like an early disco. In the ceiling is a large glass window to allow in positive energy rays from space, which then pass down through a series of pyramidal prisms before being focused into a magical 30-foot crystal which floats below in the satellite's Operations Room, 'apparently defying the known laws of gravity'. From here, the cosmic rays are beamed down to Earth and absorbed by do-gooders, amplifying their benevolence many times over, and adding to the peace of the world. Unfortunately, however, mankind was largely unaware of Satellite Number Three's presence because it was 'totally invisible to physical human eyes' and to radar, thus meaning that only advanced adepts and avowed peaceniks like George King himself could ever hope to see it, let alone then climb aboard the amazing satellite of love.[1]

Or, rather than peaceniks, maybe we should instead label the Space-Brothers and their human friends more accurately as being beatniks, a term coined in 1958 to describe those new angry (or often simply stoned) young men who rejected mainstream society in terms of its morals, rules and traditional work ethic.[2] Some beatniks were very interested in the Contactee movement. The author, occultist, homosexual, drug abuser and general all-round lunatic William S. Burroughs (1914–97), for example, was so taken with accounts of people who claimed to have enjoyed sexual congress with aliens that he took to mowing obscene designs into his lawn in the shape of giant erect penises, hoping to lure man-hungry gay space rapists down from their craft and into his bedroom.[3] So interested did Burroughs become in the topic that in the 1980s he was asked to review a book about the Contactee movement, called *In Advance of the Landing*, and obliged by praising both it and the Contactees' outlook on life. 'These eccentric individuals,' he wrote, 'may be tuning in, with faulty radios, to a universal message: we must be ready at any time to make the leap into space.'[4]

The long-haired Space-Brothers anticipated both the beatniks and the later hippie movement in terms not only of their bizarre dress sense and fear of hair clippers, but in many other aspects of their New Age lifestyles, too – once George King moved to California, he will have been onto an instant winner by talking about the alleged 'healing power' of the magical crystals at work in Satellite Number Three, for instance. If Frank E. Stranges had thought on, he could have made a fortune selling crystals with an alien connection to them himself, at least if you believe his tale – unique, so far as I am aware – of once having been caught short in need of a shit whilst aboard Valiant Thor's flying saucer. Stranges' account is so perfect that it deserves to be reproduced in full:

> I went into the bathroom and was embarrassed to note the obvious absence of toilet tissue. Then it happened. I heard a voice within my mind which I immediately recognised as belonging to Val. He said, "Frank, look to your right. You will find three buttons. Push the first, then the second, and then the third, in that order." I could hear Teel's laugh (Teel was a pretty space lady) as I proceeded to press the first button. The sensation was that of a rapid warm wind similar to a jet of air, blowing beneath the seat. The process entirely crystallised my waste-matter and caused it to drop from me. Then the second button was another jet-blast of a different pressure and temperature. Finally, the third button. This produced a pleasant, fragrant substance that made me feel as though I had been washed, cleaned, powdered and perfumed. As I came out of the bathroom ... I had a strong feeling that everybody in the room knew what had happened. Suddenly, Teel broke the silence by saying, "Well, do you want to take one home with you?"[5]

Had Stranges indeed brought home just such a crystallised turd with him to show the world, he would surely have become a true New Age legend. Hand the item over to a competent jeweller, and you might easily get a few dozen pairs of 'lucky' crystal earrings or 'healing' crystal amulets out of the thing. In California, such things would certainly have found buyers.

## Brave New World

Something strange had happened out in the American south-west. Just as the angry, jealous God of the Old Testament, who had once appeared as a pillar of fire in the desert, had ultimately given birth to His direct opposite in terms of the long-haired, unshaven, sandal-wearing proto-Commie peacenik and friend of sinners Jesus Christ, so it seemed that the mushroom cloud-haunted deserts of the southern USA had inadvertently released Woodstock out onto the world alongside Armageddon – radiation does give birth to unexpected mutants, after all. The phallic ICBM was increasingly to be combated by the circular *mandala* logo of the CND as the '50s shaded into the '60s, with protesting youths sticking flowers down the gun barrels of soldiers on university campuses, but those who had been carefully reading their Contactee literature might have been able to predict the whole Flower-Power phenomenon several years beforehand. During one of his many trips into outer space, it seems that Orfeo Angelucci had once met none other than Jesus Christ (a sun-spirit, apparently), who told him that 'This is the beginning of the New Age'.[6] Indeed it was.

Presumably Angelucci's own journeys actually took place within inner space, not outer space, but expanding your consciousness was always one of the main aims of the beatnik generation – it's just that Angelucci didn't need any drink or drugs to become a psychonaut. Or did he? Angelucci had long been a sickly person, and on the night of his first encounter with the Space-Brothers had been feeling very ill indeed, something which made his drive home from the Lockheed factory at 12.30 a.m. rather unpleasant, with his eyes glazing over and the sounds from passing traffic becoming 'oddly muffled' as he slipped into an altered state of consciousness. Eventually, Angelucci stopped his car somewhere off the beaten track, whereupon he saw two small luminescent green balls which descended down from the sky, accompanied by a voice which gave him a strange instruction: 'Drink from the crystal cup you will find on the fender of your car, Orfeo.' The substance must have been strong stuff, as a large, three-dimensional TV screen then appeared floating in mid-air, showing the faces of an alien man and woman described as being 'the ultimate of perfection'.[7] Whilst this was evidently a genuine spontaneous hallucinatory experience – few people have accused Angelucci of entering this field for the money, unlike certain other Contactees – it does have definite echoes of a drug trip about it. The presence of the crystal cup is especially interesting as a kind of symbolic signal that Angelucci was about to see something weird; drink the draught from the magic cup, and you will have strange visions, Orfeo. Just as odd was his second taste of this nectar, which occurred in December 1954 when Angelucci was approached in a California diner by a stranger calling himself Adam, who claimed to have met the Space-Brothers himself. Partway through their conversation, Adam, who said he was a physician, produced a little white pill and told Angelucci to slip it into his glass of water. As the liquid bubbled and changed to a lovely golden hue, the Contactee took a drink.[8] After only two sips, he entered into 'a more exalted state' – that is to say, Angelucci got high:

> No longer was I in Tiny's Café in Twentynine Palms. It had been transformed into a cosy retreat on some radiant star-system. Though everything remained in its same position, added beauty and meaning were given to the things and people present there ... Were my ears deceiving me? Was that music I heard coming from the direction of the glass? ... Slowly I ... looked at the glass and was held in amazement. A miniature young woman was dancing in the nectar! Her golden-blonde beauty was as arresting as the graceful thrusts of her dancing body. Her feet were so light and responsive that the music itself seemed to emanate from them.[9]

One of the Space-Brothers' core messages to Angelucci was that every person on Earth has 'a spiritual, unknown self which transcends the material world and consciousness',[10] something which is perfectly true – it's called the unconscious, and one of the most direct paths towards encountering its seemingly inexhaustible storehouse of imagery is through psychoactive drugs and hallucinogens such as LSD, with California being an early centre of experimentation with such substances. Most famously, there were the trials made with LSD, psilocybin and mescaline by the English novelist Aldous Huxley (1894–1963), author of the classic dystopian sci-fi novel *Brave New World*. Huxley had left Europe for America in 1937, well known for his writing and his work as a peace activist, and spent most of the war years living in a small house in the Mojave Desert. Here, anticipating the prospect of Hollywood script work, he devoted himself first to meditation, and then, beginning in 1953, to experiments with psychoactive substances. The most lasting literary products of Huxley's time in California were his classic accounts of drug-taking, 1954's *The Doors of Perception* and 1956's *Heaven and Hell*, which were fast to prove major inspirations for the West Coast beat generation.[11] It was whilst staying in Los Angeles in 1953 that Huxley made his first experiments with mescaline. Like Angelucci, he took his pill dissolved in a glass of water, and quickly began to see changes in the world around him – which, just as for Angelucci, stayed somehow the same, yet different. Most famously, he spent what seemed like hours staring at his own trousers ('*Those folds ... what a labyrinth of endlessly significant complexity! And the texture of the grey flannel – how rich, how deeply, mysteriously sumptuous!*'[12]). This really was a brave new world Huxley was living in.

## Lucy in the Sky with Orthon

This is not necessarily to suggest that Angelucci had read Huxley and was simply riffing off him; it has also been proposed that the CIA might have made use of LSD doses to lull the early Contactees into a state of altered consciousness and then see what slipped out of them, to test if they really were Reds.[13] Another suggestion sometimes made is that the Contactees might have been exposed somehow to a powerful hallucinogen called DMT (DiMethylTryptamine), which occurs naturally in various plants and, in trace amounts, within the human body and brain itself; it is sometimes said to aid the sleeping mind in dreaming. Interestingly, people who take DMT often undergo encounters with entities described as being profoundly 'alien' in nature. One user, for example, described approaching a space station filled with 'android-like creatures ... like a cross between crash dummies and Empire troops from *Star Wars*',[14] whilst Terence McKenna, the noted researcher into altered states of consciousness, has described a bizarre experience he had during a trip to the Amazon rainforest during the 1960s when, under the influence of certain DMT-like jungle hallucinogens, he saw 'exactly the same saucer as appears in George Adamski's photos' flying directly overhead him and looking 'as phony as a three-dollar bill'.[15]

Terence McKenna's ideas about where precisely the human psyche goes (if indeed it goes anywhere) during such trips seem ambiguous. On the one hand, he can argue that the DMT experience 'is not one of our irrational illusions' but 'real news', one which gives us access to 'a nearby dimension' or alien world, just waiting to be explored. In this case, the aliens (who take such forms as robots, humanoids, insects, lizards and elves) people encounter when tripping would somehow be real.[16] On the other hand, McKenna also speculates that such beings may instead be 'reflections of some previously hidden and suddenly autonomous part of one's own psyche'.[17] If so, then I don't really see why you'd absolutely *need* drugs to meet them; altered states of consciousness need not be chemically induced. Sometimes, metaphorical cups full of mescaline will do just as well as real ones, if you want to see aliens. Aldous Huxley took the title for his druggy 1954 counter-culture classic from a line by the poet William Blake (1757–1827), a man notorious for having numerous bizarre, but wholly non-narcotic, visions of his own: 'If the doors of perception were cleansed everything would appear to man as it is, infinite,' Blake once wrote.[18] Blake saw angels, ghosts, fairies, Heaven and Hell through his own fully cleansed mental doors of perception; but a more modern version of Blake's infinity might well be the fictional, fairy-land version of outer space which the likes of Orfeo Angelucci have peered into.

The connection between UFOs, aliens and altered states of consciousness seems an instinctive one. The English writer and counter-cultural giant John Michell (1933–2009) first drew British beatniks' attention towards the world of ufology with his 1967 book *The Flying Saucer Vision*, which, expanding upon Jung's ideas, posited the notion that the saucers were some mystical portent of a forthcoming 'radical change in human consciousness'.[19] When LSD first hit the streets of Swinging London in 1965, one especially potent batch was known as 'flying saucers', whilst one of the hippest clubs of the day was called 'UFO', with psychedelic bands like Pink Floyd giving their albums names like *A Saucerful of Secrets*.[20] From the 1950s to the '70s, it seemed there was more than one way to give your mind a trip out of this world. Many of the early American Contactees would have fitted right in with the social milieu of peace and love unleashed by the likes of Aldous Huxley, a world of hippies, LSD, vegetarianism, pacifism and free love which, in many ways, their Space-Brother contacts had done so much to predict; the final sixty-three pages of Howard Menger's book *From Outer Space to You*, for instance, consisted not of descriptions of his lovely bra-less alien wife Connie, but a long treatise of health food advice, entitled 'A New Concept in Nutrition'. He even released an experimental record filled with so-called 'space music', an extremely unusual form of piano-based jazz he said had been composed by the Space-Brothers – far-out sounds indeed.[21]

## Bees Means Hives

Another dabbler in both drugs and alternative modes of living whose thought hit an intersection between the New Age and the Saucer Age was Gerald Heard (1889–1971), a highly erudite Cambridge graduate of Anglo-Irish origins who later found his true home in California. If Heard is known of at all by today's UFO fans, it is for his seemingly insane 1950 assertion that flying saucers were piloted not by humanoid Scandinavian-looking Space-Brothers, but by a race of super-intelligent quasi-socialist bees from Mars. To most who hear this idea articulated in isolation, Heard is instantly classified as being a nut, and his thinking dismissed out of hand. However, this situation is actually quite unfair, seeing as Heard was in truth no loon, but a genuinely intelligent and respected writer, philosopher, broadcaster and spiritual sage who mixed with many of the global intelligentsia of his day. A real freethinker – his first book, 1924's *Narcissus: An Anatomy of Clothes*, appears to contain an early and forgotten version of what we might now call 'meme theory'[22] – Heard was unafraid to put forward bizarre or unpopular ideas, for good or ill.

Heard's initial literary foray into bee-related matters occurred in 1941 with the publication of his best-selling detective novel *A Taste for Honey*, which was both the first true full-length Sherlock Holmes pastiche and the first ever entry in the whole 'killer-bee' genre. Focusing around the actions of an evil beekeeper who has managed to train his swarms to murder selected victims, the book, published under the name H. F. Heard, was the basis for the obscure 1966 British movie *The Deadly Bees*.[23] Heard's later incursion into insect-based narratives, however, came in 1950 with *The Riddle of the Flying Saucers: Is Another World Watching?*, which holds a place in the record books as the first full-length non-fiction book about the subject published by a British author. The title was serialised in the *Sunday Express* throughout October 1950 as part of a circulation war, with the idea that alien bees might be watching over us from afar being considered sensational enough to be splashed all over the front page.[24] We can see from this that Heard's ideas got a generous enough public airing considering their incredible weirdness, but how seriously did he intend them to be taken? He apparently wrote the book in its entirety in a mere three weeks,[25] and it does contain a sprinkling of humour in parts, with comic phrases like 'bee-masters of the planet' and 'Why in the name of Space and Solids did they not come before?' surely being deliberately intended to provoke smiles in the reader.[26] Furthermore, in later life Heard abandoned his whole thesis in favour of endorsing Jung's ideas about the saucers being a kind of 'modern myth', and he never wrote another word upon the subject.[27] However, considered overall the book does not quite deserve to be described as the mad rubbish it is so often dismissed as.

It is worth reminding ourselves that, at the time Heard was writing, it was still not absolutely certain that Mars was a totally dead planet – indeed, so culturally pervasive had the idea of Martian life become that Heard himself felt moved to comment that the very notion was 'a little vulgar'.[28] Even several years after Heard's book had been published such notions were still being advanced. In 1958 the Harvard astronomer William Sinton (1925–2004), for example, claimed to have proved via spectroscopic analysis that there was sugar on Mars, a substance he presumed must have been produced by vegetation.[29] Sinton's claims would surely have been welcomed by any remaining enthusiasts who still clung doggedly to Heard's 1950 hypothesis at this late date, seeing as he had already speculated himself that the Martian bees, having access to various advanced (though presumably tiny) items of super-technology, might well have got around the virtually flower-free conditions of their desert homeworld by creating a form of artificial sugar from some unspecified 'synthetic substance' which would in effect be a new industrial form of chlorophyll. By mixing this with water and exposing it to air and sunlight, the natural chemical processes of certain plants could then be imitated on a vast scale, said Heard, creating huge supplies of sugar for the bees to eat virtually on-demand.[30] As for the whole idea that the saucer pilots must have been bees in the first place, it seems that Heard drew his inspiration from a report in the *Los Angeles Times* regarding an astronomer named Dr Gerard P. Kuiper (1905–73), who in March 1950 had made an idle remark to the effect that, Mars being such a barren planet, the only advanced beings who could possibly have survived up there must have been a breed of intelligent insects. Kuiper was scathing about the whole contemporary UFO-mania, and had meant this comment sarcastically, the implication being 'How the Hell could an insect fly a spaceship?', but Heard does not appear to have realised this, and reported Kuiper's words in his book seemingly at face value.[31]

Then, Heard quickly turned Kuiper's real opinion on its head, arguing that in fact bees made *ideal* spaceship pilots, not terrible ones. 'We need not bring their physical weakness against a set of super-bees, provided that their knowledge has, in the Baconian phrase, been turned by them into power' through their mastery over science, says Heard.[32] Examining recent Press reports of flying saucers, Heard concludes from tales of their impossibly speedy gravity-defying manoeuvres that the bees' spaceships are able to accelerate to immense speeds from a complete standing start. Put a humanoid in such a craft and they would be crushed by the sheer power of the g-forces involved, splattering against the walls as soon as they so much as started the engine. Due to their small size and the 'amazingly tough' structure of bees' chitinous bodies, though, Heard thinks that they alone would be able to survive the dangers of such accelerative force. Also, the way that bees seem to defy gravity themselves when flying about on Earth might have provided Mars' super-bees with the initial design inspiration for their later creation of the saucers, he proposes![33] Moreover, Heard reasons that the bees, as undisputed Lords of Mars, would have escaped from the usual pressures of natural selection, and evolved into what can only be described as a species of living jewels, with bodies every bit as hard and polished as their fantastic mega-minds, becoming:

A creature with eyes like brilliant cut diamonds, with a head of sapphire, a thorax of emerald, an abdomen of ruby, wings like opal, legs like topaz – such a body would be worthy of this "super-mind" … [In their presence] it is we who would feel shabby and ashamed, and maybe, with our clammy, putty-coloured bodies, repulsive![34]

Compared to the Martian super-bees, it was us pathetic humans – with our skins 'mainly the tint of a toadstool'[35] – who now seemed the lesser breed; and that was the point. Reading between the lines, it becomes apparent that Heard was hoping that mankind would one day transform into a species of god-like humanoid hive dwellers itself, too.

## Hive-Minds

Heard didn't have a great opinion of the supposedly 'advanced' state of Western civilisation. A pacifist campaigner just like his good friend Aldous

Huxley, in whose company he had left England for America in 1937, Heard in his book eagerly subscribes to the Contactee-like view that the saucers were now visiting Earth in such numbers due to our development of nuclear weapons. In January 1950, astronomers in Japan reported an unexplained sighting of what looked like a gigantic explosion on the Red Planet, in which a huge plume of dust some 40–60 miles high and 700–900 miles in diameter billowed above the Martian surface. This coinciding with debates in the US about whether or not to build an H-bomb, some drew connections between the two events, with Heard himself hoping that the blast did not signify that the bees were not quite as peaceful and spiritually advanced as he had previously guessed. Perhaps in pursuing our own Cold War against Russia down here on Earth, the West had inadvertently fired the starting trigger on a second Cold War front with the insects of Mars, too?[36] Recalling the 1959 event at White Sands Missile Range in which one of Wernher von Braun's test missiles had supposedly been deflected off-course by aliens, Heard proposed that maybe a hovering mothership full of bees had shot it down as an imminent threat to their vehicle.[37] Might such accidental events have convinced the bees we were hostile?

Phobos and Deimos, the twin moons of Mars, may well have served as rapid-launch hives for the bees' saucers in case of perceived military threat, proposed Heard.[38] All those years humanity had wasted staring at the Martian canals, (to which he apparently gave credence[39]) laments Heard, when we really should have been examining the bees' preparations for future invasion![40] Heard uses the researches of the noted American anomalist Charles Fort (1874–1932) – the first man to propose that strange lights seen in the sky might be alien spacecraft, and whose books contained many accounts of astronomical anomalies – as evidence that the bees had been monitoring us silently from above for 150 years, quietly assessing whether or not we were likely to be any threat to them.[41] Following America's dropping of atom bombs on Japan in 1945, it appeared to Heard that Mars' monitoring of Earth had stepped up a gear. Maybe the bees were worried we would blow up the entire planet through our foolishness, or even make the sun itself begin giving out harmful rays, or explode somehow? Or, worse yet, maybe the dust from our ruined Earth would float into space and hover in a big belt around Mars, blocking out the sun's light and thereby impeding alien honey production? The potential reasons for the bees to fear and hate us were endless![42] As Heard pithily put it, 'Sky-suicide is no private matter.'[43]

There was still hope, however, because Heard, being a one-time science correspondent for the BBC, was intimately acquainted with the researches of the Austrian entomologist Karl von Frisch (1886–1982), who in the 1940s had discovered that Earth bees can communicate with one another via the medium of dance – bees use their so-called 'waggle dance' to tell one another in great detail where good flowers for harvesting pollen are located.[44] Of course, Earth bees tend only to talk to one another about mundane matters like honey, but this problem was not insurmountable. Might it not be possible one day for humans to learn to speak their dance language and then steer the conversation around to topics of mutual interest, by 'seeking for an opening for negotiation when things are quiet' in the bees' world? To 'break into a bee's mind' in this way would be 'a greater discovery than breaking the atom', says Heard, because, by giving them offerings of pollen and telling them they are our friends, we could train them up to be interplanetary diplomats who could 'act as go-betweens' betwixt Earth and Mars, thereby helping avert a truly catastrophic outbreak of interplanetary warfare.[45] Heard explained himself thus:

> If we could get into touch with [Earth bees] here and now (and there seems no reason why we should not) … then they might be able to act as invaluable translators and interpreters, when and if "Bees" of a still more advanced breed might swarm upon us. Our bees might help to bridge the gap between us and those "others". They could indicate to them that, in spite of our unprepossessing appearance … we are … capable, if treated kindly … of reasonable behaviour.[46]

The Martian bees may well look at these terrestrial bee-diplomats and see in them earlier versions of themselves, much like human beings observing chimps. Whilst most scientists ponder Earth bees' way of life and say that it has become 'fossilised' in the hive, and incapable of any further change, Heard is not so sure. He thinks bees could have evolved the gift of dance language relatively recently, and speculates that bee civilisation on Earth might even be considered capable of intellectual innovation. After all, when bees evolved their waggle dance to tell each other about where the best pollen stores were, then were they not in a sense also inventing maps?[47] If bees on Earth have already become cartographers, then what else might the future hold for them? On Mars, bees may still live in hives, but probably build them from more advanced materials than Earth bees do, like metal or glass, and so too might the Earth bees one day. Being able to build spaceships, Martian bees have clearly already had their own Industrial Revolution, and there was no reason why terrestrial bees should not do the same.[48] Their minds may have been tiny, explains Heard, but this is of no matter seeing as 'Some idiots have very large brains'.[49] It is not individually that bees' brainpower should be measured, but collectively:

> We know that bees have come a long way … It is pretty certain that they started with a far simpler way of life than now is theirs; as we started without gear and goods and plant and tools and cities and transport. It is generally conceded that the bees, which have cities and an hierarchical society, came up from solitary forms … [However, since starting to live collectively] they have built up cities and the organisation

of cities, control of population, supply, distribution of power, and order of succession in a manner so masterly that beside it our own efforts look amateurish and dangerously incompetent.[50]

Wouldn't it be wonderful if mankind could come together, stop pursuing our silly Cold Wars, and do the same? To Gerald Heard, there was a distinct possibility that this impossible-sounding dream may one day come to pass ...

## A California State of Mind

Bees, like Martians, have long been used for the purposes of political argument. Notably, the presence of a Queen apart, they seem like perfect little Communists, having collectivised their agriculture and adopted a form of communal living in which the notion of private property has no clear place. Each beehive, it might be said, is its own perfect little Soviet, slotting in seamlessly to the greater collectivised whole of bee-kind. Some disapproving writers, such as Tickner Edwardes (1865–1944), author of 1908's *The Lore of the Honey-Bee*, have despised the beehive as 'a living example, a perfect object lesson in what socialism, carried out to its last and sternest conclusions, must mean'. Bees could act in no way other than that in which they were pre-programmed to do so, meaning that for Tickner 'every worker-bee is herself the state in miniature, all propensities alien to the pure collective spirit having long ago been bred out of her' by life in the Communist hive.[51] Early socialist newspapers in both Britain and France were called *The Bee-Hive* and *The People's Bee-Hive* respectively, with the hive being adopted as a symbol of left-wing utopia because it seemed to embody a miraculous expression of collective solidarity, with everyone working together in order to ensure a golden future for all (literally, when you consider all the honey).[52]

It would appear Gerald Heard would have agreed with this latter viewpoint, not the former. For such an independent intellect, Heard had a curious love for experiments within the field of communal living. Sometimes dubbed 'the first hippie', Heard was certainly a New Age pioneer. After meeting the Hindu sage Swami Prabhavananda (1893–1976) in Hollywood in 1939 Heard became converted to the Vedanta branch of Hinduism, and in 1941 set up his own spiritual commune, Trabuco College, located in a remote area south of Los Angeles. Here, Heard meditated (for six hours a day!) in the company of students who gathered there to pray, engage in the comparative study of religions, and hear lectures from visiting speakers like Aldous Huxley, who had also been turned on to Hinduism by Heard. What George Adamski might have liked to have done on the slopes of Mount Palomar, Heard did for real, in a sincere and well-informed fashion. Sadly, attracting enough similarly sincere students to Trabuco for the place to be financially viable proved a problem, and the venture collapsed in 1947. In years to come, Heard lectured at many of the top US universities, from Harvard to Princeton, wrote numerous well-received books about spirituality and philosophy (and one poorly received one about space bees), as well as a number of detective novels and ghost stories, and became a TV and radio personality with a reputation as a well-informed generalist who knew plenty about both science and the humanities, rather than a narrow specialist in any one field. 'The universe is my hobby', Heard once said, and he meant it, too.

In the 1950s he also imitated Huxley by experimenting with LSD, later meeting with pioneer psychonaut Timothy Leary (1920–96), and praising the hallucinogen as an excellent way to enlarge a person's mind, dissolving barriers between oneself and the world, and stimulating mystical experiences. This was in some ways a continuation of his previous life in England, where he had been intimately involved with psychical research, being convinced that each man's individual mind shared within the life of some greater, collective one, which in itself then merged with the universe and thus also with the mind of God – another resurfacing of the old microcosm-macrocosm idea. In later life, he would advise young people looking for meaning in their lives to study the paranormal as an initial means to open themselves up to the possibility of mystical and spiritual experience. A member of the Society for Psychical Research between 1932 and 1942, Heard penned several articles about the history of mediumship. He felt it was mankind's destiny to evolve towards a state where, by developing more acute psychic powers, it was possible to gain access to the collective mind of our species, speaking of a future being called 'Leptoid Man' (from the Greek *lepsis*, 'to leap') who would enjoy a much more advanced state of consciousness than we did at present.[53]

## The Great Leap Forward

It was in 1929 that the wider world first began to take real notice of Gerald Heard, following the publication of his award-winning book *The Ascent of Humanity*, a complex meditation upon the future evolution of human consciousness. Believing optimistically that progress in science and technology went along with progress in society (an argument he was later to modify ...), Heard felt that the path of progress was shaped like a sort of spiral, growing ever-narrower as it reached the top, at which point mankind would achieve a state of 'complete co-consciousness' with one another and Creation. In his 1936 book *Exploring the Stratosphere*, Heard was to illustrate this idea with an internal shot, taken from below, of a rocket-launching tower constructed by the American rocket pioneer Robert H. Goddard (1882–1945), a man known as the 'angel-shooter' for his desire to conquer the heavens.[54] Arguing that the mind of man was somehow like Goddard's rockets, with our ascent towards a higher plane paralleling that of the rocket's own upwards ascent into space, Heard pointed out that the ascending diamond-shaped struts of Goddard's launch-tower when viewed from below, as in his photograph, mirrored the upwards-tapering shape of his own imaginary spiral of human progress, as outlined in *The Ascent of Humanity*. Man and rocket were in some sense a part of the same thing; just as newer rockets could fly higher and higher as their design evolved, so could man's mind rise higher and higher as its own design evolved, too. This rocket tower of Goddard's, incidentally, was located amidst the desert sands of Roswell, New Mexico – read into that what you will.[55]

Another model Heard may have had in mind when writing his 1929 book, however, was a bizarre plan hatched by the Russian architect Vladimir Tatlin (1885–1953) to build the Soviet Union its own Communist version of the Eiffel Tower. Never actually built, Tatlin's *Monument to the Third International* did make an appearance in 1920 as a large-scale 20-foot-tall model, and looked just like Heard's imagined spiral of human consciousness embodied. Made of metal and glass, it had a sealed, telescope-like interior, which was intended to rotate in line with the heavens, and a gigantic diagonal strut resembling a launch pad running up its side, as if to launch the rocket of the proletariat up into the socialist Heaven of tomorrow, powered by the fuel of pure, concentrated dialectical materialism.[56] In these analogies, the rocket of the future was very much a multi-stage device like those first conceived of by Konstantin Tsiolkovsky. Capitalism was like the initial booster rocket which, in Marx's dialectical scheme of history, allowed Communism later to blast off from it, heading forever 'onward and upward'. Once the capitalist booster rocket had done its work in getting Communism up and off the ground, however, it quickly became burnt out and spent, detaching from the main rocket body and being allowed to fall away back down into oblivion.[57] In one of the most hubristic mental projections of all time, such Communist giants as Marx, Engels and Lenin had earlier reclaimed Newtonian physics itself as being a dialectical force – just as Newton had demonstrated how, once the planets had been placed into motion, they would keep on moving in orbit by themselves, so Marx had allegedly shown how, once mankind had given himself a shove forwards, he would keep on moving in that same positive direction forever more, like a self-propelled evolutionary rocket through space, albeit upon a somewhat spiral path. As Tsiolkovsky had once calculated the upwards trajectory of Russia's future rockets, so Marx had calculated the upwards trajectory of future Russian man.[58]

## I Want to Be a Busy, Busy Bee

Whilst Heard himself was no Communist, he too appears to have wished to jettison many elements of the capitalist world order behind him once mankind had embarked upon the rocket journey towards a higher consciousness. And this, I think, is where the bees came in. It should be obvious that Heard was an intelligent man, whom you would not, under ordinary circumstances, expect to believe in the existence of bees on Mars. Indeed, the more you read of Heard's UFO book, the more you realise that its central premise is built upon exceedingly flimsy evidence. Might the advanced Martian bee-civilisation he posits, therefore, be best thought of as some kind of playful 'thought-experiment', or *jeux d'esprit* in which Heard imagines, using an ostensibly factual framework, the basic kind of society which he hopes his beloved Leptoid Man might one day achieve? He later quite readily endorsed Jung's ideas in favour of his own, after all, and in 1947 had imitated Aldous Huxley in writing his own novel about a dystopian future, named *Doppelgangers*; maybe *The Riddle of the Flying Saucers* was his own disguised attempt at writing a contrasting piece of utopian fiction? Living within a giant alien beehive might not have sounded much like utopia to most people, but an examination of Heard's earlier writing from his days in England reveals an apparent desire for humanity to develop a kind of hive-mind to it. In companion books to *The Ascent of Humanity*, like 1931's *The Emergence of Man*, Heard argued that early mankind had once had no true sense of individual selfhood to it, instead enjoying a sort of tribal 'co-consciousness' in which, like bees in a hive, everyone felt themselves to be not autonomous mental units, but inseparable parts of a larger whole. He talks of cavemen being in a state of 'constant telepathic communication with the rest of the group' and participating within a collective prehistoric unconscious. 'The first human unit is the group, not the individual', he wrote. For Heard, during Palaeolithic times cavemen did not even draw a clear distinction between their group-mind and the external world of dead matter, engaging with reality through a kind of *participation mystique* in which drawing pictures of animals being killed with spears would magically lead to identical successful hunting expeditions occurring in reality.[59]

However, what Heard saw as being the excessive state of individuality achieved by modern-day Western capitalist man was not wholly to be despised. Drawing pictures of dead mammoths doesn't really lead to a nice big meal that night, and, if humanity was to survive and evolve any further, then new technologies, inventions and modes of living were needed, which could only be developed by those born with a fuller sense of self to them; Earth bees have conspicuously failed to invent the wheel, for example. The lesson of Darwin was that if creatures didn't change, then they died – Heard's example was that of the dinosaurs who grew huge and strong, but were ultimately outwitted by the more cunning rodent-like mammals who became our ancestors.[60] Nowadays, however, the individualised mind of man was growing ever closer to solving all the problems of day-to-day living in terms of food, hygiene, medicine and so forth, but equally had grown selfish and warlike, with the capacity to kill off half the race in huge global conflicts. The solution was for us to reverse our individualistic path of evolution, which was merely the 'growth-pang' of something far greater, entering back into a modified version of that same hive-like 'superconsciousness' which our prehistoric ancestors had once enjoyed. It turned out that, in order to go forwards, sometimes mankind actually had to go backwards – tracing out a complicated spiral path to the heavens like one of Goddard's rocket towers, or Tatlin's *Monument to the Third International*.[61] In short, then, just like fabled Music Hall comedian Arthur Askey (1900–82), I think there was a part of Heard's psyche which made him want to become a busy, busy bee one day himself!

Ultimately, one of Heard's main perceptions about the nature of the universe was that mind and matter were mere aspects of one another – the same conclusion drawn by men like Edgar Mitchell and various other mystics down the ages who have found the idea of a cosmic parallel between microcosm and macrocosm appealing to them. In his 1932 book *This Surprising World*, Heard opined that mankind's 'essential nature' lies in his mind, with his body being 'only its projection'. In a similar fashion, the universe's essential nature was also a mind – that of God – with matter being only its projection, a viewpoint which may have been familiar to the old Neoplatonist philosophers of Greece and Rome.[62] Like Jung with his talk of UFO *mandalas*, Heard spoke of the need for each individual, and wider society itself, 'to re-mend the fissure in his own psyche and so see himself and his community, it and Life, and Life and the universe as one'.[63] The way ahead was to merge the mind of man further with the God-ensouled matter of the universe by wiping away the barriers between it through mystical, ecstatic experience, thereby allowing man and God to enjoy some form of intercourse with one another. But what might happen if some deluded souls chose to take such an instruction to enjoy intercourse with the universe literally?

# 13

# Venus Has a Penis: The Coital Cosmology of Charles Fourier

Surely the most bizarre post-Enlightenment comeback for the idea of living – and indeed loving – stars and planets came from the pen of a very odd Frenchman named Charles Fourier (1772–1837). Fourier was one of the first true prophets of the Left, his particular brand of thinking eventually being dubbed 'Utopian Socialism' by the movement's later guiding lights, Marx and Engels, twin authors of *The Communist Manifesto*. Here, they described such early socialists as Fourier, writing at a time of 'universal excitement' about the potential of the working class, of peddling 'fantastic pictures of future society', mere utopian 'castles in the air' which only those deluded souls with a 'fanatical and superstitious' cast of mind could possibly ever believe in. The heart of a man like Fourier was in the right place, Marx and Engels implied, but his head had simply gone ga-ga – either that, or his writings were really just one big joke.[1] So strange was his vision of the socialist New Jerusalem that during the twentieth century he was adopted by the Surrealist André Breton (1896–1966) and his fellow travellers as a kind of mascot of what might be termed 'the politics of the impossible', with Breton's long laudatory poem *Ode to Charles Fourier* being released in 1947.[2]

The child of a linen draper from Besançon, Fourier's father intended him to follow in the family trade, but Charles found the prospect of a life spent in commerce to be a most disagreeable one. His youthful habit of telling customers at his father's shop the wholesale price of his cloths, thus allowing them to work out how much they were being marked up for sale, says much about his inherent unsuitability for such work. Nonetheless, he was obliged by financial necessity to pursue such a life, which he did for many years in both Lyons and Marseilles, first as a private businessman, then as a clerk for a large wholesalers firm, before finally finding a role as a commercial broker. From 1799 onwards, however, he began to formulate his theories for a new, post-capitalist society, writing manuscripts during his spare hours. In 1814, Fourier retired to live with his sister in Belley, devoting himself to writing full-time. He died in 1837, following an inflammation of the bowels, but his ideas lived on long after him.[3] Fourier was one of those overly optimistic thinkers who aim to create a universal 'Theory of Everything', and it sometimes seems as if he wrote about every topic under the sun, from the various different types of cuckolded husband to a new dictionary of hyper-punctuation he claimed to have written (and then lost),[4] but he made his greatest impact in two seemingly separate-sounding areas; cosmology and social reform. His big theory was that of 'passional attraction'. This posited that human beings were naturally motivated not by desire for profit or money, as the current structure of society suggested, but by natural passions like love and the desire for variety, alongside the five 'sensuous passions' of sight, hearing, taste, smell and touch.[5]

## Let Them Eat Cake

Declaring himself 'The Messiah of Reason', Fourier proposed creating a brand-new society in which the true passions of mankind, rather than being suppressed, would be indulged. For example, we all know that gluttony is a sin, but for Fourier a desire to stuff your face with cake was perfectly natural and should be indulged, not discouraged. 'Gluttony is a source of wisdom, insight and social accord', he once wrote. 'It was in [God's] power to give [people] a liking for dry bread and water', but instead he made all children born with 'a liking for dainties'. So in this case, 'the question ... is who is wrong, God or morality?' Instead of discouraging children from wanting to eat cake for breakfast, tea and supper, surely instead we should try and create a society in which such sugary treats are cheaper and more easily available than bread?[6] The solution was to do away with our current civilisation altogether, and create a brand new unit of social organisation, termed the *phalanstère*, generally rendered in English as 'phalanstery'. The term was derived from the ancient Greek word 'phalanx', denoting a type of military fighting unit in which men were grouped so tightly together that they operated in effect as one. Each phalanstery would be a self-contained commune housing exactly 1,620 members, derived from all three genders (Fourier considered children to be a separate sex). Of these, 810 would be women or girls, and 810 men or boys. This number was arrived at by Fourier somehow managing to work out that there were exactly 810 different personality types of human being in existence, thus meaning that a community of 1,620 would be an entire world in miniature.

Within the phalanstery, the currently boring and degrading world of work would be transformed into a pleasurable one of play. For example, seeing as one of mankind's main inbuilt passions was for competition, as currently expressed in the negative forms of *laissez-faire* economic warfare and actual military warfare, this natural impulse should be channelled into more positive arenas. As such, in Fourier's future world there would be regular fake 'wars' arranged between cooks from one phalanstery and cooks from another, to see who could make the best meals. In his text 'Major or Gastrosophic War', Fourier made plans for a kind of pastry-based World Cup, in which competitors from around the globe would gather together in the city of Babylon to see who could bake the best *vols-au-vent*. Brilliantly, Fourier had formulated a highly complex equation for calculating the level of what he termed 'nice tastes' in an object, namely 'Y: [K rotated 270°] : 3.4.2 : K : 3.5.4 : [K rotated 180°] : 2.3.2 : [K rotated 90°] : [Y rotated 180°]'. This formula would be used to help judge the quality of the *vols-au-vent* created by each 'army of the 12th degree' which Fourier imagines will one day be sent to Babylon by 'about sixty great empires [of phalansteries] that have each provided 10,000 men or women' to compete in the pastry war. Once installed in a gigantic 'battle kitchen', these warrior-chefs will then bake their *vols-au-vent* before submitting them to the judgement of a 'great gastrophical Sanhedrin', defined by Fourier as being 'a High Jury which functions as an ecumenical council in this matter'. Battle is joined between the different armies of chefs for around six months or so, at the end of which the winners get showered with 300,000 bottles of champagne and receive a specially engraved gold medal for their efforts, whilst the cheers of the multitude 'echo far off in the caves of the mountains of the Euphrates'. Thus, true warfare between nations will be replaced with the harmless mock warfare of play – something which (in the West, at least) has actually since happened, albeit with competitive sports such as football and rugby, not pastry baking. Fourier imagined three main kinds of future fake warfare; gastronomic, industrial or engineering-related, and love-based, the last of which would presumably have gained the largest viewing figures.[7]

## Workers' Playtime

Furthermore, seeing as all 810 different personality types would be present in every phalanstery, each person would be allowed to keep on swapping jobs for the sake of variety, like a butterfly flitting from flower to flower. Either that, or they would be 'scientifically' assigned the job towards which they were most suited; most famously, the 'little hordes' of small children in each phalanstery would be given the job of cleaning toilets, seeing as kids everywhere enjoy rolling around in mud and getting dirty, as well as thinking bums and turds to be inherently funny, and so would find the task of wiping up everyone's else's shit to be great japes. In fact, so pleasing and play-like would everyone's work become that labour would be transformed into a form of surrogate sex;

Fourier felt that, just as with gluttony, all sexual desires were natural things which should be indulged in with absolute freedom. The apparent coiner of the word 'feminism', Fourier – a lifelong bachelor himself, and self-proclaimed 'protector of lesbians' – thought the entire institution of marriage should be abolished, and women set free from family bonds to take at least four lovers each. Sexually speaking, anything would go in the phalansteries; S&M, incest, bestiality, sapphism, homosexuality, group-sex, paedophilia, sodomy, holding hands, whatever man's motive passions led him towards. Seeing as the laws of passional attraction drew people on to do such things, to prevent them would be as unnatural as stopping a ripe apple falling from its branch, and thus an actual breach of the laws of gravity. In his text *La Nouveau Monde Amoureux* – which proved so scandalous it was not actually published until as late as 1967 – Fourier even proposed the idea of a so-called 'sexual minimum' being created, a sort of amorous welfare-state safety net in which each citizen would be guaranteed a certain amount of intercourse by the State. Those who found themselves jilted would soon be left satisfied by bands of wandering 'fairies', he suggested, gangs of public sector prostitutes who would seduce the lovelorn upon sight, giving them all the intimate attention they required.

This wonderful new 'Age of Harmony', as it was dubbed, would last for 80,000 years and have hugely beneficial effects upon human health, said Fourier. In his new Harmonic paradise, it became obvious that humans would evolve into sexual supermen, growing to stand seven feet tall and living to the age of 144, in a world wholly free of illness, misery and pain. Our teeth would become replaceable, and our lungs grow such capacity that we would become able to breathe underwater and thus cultivate the seabed. The globe's population would swell to some three billion, with 37 million poets equal to Homer, 37 million physicists equal to Newton, and 37 million dramatists equal to Molière living upon it at all times (although these were only 'approximate estimates').[8] Amazingly, Fourier's scheme won a number of supporters. Much of his writing actually centred upon more sensible topics like agricultural reform, and did not seem quite as mad as talk of erotic childhood toilet cleaning or giant international pastry wars. Following the publication of his book *The New Industrial and Societary World* in 1829, an abridged condensation of his ideas into more palatable form, Baron Guillaume Capelle (1775–1843), the Minister of Public Works, was planning to establish and fund an experimental version of one of Fourier's communes under the patronage of the French Government. In 1830 revolution broke out in Paris, however, putting a stop to such plans. Nonetheless, Fourier's ideas had now captured the imagination of the public, with several newspapers devoted to spreading his gospel being established, the most notable of which was titled *La Phalange*. Inspired, some disciples tried to put their mentor's ideas into practice themselves, and several

Fourierist communes were actually established, largely in America, though all failed in the end.[9]

## How Do You Like Them Apples?

The 1830s and 1840s, being a time of revolutionary fervour across mainland Europe, saw a great public appetite for visions of a new social order, something which gave Fourier's notions a real boost. The basic outline of his unrealistic utopia had first been laid out by Fourier decades earlier, in an anonymously attributed 400-page 1808 publication called *The Theory of the Four Movements*. This work was quickly withdrawn from sale by Fourier, however, once he got wind of the mocking attitude that its insane contents had engendered in most of its readers.[10] Surprisingly, the vast majority of the ridicule heaped upon Fourier's book centred not upon his weird plans for phalansteries, but upon his even more bizarre ideas about the universe. What must be remembered about Fourier is that he felt his idea of 'passional attraction' was a kind of scientific continuation of the work of Newton upon gravitational attraction. 'I ... have surely completed the task that the Newtonians began and left unfinished', he wrote proudly, whilst never missing an opportunity to belittle his long-dead rival. Whilst 'as a mathematician Newton did all that we had a right to expect from him,' Fourier loftily declared, he had nonetheless commenced his studies 'at the wrong end of the subject' by focusing his calculations upon 'a few secondary branches of Nature's laws' like gravity and optics rather than, for example, the uncontrollable human desire for cake and sex. Because of Newton's wrong-headedness, the current pseudoscience of astronomy could 'only explain the EFFECTS and not the CAUSES' of planetary movements and orbits in space.[11]

In a haunting parable, 'The Four Apples', Fourier explained how, throughout the long march of human history, there had been four separate apples of immense importance to mankind. First of all, there were the two bad apples, the one with which the Serpent had tempted Eve, leading to mankind's expulsion from Eden, and the golden apple which Paris had given to Aphrodite in Homer's *Iliad*, thus causing the Trojan War. After this, however, came two counterbalancing good apples, the one which is alleged to have fallen upon Sir Isaac Newton's head giving him the idea for his theory of gravity, and another such fruit which was encountered by Fourier himself sometime during the 1790s. Whilst walking through Paris one day, Fourier stopped to buy an apple from a vendor. Asking its price, he was shocked to find each fruit cost the equivalent of fourpence, whereas back in the sticks he could get a dozen apples for a halfpenny. Observing the 'extortions of commerce' which could make an apple cost ninety-six times as much in Paris as where it was actually grown in the countryside was, said Fourier, the final straw which made him realise the corrupt nature of the entire capitalist system, and led him on to developing the idea of passional attraction – a momentous realisation commemorated today by a large metal sculpture of an apple created in 2011, which rests atop a multi-coloured glass cube near Paris' Place Clichy.[12] Proclaiming that 'the Newtonians had only half-explained the laws of the branch of movement', and had 'grasped a mere fragment' of the laws which governed the universe, in his 1808 book Fourier suggested that, due to their complexity, his words 'must be read at least twice, and preferably three times', if you wanted to understand them properly. His tables and diagrams, too, had to be 'read and re-read' until fully memorised; those indolent fools who were not willing to do so should simply 'shut the book rather than continue reading'.[13] This was sage advice. Many readers may well have seen fit to read Fourier's words twice, or even dozens of times; not because they were necessarily that hard to grasp, but because they were so unbelievably, marvellously weird.

## Star-Crossed Lovers

Fourier thought that Newton was fine so far as he went, but that the major attractive force keeping planets in orbit around their suns was not really gravity, but passion. Just as mankind was governed and moved by his own motive passions, so were the heavens – yet another revival of the old macrocosm-microcosm idea. What this meant was that the planets, moons and stars were in some sense living animals, with senses like sight, touch and taste, and who were seemingly obsessed with having sex with one another; most notoriously, Fourier declared that eclipses were caused by the sun engaging in a 'conjugal embrace' of the moon, an idea famously satirised by the French cartoonist J. J. Grandville (1803–47), but one which may have found some favour amongst medieval alchemists.[14] In Fourier's own words, 'A planet is a being which has two souls and two sexes, and which procreates like animal or vegetable beings by the meeting of the two generative substances' which are emitted from their two poles.[15] By this, Fourier meant that the North Pole of every planet was male, and the South Pole of every planet female. Each pole was actually a giant genital, emitting a sort of subtle, airy, sperm-like substance of either male or female quality, termed 'aroma'. Throughout his work, Fourier is constantly talking about planets emitting aromas onto one another in order to turn each other on, mate and generate life. This sounds initially like they are broadcasting arousing smells out into space like some bizarre kind of penile or vaginal farts, but actually these aromas are better thought of as being a virtually incorporeal 'fluid' which connects the planets together into their own little solar systems throughout the universe. Basically, such 'aromal fields' were an erotic form of gravity, making planets orbit around their larger suns, or moons around their parent-planets, in much the same way that love-struck teenagers might follow the object of their affections around everywhere they go. Sometimes unused stores of this subtle semen could be seen being spurted messily out from a planet's pole in the form of what we on Earth would call the *aurora borealis* and *aurora australis*, the Northern and Southern Lights. Seeing as each planet had a male and female pole, they could be thought of as having sex lives similar to those of many plants, which also commonly posses both male and female generative organs within themselves.

However, should a planet's North and South Pole fancy trying something different from pleasing one another, they could also choose to emit their aromas out into space, seeking out the polar genitals of other planets, too; so Jupiter's male North Pole could have aromal sex with Venus' female South Pole, and *vice versa*. These unions would then produce 'children', in terms of the animals, plants and minerals found on each world. Seeing as many of these cosmic kids were produced via interplanetary gang-bangs involving more than two partners, however, working out the precise parentage of each creature or substance on any given world could sometimes be a problem complex enough to defeat even Jeremy Kyle. Here on Earth, a kind of flowering Mexican plant called the tuberose, for example, apparently had three parents, Earth's southern polar 'vagina', the northern polar 'penis' of Uranus, and the southern polar genitals of the sun.[16] At no point does Fourier provide any actual *proof* of any of this being true, you understand, he simply presents it all on the page in a completely matter-of-fact way whilst airily informing his readers that, should they desire to see the indisputable evidence of the reality of his ideas, they will have to wait some thirteen years until 1821, when he estimates he will finally have had enough time to write it all down properly. 'So be patient until 1821!' Fourier advises his readers, unrealistically.[17]

## Sexual Healing

Whilst Fourier taught that good things will come to those who wait, the Earth during his own day was having its patience tested to the very extreme. Many of the children produced through the Earth's past adventures in galactic sex had gone horribly wrong, with unpleasant animals such as rattlesnakes, sea monsters and bedbugs, and horrible illnesses like leprosy, rabies and syphilis infesting the planet. Fourier theorised that each planet, star and moon went through several stages in its lifespan – including a planetary puberty in which each is 'shaken by a series of quakes and volcanic eruptions which would make possible the release of noxious fluids buried underground', a process analogous to the angry eruption of spots and pimples on the face of a greasy adolescent.[18] With this in mind, Fourier chose to explain these deformed, Caliban-style children of the Earth as reflecting the fact that both it and its highest inhabitants, human beings, were still stuck in an early and beastly stage of development. Given that the macrocosm is but a reflection of the microcosm, it was no wonder that the ugly race we had become was now mirrored in the unpleasant fauna, flora and diseases currently prevailing across our planet. By stupidly continuing to live within an exploitative capitalist society in which apples were routinely overpriced, men had thrown planet Earth out of its intended orbit, meaning its aromal fluids were not mixing with those of its other planetary suitors correctly. Because Earth happened by chance to occupy an absolutely central position in the universe, our sexual sickness had also begun to spread out to other worlds, too, like an interplanetary AIDS virus. Earth's spermal aromas were becoming noxious, which had led to the death of our only orbiting moon, and given the sun a 'slow fever or consumption', as could be diagnosed by the increasing appearance of sunspots. Whilst a 'rescue column' of some 102 planets was on its way to our solar system hoping to save us from sexual destruction, it had set out on its 'forced march' back in the days of Julius Caesar and still had not arrived. The only solution was for everyone to start living within phalansteries, as aliens did on other planets, and engaging in bizarre but fulfilling sex acts with one another. With mankind's correct passional attractions re-established in this erotic workers' paradise, the Earth would shift back towards aromal equilibrium, and our globe's sexual gravity return to normal. Our pale and white-skinned 'corpse moon' would then fly away to the Milky Way and dissolve into blissful nothingness, being replaced by five much fresher, nubile and up-for-it young ones, after which the planets would return to their proper Harmony and realign in a kind of daisy-chain to better facilitate intercourse. Welcomed back to the celestial orgy, all the other planets would quickly gather back around Earth, and start shooting off their aromal fluids all over our grateful sphere like doggers in a pub car park.[19]

Showered in space sperm, the scene would be set for the Earth to become an Edenic paradise, with our orb undergoing an all-encompassing planetary orgasm lasting some 80,000 years. The effects of this giant orgasm, Fourier explained in *The Theory of the Four Movements*, would be extreme indeed. First of all, the *aurora borealis* will begin to change into something called the 'Northern Crown'. Currently, says Fourier, our North Pole is in 'violent upheaval with the need to create', but cannot, seeing as the Earth's sexual gravity is all wrong. The present erratic appearance of the *aurora borealis* is sort of like the North Pole prematurely ejaculating small dribbly bits of sperm in sex-starved frustration, he says, this 'useless effusion of creative fluid' being unable to join up with the aromas of either the South Pole or other planets. Ultimately, though, as people begin to live in phalansteries and develop Fourier's passional utopia ever further, the *aurora borealis* will become more and more active, recover its potency, and 'broaden out into a [permanent] ring or crown' which will emit not only light, as now, but also heat. If you want to visualise this phenomenon, said Fourier, then you should simply think of the rings around Saturn, which are also made of glowing aromal sperm – as, indeed, are our immortal souls.[20]

Fourier believed in a complex system of reincarnation, with the souls of the dead inhabiting the aromal crowns, bubbles and rings which orbited their home planets whilst they awaited delivery of their new physical bodies. When mediums and psychics talked to the dead or travelled in their so-called 'astral-bodies', they were really swimming through a sea of spiritual sperm. To Fourier, each human (or alien) soul was but a fragment of its parent planetary soul or life-force, which flowed through the planet in a way analogous to the circulation of human blood, thus meaning that during life we were all simply solid fleshly embodiments of each planet's aromal fluids; so the new Northern Crown was in many ways the warm, spermy Heaven we would all float up to when we died. Nobody could remember their past lives, but for Fourier all those currently occupying a low station in life were once kings, princes and rich men, and *vice versa*. We lived for exactly twice as long as sperm-spirits as we did as corporeal people, with someone who dies aged twenty orbiting Earth as an aromal ghost for forty years, but a centenarian taking two centuries' worth of holiday in space. Furthermore, as living creatures, planets, too, would all ultimately die and their souls be reincarnated in baby comets, taking all of our aromal souls away with them to be reborn anew elsewhere in space. Currently, though, the Earth had been placed into a kind of 'interplanetary quarantine' due to its toxic, rotten sperm, making it all the more vital that we quickly managed to usher in the Age of Harmony.[21]

## Secret Lemonade Thinker

When the Age of Harmony arrives, the new halo of solidified sperm circling the male North Pole will become so hot that it eventually leads to a kind of benign global warming taking place, in which the Arctic grows as warm as Spain or Sicily, and frozen places like northern Canada and Siberia see a rise of six to twelve degrees in average temperature, meaning they can finally be farmed profitably by mankind. Only warm winds will blow from the North Pole from hereon in, putting an end to bad weather, producing an 'even temperature' and halcyon days with 'gentle and unruffled skies' almost everywhere, making it possible for farmers to have two harvests per annum instead of one, with summer lasting for eight months. Orange groves will flourish in Warsaw, vineyards in Moscow, and St Petersburg be transformed into a Mediterranean-style resort along the lines of Nice or Montpellier.[22] Sun-seekers needn't worry about being burned to a crisp, either, as in the future Harmonic Age suntans will make a person white, not brown, and thus eventually 'the inhabitant of Senegal will be whiter than the Swede' – excellent news for racists![23] Even better, should you go for a dip whilst holidaying in one of these new Russian Riviera resorts, you will end up swimming not in ordinary sea water, but in lemonade! And bathers don't need to worry about being eaten by sharks or swallowed by whales, either, because all the monsters of the deep, once exposed to these new lemonade seas, will either die or undergo a pleasant reform of character. According to Fourier, the heat, light and sexual aromas emitted from the new North Pole will:

> … change the taste of the sea and disperse or precipitate bituminous particles by spreading a boreal citric acid. In combination with [sea-] salt, this liquid will give the sea a flavour of … lemonade … It will then be easy to remove the saline and citric particles from the water and render it drinkable, which will make it unnecessary for ships to be provisioned with barrels of water. This breaking down of sea water by the boreal liquid is a necessary preliminary to the development of new sea creatures, which will provide a host of amphibious servants to pull ships and help in fisheries, replacing the ghastly legions of sea monsters … The sudden death of all of them will rid the Ocean of these vile creatures, images of the [current] intensity of our [bloodthirsty] passions … Death will strike them all at the same moment![24]

Sadly, Earth's lakes and inland seas, not being connected to the world's oceans, will receive only 'the subtlest flavourings' of lemonade, sourced direct from the aroma-infused atmosphere. However, there will be so little of it present initially that the fish in these lakes will not immediately die like the sea monsters, but gradually, as generation succeeds generation, become used to living in weak lemonade, growing ever-stronger with its taste until they become a race of super-fish as the level of lemonade content around them grows 'so slowly and so imperceptibly' with each passing year. With this in mind, Fourier recommends that mankind, as soon as the Northern Crown becomes visible, should 'set in motion the same procedure for sea creatures as Noah did for those of the land', and gather up quantities of all the useful sea life – those creatures 'which do not attack divers', like herring, mackerel and shellfish – and transplant them into large salt-water lakes. Here, the friendly fish should be left to develop and accustom themselves to conditions of mild lemonade until the point where it becomes possible to restock our fully fizzed oceans with them once more.[25]

## The World Turned Upside-Down

Fourier tried to give evidence for these mad assertions by reference to certain allegedly inarguable facts about Earth's geography. If Earth was not meant to have a Northern Crown, said Fourier, then God's design of our planet's continents was nothing less than 'a sign of ineptitude' upon His part. There is a greater volume of landmass in the northern hemisphere than the south, yet so long as the North Pole remained frozen, most of this, like Siberia and the Arctic, was virtually uninhabitable for man, which seemed like rather a waste. Furthermore, when examined on a globe, northern landmasses appeared to gather around the frozen north in a big circle, like travellers around a campfire. If God hadn't meant for the northern continents to huddle around a future Northern Crown in this way, said Fourier, then this arrangement would make Him 'look ridiculous'. Earth's current geography, he said, was 'not random', and all of its apparently negative disadvantages in the present day would be changed into positive advantages, come the future Age of Harmony.[26]

In actual fact, God – who was really the impersonal laws of mathematics and physics[27] – had designed our planet to look a lot flashier than it does today, as the Harmonic Man of the future would one day see. The sun, Fourier claimed, was not really a big ball of fire at all. It was an ordinary planet, with humanoid inhabitants termed Solarians, surrounded by a big bubble of aromal love-light of the same kind which formed the rings of Saturn, and which might prove so erotic to other planets that it could well cause strange new worlds to invade its orbit in search of rumpy-pumpy. According to Fourier, our current solar system (or 'vortex', as he had it) was severely under-populated, due to the Earth's rotten sperm, resembling nothing but 'the fleeing remnants of a regiment destroyed in battle.' Many other vortexes had between 400 and 500 planets, suns, moons and other such bodies within them, he said, 'all equipped with rings, crowns, polar caps and other ornaments' which improved the appearance of the night sky no end. Fourier proposed that one day a giant spherical comet the size of Jupiter might enter into our vortex, become aroused by the sun's aromas, enter into its orbit, pass through puberty and become a fully grown planetary concubine.

At this point the huge comet's own sexual gravity would kick in, causing a besotted Earth to leave its current orbit and become 'one of the moons of the intruder', together with a similarly adoring Venus and Mars. Bathed in sperm from the sun and its newly captured planetary moons, this young 'vice-sun', as Fourier termed it, would 'soon become the richest and most fertile planet in the whole vortex', growing double rings of orgasmic joy around its equator, or double crowns at both its poles, due to the constant aromal orgy in which it would be engaged. This development, said Fourier, was 'highly probable' to occur, although he admitted that there was an alternative scenario possible, in which an invasion fleet of 300 to 400 smaller comets 'might suddenly turn up' nearby and begin orbiting around our sun, swallowing its sperm like there was no tomorrow, and feeding off it. On the other hand, it was always possible that several sexually sympathetic stars of the Milky Way, feeling our vortex's present loneliness, might well detach themselves from their galactic parent and fly towards us, creating 'dazzling legions of hypermoons'.[28] With our own solar system given a dose of cosmic Viagra in this way, all the other solar systems would soon follow suit, due to our position as the 'pivotal hinge' of the entire 'sidereal apple'. As all the galactic suns, currently 'heaped up at random' like piles of fruit began to mate and reproduce, filling up all the empty spaces which presently lay 'without proportion or utility' in the dark night sky, said Fourier, Harmony would suffuse the current interstellar desert, with thick jets of planetary sperm criss-crossing the blackness 'like bullets on a field of battle'.[29] The perverse possibilities were endless!

## The Lion Leaps Tonight

Was Fourier just a madman? Some acolytes have answered 'No!' suggesting that he only placed such laughable speculations in his 1808 manifesto in a misguided attempt to gain more public attention for his (supposedly) more sensible ideas about phalansteries.[30] After all, at the beginning of *The Theory of the Four Movements*, Fourier did specifically write that 'a subject as elevated and extensive as this will not interest the majority of readers, but it will be interspersed with enough curious detail to compensate for some dry passages.'[31] As a PR strategy this is a risky one, however, akin to Marx and Engels creating an appendix to *The Communist Manifesto* in which they patiently explain that, under Communism, people will one day begin excreting chocolate ice cream. Alternatively, some Fourier fans have proposed that their hero meant all this nonsense metaphorically, encouraging his readers to 'think the unthinkable' when it came to remaking capitalist society – as Marx and Engels later did when imagining the possibility of a world without money or private property.[32] Take one of Fourier's most notorious predictions, the idea that, as more and more interstellar sperm spreads across our rejuvenated planet, the wild beasts which currently threaten mankind will mutate into our 'useful servants'. In a bizarre 1822 text, 'Detail of a Creation of the Hypo-Major Keyboard', Fourier explained that, 'through an effort of counter-moulding', the planets, stars and moons whose aromas had in the past given birth to unpleasant predators would soon create new races of anti-predators to do our bidding. Fourier gave a partial list of these wonderful anti-animals which would one day inhabit our lemonade waters:

- Anti-whales that tow vessels through calm waters.
- Anti-sharks that help track down fish.
- Anti-hippopotamuses that tug our boats upriver.
- Anti-crocodiles, or 'river-collaborators'.
- Anti-seals, or 'sea-steeds'.

Even more fantastic will be what will happen to Earth's big cats, which will have their characters reformed to such an extent that they voluntarily submit to act as a kind of free universal taxi service. One day, says Fourier, there will be:

> ... an effort of [aromal] counter-moulding, through which [the planet] who gave us the lion will give us as a counter-mould a superb and docile quadruped, an elastic carrier, the ANTI-LION: the kind of post-animal that would allow a rider ... to lunch in Paris, dine in Lyons, and sup in Marseilles ... The anti-lion, anti-tiger, and anti-leopard will be thrice the size of the current versions. Thus an anti-lion will easily cover eight yards with one bound, and the rider, on the back of his charger, will be as comfortably installed as if in a well-suspended berlin [a type of luxury horse-drawn carriage]. It will truly be a pleasure to inhabit this world once we have such servants to enjoy![33]

In addition, these 'elastic carriers' would be able to leap over any ditches or other obstacles they came across, allowing them to cut down on journey times even further. Traditionally, most people have just laughed at this description, but Fourier's English translator Hugh Doherty, writing in 1851, preferred to say that while 'the conception [was] fanciful', nonetheless Fourier's basic 'intuition was correct'. Railways and steam trains, argued Doherty, 'were not invented when he wrote his theory', but in reality 'the locomotive-engine is a panting lion, fifty or a hundred times more powerful and docile than the animal imagined by Fourier'. According to this rather generous viewpoint, just like Native American Indians labelling steam trains as being a new breed of 'iron horse', Fourier had simply got his imagery muddled, imagining a 'lion train' instead. In a confused way, 'knowing that man would want to travel rapidly from one region to another when refinement became general', Fourier had, like the genius he was, imagined super-fast travel across entire countries taking place years before such wonders actually came to pass. It was just that, unfortunately, he had accidentally envisioned this occurring via a series of wholly imaginary mutant sperm-animals, instead of via machines. As further proof of this argument, Doherty proposed that his mentor's equally mad-sounding idea of huge 'anti-condors' picking people up and flying them from city to city at incredible speeds was simply an early anticipation of hot-air ballooning. Taking this logic to its extreme, you could almost say that Fourier invented Concorde![34]

## Some Men Have a Fifth Limb

Such arguments are deeply desperate, and most attempts made to reclaim Fourier from the charge of loopiness seem unconvincing, to say the least. His acolytes' most effective way of trying to make Fourier seem less of a loon was simply to try and cover up some of his more ludicrous writings, something made easier by the fact that several of his most absurd essays existed only in manuscript form at Fourier's death.[35] For example, it was not until 1992 with the publication of Jonathan Beecher's (b. 1937) study *Charles Fourier: The Visionary and His World* that we learnt of the existence of a wonderful new human feature termed the *archibras*, a sort of fifth limb/ tail which mankind would eventually develop due to being sprayed with interstellar sperm. According to Fourier:

> The Harmonian Arm or *archibras* is a veritable tail, a tail of immense length and with 144 vertebrae ... [It] terminates with a very small elongated hand, a hand as strong as the claws of an eagle or a crab ... When a man is swimming, the *archibras* will help him move as fast as a fish. It can stretch to the bottom of the water, carrying his fish nets ... With its help a man can reach a branch twelve feet high ... pick fruit at the very top of the tree and put it in a basket ... It [also] serves as a whip ... It can be used to tame a wild horse; the rider can tie up the horse's legs with his *archibras* ... in the playing of musical instruments it doubles a person's manual faculties.[36]

It was no wonder Fourier-fanciers didn't want such comical fantasies released, but the various Fourier-themed newspapers which sprang up had to be filled somehow, and so many of his odd ideas did in fact seep out into the public realm, with such important essays as *Melons Which Never Deceive* being printed. Here, Fourier set out to explain how the (allegedly) well-known old dictum that 'melons are as hard to know as women and friends' will one day be proved false. Currently, says Fourier, melons have a hard external husk, so it is tricky when buying one to know whether it will be nice and firm, or horrible and squishy, inside. 'Blunders are so frequent' in this matter, according to Fourier, that every day 'we joke about the one who carries a melon'. Because the microcosm reflects the macrocosm in Fourier's philosophy, however, it turns out that this state of affairs has been intentionally brought about by the watching planets, who deliberately combined their sexual aromas in order to engender a fruit which reflected the rotten soul of corrupt mankind, a lying 'legion of double-dealers' who deserve to be paid in their own true currency, 'which is falsity'. Thus, whilst Fourier does admit that the treacherous melon has other qualities (which alas 'there is no time to mention') he claims that the primary purpose of its existence is that of cosmic irony, with the current 'traps of the melon' being Nature's cruellest joke on man. However, come the time when phalansteries have been established, they will apparently contain within them a group of specially trained workers called 'melonists', who will gain semi-orgasmic satisfaction from participating in a highly complex melon assessment and distribution system of Fourier's own devising, which will enable the right kind of fruit to go to the right kind of diner without fail; squishy melons will go to feed the cats and horses, for example, whereas the very firmest melons will be laid upon the tables of an elite group of highly advanced melon-eaters, known as 'The Command'. In this way, says Fourier, at long last 'not a man, *not a cat*, can be deceived about the melon'.[37]

## Love Makes the World Go Round

Such embarrassing reveries may sound thoroughly deserving of suppression, but the groundwork for Fourier's theories about the 'harmonic irony' of melons had already been laid down in print as long ago as 1808, in *The Theory of the Four Movements*. Here, when explaining how God was really mathematics, he had also laid down a pseudo-scientific explanation as to why it was that all things born under (and on) the sun had a symbolic component to them. First of all, Fourier explained how the laws of both Newtonian physics and passional attraction – that is to say, the depersonalised manifestations of God Himself – were 'co-ordinated with mathematics' because otherwise 'there would be no Harmony in Nature, and God would be un-Just', a rather Greek-sounding statement. To Fourier, Nature was composed of three basic things. Firstly, there was *God* or *Spirit*, the 'motive principle' behind that which moved – the force of momentum, for instance. Secondly, there was *Matter*, that which was moved – a ball being dropped, maybe. Finally, there was *Justice* or *Mathematics*, defined as 'the principle governing movement' – the laws of physics which made momentum act upon a dropped ball in the precise way that it did, say. The interaction of these three principles, said Fourier, meant that the universe was never arbitrary in the way things worked; if you dropped a ball to the floor a hundred times, it would never get bored and float upwards instead on your ninety-ninth go for the sake of variety. To him, this proved that 'God must be in accord with mathematics as He moves or modifies matter', and so must actually *be* maths, in some way.

Now, consider the orbit of planets. Due to the interaction of Spirit, Matter and Justice, their orbits must describe a shape, generally some form of ellipse; the laws of the universe have ordained that they have to move, and all things move in a pattern of some kind, these patterns being classifiable under the heading of 'geometry'. As Fourier says that the planets are living things, guided by their passions, this must mean that the laws of geometry apply equally to inanimate and animate matter. Throw a dart, and it will describe an arc, which could be drawn down on paper as a kind of parabola. Maybe the passional emotions could also be represented in this way, as geometrical shapes or figures descriptive of the laws which govern their own movement? For example, Fourier said that love had the quality of an ellipse, which would make sense as lovelorn planets and moons orbited one another in an elliptical fashion whilst exchanging aromal fluids. 'All passions,' said Fourier, 'produce effects in men and animals which are geometrically governed by God.' Human emotions thus had an inherent geometrical basis, which meant that the planets could, by making things grow in certain geometric shapes through their aromal influence, make them symbolic of mankind himself – and, contrariwise, that mankind, by giving way to certain passions, could alter the very geometry of the universe itself, thus meaning that love was potentially stronger than gravity. As Fourier put it, 'This means that the properties of an animal, a vegetable, a mineral, or even a series of stars, represent an effect of the human passions' upon the external world and vice versa. Or, in other words, the microcosm was inherently reflective of the macrocosm.[38]

Ever the systematiser, Fourier created a table linking each of the twelve main human passions to its own associated colour, element, mathematical function, mineral, musical tone, and so on. Here are the entries in this table for 'love' and 'ambition':

| Musical Note | Human Passion | Planet | Colour | Mathematical Function | Geometrical Figure |
|---|---|---|---|---|---|
| Mi | Love | Venus | Azure | Division | Ellipse |
| Sol | Ambition | Sun | Red | Multiplication | Hyperbola |

Examining such a table, you can see Fourier's mind at work; presumably, the sun is the 'planet' (he makes no consistent distinction between stars and planets) of ambition because it seeks to outshine all others, and to multiply something mathematically can make a small but pushy number very large instantaneously, transforming two into two million simply by virtue of the magic formula 2 x 1,000,000. Love, meanwhile, is naturally associated with Venus, named after the Roman goddess of *amore*, and prefers division to multiplication, with lovers dividing their personalities between one another and thus paradoxically creating something larger than the sum of their parts. Clearly, all this is an attempt by Fourier to impose a sense of meaning upon an otherwise empty and meaningless universe. As he admitted, 'Without analogy, Nature is no more than a vast patch of brambles.'[39]

## How the Elephant Got His Anus

Fourier called his new theory of correspondences 'the most pleasing of the sciences', and he was right. Not only does it 'give a soul to all Nature', it also transforms the world into a gigantic symbolist picture book where, for example, the clever planets deliberately gave the lion small ears, 'as if trimmed by scissors', in order to make the King of the Jungle represent most human kings, whose fearful courtiers are reluctant to tell truth to power, thus meaning that monarchs 'are *morally* deprived of the use of their ears'. Donkeys, however, with their big ears, stood for the ordinary peasants, whom nobody cared about openly insulting in public.[40] The 130 different types of poisonous snake which Fourier felt dwelled on Earth, meanwhile, stood in for the 130 different types of human 'calumny and perfidy' which he claimed to have identified.[41] Fourier also pronounced that there were twelve distinctive types of bird heads in existence, each of which stood for the twelve different grades of human intelligence or stupidity, running from moron to genius. Since birds could fly higher than any other animals, and human reasoning could rise higher than that of mere beasts, it was appropriate indeed that the interior of our heads and the exterior of theirs matched up like so.[42] Some birds also represented different types of human lovers. Male ducks, thought Fourier, were very quiet and affected with 'loss of voice' by comparison to females, making them analogous to hen-pecked husbands who dared not answer back their wives:

> When he wishes to woo his noisy partner, the [male] duck presents himself humbly, bowing his head and bending his knees like a submissive but happy husband lulled by illusions. As a token of this, the [male] duck's head is dipped in a glistening green, the colour of illusion.[43]

The planets had made particularly good use of erotic geometry when constructing the elephant, whose huge frame contained many curious lessons to be read. The reason elephants were often covered in mud was because this provided a moral example to humanity, corresponding with 'the image of the virtuous man who chooses the path of poverty' rather than ill-gotten riches. His trunk, meanwhile, is so silly-looking because, as the most useful and productive tool of the elephant's anatomy, it represents the fate of ordinary workers in human society, who perform all the hard graft for little reward but mockery. His big fat bum, also, had been deliberately made 'laughable' in appearance by the planets, thus standing 'as an emblem of virtue's fate' as people laugh behind virtuous men's backs just like they laugh at the wobbly bums of the obese as they pass by. Likewise, the 'extreme smallness' of the animal's eyes denote the 'narrow views of the virtuous man', who doesn't go around greedily eyeing things up for his own enrichment, whilst the 'immense mass and flattened form' of the elephant's huge ears depict the fact that everywhere men of virtue hear hypocritical words being spoken by others. Furthermore, by contrasting one geometrically symbolic animal with another, additional lessons could be learned. Thus, the evil twin of the 'noble' elephant was the 'base' dog. Whilst elephants made love in a 'decent and faithful' way, Fourier observed, dogs did so in a 'scandalous and criminal fashion'. Whilst elephants were responsible parents, giving birth to few children but raising them properly, dogs were happy to 'engender anthills, litters of eleven ... veritable heaps' of puppies, 'three quarters of which will perish by the knife, the tooth, or starvation' just to satisfy their incontinent desires. By comparison to the pachyderm, the dog was an emblem of the non-virtuous man, false in love and friendship, lustful and a truly terrible parent.[44]

## The Aroma of Uranus

The planets had also seen fit to ensure that all members of the vegetable kingdom expressed some kind of metaphor within themselves too, one which was frequently sexual in nature. Cabbages, for example, were 'hieroglyphs' of secret lovers like Romeo and Juliet, who had cause to hide their passions away from a disapproving world. Many species of cabbage had flowers, like the blooms such clandestine lovers exchanged with one another, but these were usually hidden away deep beneath puffy and ugly thick green leaves, for fear of being discovered. Cauliflowers, by contrast, were shameless vegetable sluts, their interstellar geometry depicting 'the delights of emancipated youth' whose many easily visible white flowers were like young flesh openly placed upon display.[45] Plums were a particular botanical obsession of Fourier's. The children of Uranus and her moons, they were considered to be especially erotic in nature by him, the very term 'plums' still being used as a metaphor for testicles even today. Fourier enjoyed their symbolic import so much that he even created an 'aromal catechism of plums', in which he laid out a bizarre series of doctrinal Catholic-style questions and answers to be recited by his disciples along the following lines:

Q. *Who created the Apricot, the pivotal fruit among plums?*
A. *Herschel [Uranus], the Cardinal [planet] of Love (shedding the pivotal aroma of matronage).*

Each different type of plum, seeded by the aroma of a different aspect of Uranus upon the Earth, possessed a 'dominant aroma' which gave it its hidden, cosmic meaning. The Reine-Claude plum, child of the moon Hebe, stood for 'fidelity', for example, whereas the Golden Drop Plum, child of the moon Cleopatra, embodied 'coquetry'. Having laid all this out, however, Fourier admits that some more sceptical readers might ask how he knows all this to be true. Was he perhaps present at 'the council of amorous allegories held by these gallant planets' before the creation of plums took place? Fourier admits that he was not, but protests that, as Newton's rightful successor, he ought to know. Indeed, to prove that he *did* know his onions (and cabbages and plums), Fourier went on to pen a long and elaborate tract called *Cosmogony*, which lay unfinished at his death, but which contained by far the most detailed outline of his thinking – including the catechism printed above – ever provided.[46]

It is this text, several thousand words long, which provides the clinching proof that Fourier really did mean what he wrote. Apparently possessed by an uncontrollable desire to reunite man with Nature, he came up with a new version of Pythagoras' idea of the Music of the Spheres, deeming the entire universe (or 'polyverse', as he had it) to be an 'orchestra' in which only human beings were currently playing their (sexual) instruments out of tune; 'the derangement of one of the keys hinders the play of all the others', he said. Considering existence to be one gigantic Great Chain of Being, with mankind 'an inferior link' and God right at the top, Fourier devised a kind of musical scale, called the 'First Octave', which laid out the structure of the Music of the Spheres by drawing correspondence between the rising sequence of musical notes and ever larger links in the cosmic chain leading up towards the Deity:

| | | |
|---|---|---|
| *Ut* | Monoverse | a Human Couple |
| *Re* | Biverse | a Planet |
| *Mi* | Triverse | a Universe [Fourier's term for 'galaxy'] |
| *Fa* | Quatriverse | 1,000,000 Universes |
| *Sol* | Quintiverse | 1,000,000,000,000 Universes |
| *La* | Sextiverse | 1,000,000,000,000,000,000,000,000 Universes |
| *Si* | Septiverse | 1 followed by 48 zeros Universes |
| *Ut* | Octiverse | 1 followed by 96 zeros Universes |

A similar octave would then repeat itself for the next eight orders of galaxy, and so on *ad infinitum*, from Centiuniverse to Milliuniverse, until we eventually reached the final note, that of God playing with Himself and humming in holy ecstasy. However, the first and smallest note in the scale, that of a pair of human lovers, was currently badly out of tune, as we have already seen. Thus, for as long as mankind was unable to finger his pink oboe correctly, the entire galactic orchestra would forever fail in its attempt to play the orgiastic cosmic symphony. Therefore, Fourier proclaimed his thinking to be a form of 'medical science' aimed at healing first of all our stricken planet, and then the rest of Creation, too, thus allowing the Music of the Spheres to be properly re-tuned. Once this has happened, Fourier explains, the planets will again be able to exchange sexual aromas with one another correctly. This will then disprove the greatest slander which mainstream astronomers have put out against the planets, namely that they are 'lazy', and nothing but 'great inert bodies passing eternity in promenading up and down' in their orbits through the night sky 'like our idle gentry'. In fact, the planets were anything but idle, having no greater pleasure than to emit their aromas down onto each other, thereby creating new plants, animals and minerals. Indeed, according to Fourier, the planets actually engaged in giant interstellar masturbatory contests with one another lasting for some 600 years apiece, 'a struggle of ambition [and] self-love, in which each displays its ability in competition', the erotic equivalent of mankind's forthcoming global pastry wars. In times past, treating the

Earth like the helpless Rich Tea in a galactic game of soggy biscuit, planets had gushed their sperm down onto our planet in order to see what wonders they could seed there through their aromal 'germs'. Then, says Fourier, they gathered around and judged the results of their onanistic labours; during one bout of sperm-based combat, for instance, Saturn accidentally created fleas, nasty little creatures which meant it 'had to undergo censure' from its disappointed planetary friends. Spermy seeds spilled onto other planets in this way lie in the ground for a while and then sprout up all at once, said Fourier, like skeletons from the Hydra's teeth; thus, when Jupiter wanked the ox into existence, its aromal sperm seeped into the ground and made Mother Earth pregnant with these creatures, their foetuses 'being elaborated in the bosom of the planet', before they later burst out up from the soil like mobile fungi, forming the first ever herds of wild oxen.

## A Rose by Any Other Name ...

The planets would find life 'very irksome' without these contests, apparently, and it was a good job they did have them, otherwise the universe would be almost lifeless. Through close study, Fourier managed to determine that the elephant, oak and diamond were children of the sun; the horse, lily and ruby offspring of Saturn; cows, jonquils, and topazes kin of Jupiter, and dogs, violets and opals called the Earth their parent.[47] However, as to 'what star has made us a present of the toad', Fourier admitted he did not know, though he hinted that 'my suspicions rest on Mars.' In his unfinished manuscript, Fourier provides much more detail about his theory of the aromas, which explains how he was able to determine the parentage of these interplanetary creations so precisely. Apparently, each planet has twelve governing passions, just like humans, which correspond to their twelve aromas, and which are then 'susceptible to combination without number'. Sometimes the celestial bodies simply emit these aromas out across space onto one another through 'jets or *fusées*' which travel at the speed of light. At other times, they rely upon comets to act as an 'aromal troop', becoming covered in space sperm and dripping it from planet to planet as they pass by, like bees laden with pollen. Apparently, every star and planet 'imbibes ... various juices' from such flitting bee-comets but also 'sheds upon them' others of its own.

Meanwhile, every planet or star has its own unique 'dominant aroma', which corresponds to its basic character or personality. Whilst this aroma mixes with the aromas of other planets and stars to create life, it is nonetheless possible to detect, in any animal, plant or mineral, which planet's dominant aroma prevails in the space sperm which combined to make it up, just so long as you have a well-trained nose for the task. So, for example, you can sniff flowers and work out which planet's dominant sperm they most remind you of; according to Fourier, jonquils smell of Jupiter, violets smell of Earth, and roses smell like the obvious product of the genitals of Mercury. Sometimes the planets even sire 'planetons', or 'baby planets', who are nursed and grow in the Milky Way before fleeing forth in 'swarms of comets', like hordes of tadpoles escaping their frogspawn in a pond, wandering around the universe before eventually growing up into full-blown adult planets. Once it reaches true maturity, like a fine wine, the quality of a planet's aromal creations will begin to improve; however, any aromas from a mature and healthy planet, mixing with Earth's own currently unhealthy and immature ones, will only produce freakish creatures like snakes or spiders, or worthless minerals like basalt, which are no use to anyone, being but the 'mistakes' of the universe. Consequently, says Fourier, Earth's precious metals and jewels are running out, with diamond and gold-mines being rapidly exhausted. So poorly equipped are we with such shiny substances today that most people eat with only wooden spoons; come the Age of Harmony, however, the planets will shower Earth with the spermy seeds of precious metals, meaning everyone, not only the rich, shall one day sup from a silver spoon. The night sky is filled with bright, shining suns; as microcosm corresponds with macrocosm, man also 'ought to be clothed and adorned like the universe', with big sparkly diamonds and rubies pinned all over us, thus becoming 'clothed with stars and dwelling in splendour'. When our planet has recovered its Harmony, the other planets' acts of musical masturbation will turn iron to gold, rock to pearl and sand to silver, with none of these treasures being buried deep in the ground, as now, but easily accessible upon the surface of the soil, like pebbles.

## Fire in Their Eyes

Furthermore, other planets' aromas, mixing with our own Earth's rejuvenated Harmonic ones, will give rise to the seeding of brand-new mineral substances in the ground, enabling mankind to perform all manner of wonders. Chief amongst these novel minerals will be certain 'paste-matters of new creation' from which we will be able to create a fresh form of glass for use in telescopes so powerful we can use them to communicate with aliens. In a chapter of his book *The Passions of the Human Soul*, dealing with what he terms the 'POSITIVE UNITARY EYE, OR TRANSPARENTIAL DIAPHANIC VISION, CO-IGNEOUS, HOMOGENOUS WITH FIRE', Fourier explains how, come the Harmonic Age, human eyes will evolve the amazing power of 'diaphanic vision', allowing us to see through flames and peer straight into the sun, where we will be able to see people waving down at us. When combined with the new super-glass in tomorrow's telescopes and microscopes, our new fire-piercing eyes will also become capable of x-ray vision whenever an object is illuminated by the sun or flames. So, beam sunlight or fire onto a person's stomach, then examine it beneath a super-microscope and you can see what ails their organs or, during pregnancy, deduce what gender the baby is.[48] Meanwhile, the crystalline shell of solidified space sperm which surrounds all planets and stars can also be used as a kind of mirror or 'celestial magic lantern' by the super-eyed Harmonians of tomorrow. By aiming telescopes up at this newly visible mirror, you will be able to see the reflection of things taking place miles away below on Earth, such as fleets setting sail from distant harbours, or people having sex in bushes. Sailors could then be equipped with giant wheels on which the letters of the alphabet would be written; by spinning them, they could identify themselves, give estimates of arrival times, or request provisions. The only problem Fourier could see with this scheme was that some fans of sex in bushes might object to being spied upon, but he countered that, come the Harmonian Age, 'morals will no longer be the same' with everyone shagging in the shrubbery anyway, so it wouldn't really matter.[49]

Remarkably, it also turns out that our present unevolved eyes are 'subversive' in nature, and have cunningly tricked us into thinking that the moons, stars and planets are 'twenty [or] thirty times farther off than they really are', something Fourier tries to prove by claiming that, if the celestial bodies are really as far apart as we currently think, then their sperm would surely go stale in transit and 'lose all its intensity, as happens with wine too long in bottle', thus rendering it useless by the time it achieved splash-down. Planets apparently use their sperm to send each other telegraphic messages (stroking it out via a dot-dot-dash pattern, maybe?), requesting emergency supplies of tasty aromal fluid to feed off. If they were really as far apart as today's idiotic astronomers claim, then these sperm-starved planets would simply die of hunger, Fourier says, like giant interstellar prostitutes deprived of passing custom. (Fourier even tries to blame poor harvests on Earth upon slow-travelling aromal fluid; when the planets don't eat sperm, we don't eat bread.)[50] Thus, when we get our new super-eyes and super-telescopes, we will find to our surprise that the planets are much nearer than we thought, allowing us to view them and their inhabitants closely. 'We shall see their fields, their animals, plants, buildings and individual movements as distinctly as we see the passengers from our windows'; even 'the dwarfs inhabiting Juno' will appear crystal clear. To give us an even better close-up of other worlds, Fourier further recommends 'supplying all the great observatories with albinos', whom he assumes for some reason to have superb powers of eyesight (in fact, it is the opposite).[51] Apparently, our much more Harmonic alien neighbours have already been spying on us through their own super-telescopes for years and laughing; they particularly enjoyed the Battle of Austerlitz.[52]

Ultimately, our new eyes and lenses will allow us to access something called the 'EXTRAMUNDANE PLANETARY TELEGRAPH', with us being taught a new Harmonic Language by the inhabitants of Mercury (Mercury being the messenger of the gods), a strange but beautiful tongue whose alphabet has thirty-two letters, and whose syntax and grammar correspond somehow with the relative position of the sun to other stars. The aliens will lay out big letters, three feet high, on their planet's surface, and teach us to speak Harmonic via a form of visual demonstration. Once we have mastered this lovely lingo, the men of Mercury will use it to teach us the full history of the universe, provide us with new forms of mathematics, explain how to build various wonderful new inventions, lay out the hidden properties of all gases and chemical elements, give us free music lessons using hitherto unknown instruments, and even lend us handy recipes for new flavours of delicious *ragout*, far superior to those currently known on Earth, even to the greatest chefs of Paris.[53] In this Age of Telegraphic Harmony our current dead satellite, 'that mummy, the moon', with its pathetic colour 'like a Dutch cheese', will finally disappear to be replaced by five others which will orbit us emitting beautiful colours and producing in the night sky 'the effect of a garden illuminated with coloured lamps'. The only downside would be that 'speculators in lamps' would most likely go out of business; but, says Fourier, this would be a small price to pay for living within the Age of Harmony.[54] All things considered, I think you'd have to agree ... but, strangely enough, we're still waiting for it to begin. I suspect we always will be.

# 14

# Turn On, Tune In, Drop Out: Talking with Aliens, Talking with Ourselves

'Sometimes I think we're alone in the universe, and sometimes I think we're not. In either case, the idea is quite staggering'; so said Arthur C. Clarke.[1] If there really are alien beings out there, however, then ought it not be possible to speak to them, albeit by means a little less ridiculous than Fourier's idea of using super-telescopes to examine giant alphabet cards laid out for us by the inhabitants of Mercury? The answer is 'Yes, but only if you don't mind talking to yourself'. Beginning in 1998, and continuing for the next seventeen years, scientists at Australia's Parkes radio telescope in New South Wales repeatedly picked up strange signals, known as perytons, on their equipment, which appeared similar in duration and frequency to fast radio-bursts of a kind known to emanate from outer space. Were they being sent a message from another galaxy? For almost two decades, these phenomena were a mystery – until in 2015 an answer to the conundrum was finally provided. A PhD student working at the site, named Emily Petroff, set up an interference monitor which enabled her to discover the signals were in fact being caused by the less than stellar source of an ordinary microwave oven in the facility's kitchen. Whenever hungry scientists opened the oven door to stop the cooking process before it was finished, it led to peryton bursts being emitted, which were then picked up on the radio telescope. There were no messages from aliens being recorded out in the Outback at all; merely proxy rumblings from the astrophysicists' stomachs.[2]

Such radio telescopes are generally located in as isolated areas as possible, so as to minimise the possible interference from sources like mobile phones and radio stations. However, no matter how remote the location, no radio telescope can avoid having to interact with possibly the most signal-distorting device of all; the human brain. Stories such as the above are surprisingly common, because the history of human attempts to communicate with other planets is a long one. Problems with this quest have been twofold, however. First of all, there was the question of how such messages were to be sent and received. Secondly, there was the equally puzzling problem of how human beings and alien races were to make themselves understood to one another, even if such signals did get through. On Earth, the language barrier between men of different nations was bad enough; but what if aliens didn't turn out to have any kind of language at all?

## Signal Failures

Initial plans for signalling to men on the moon date back to the 1820s, when the German mathematician Karl Gauss (1777–1855) decided it might be possible to let any potential lunarians know we were here using the universal laws of geometry. Probably the most famous such law is Pythagoras' Theorem, diagrammatic proof of which Gauss proposed could be laid out upon the vast plains of Siberia. A giant triangle with three squares would be outlined in green using long lines of specially planted trees and forest, with the interior of the shapes coloured in a contrasting yellow hue with huge fields of rye or wheat. Alternatively, Gauss is said to have suggested using his own 1818 invention, the heliotrope, which employed precisely aligned arrays of mirrors to reflect sunlight over great distances towards a given point, to create a bright spot somewhere on Earth so large it could be seen from the moon; in an 1822 letter, Gauss suggested that 100 mirrors thus aligned would be enough to do the job. Similar ideas were put forward in 1840 by Johann Joseph von Littrow (1781–1840), Director of the Vienna Observatory, who proposed digging a series of circular canals, 30km in diameter, in the middle of the Sahara desert, and filling them with water. Then, a layer of kerosene would be poured on top and set on fire at night. Seeing as you would never get giant perfectly circular fires in Nature, the moonmen would quickly be able to deduce that the Earth was inhabited with intelligent arsonists. The slight trouble with these proposals, however, is that – Gauss' 1822 letter discussing mirror signals apart – we have only second-hand evidence that either man ever actually made them.[3]

Such stories have been often told, but with so many variations as to make them seem almost like urban myths – some say Littrow's fiery canals were meant to be square, for instance. Are they mere legends, then, an early counterpart of the contemporary crop-circle saga in reverse? Not necessarily. Both Gauss and Littrow did genuinely believe in the probability of life on other planets, with Gauss in old age even entertaining the bizarre idea that we might one day all be reincarnated as 'tiny creatures' living on the sun.[4] Then again, Gauss and Littrow's alleged schemes may have been mocking parodies of the loony ideas of their much-mocked contemporary Franz von Paula Gruithuisen (1774–1857) of Munich University, who had become famous after claiming to have seen a series of non-existent 'colossal buildings' on the moon during the 1820s – he was another one of the astronomers targeted in Richard Adams Locke's moon hoax, discussed earlier. Gauss and Littrow thought Gruithuisen was a self-deluding idiot, an opinion which must only have been heightened when he began explaining a faint patch of illumination periodically noticed on Venus as being due to 'general festivals of fire given by the Venusians' in order to celebrate 'the time when another Alexander or Napoleon comes to supreme power' on the planet, thus marking the new 'reign of an absolute [alien] monarch'.[5] Was Littrow's idea of setting fire to canals simply a skit of this daft assertion, or a development of it into slightly more sensible form? We shall never know.

Many of the subsequent plans for interplanetary communication were clear variations upon the ideas of Gauss and Littrow. One unnamed Victorian engineer, for example, allegedly paid direct homage to the latter by uncharitably deeming the entire Sahara Desert to be 'useless', and proposing that it should be filled with 2,400 miles of canals arranged in the shape of a huge right-angled triangle, which would then be illuminated with electric lamps so as to attract the attention of any watching Pythagoreans on Mars. As a bonus, the engineer added, the pointless sandy wasteland would be irrigated and put to some use at last, at the same time as ridding Europe of its unemployed masses, who would be forcibly shipped out there to dig the canals in the first place.[6] The famed French astronomer and best-selling author Camille Flammarion (1842–1925) – of whom much more later – also proposed creating huge geometrical figures in the landscape, but then changing them in an intelligent way. By marking out a huge square, and then splitting it down into its smaller constituent triangles, for example, we could let Martians know we had been paying full attention during Maths lessons. Seeing as we could currently 'see anything on Mars that is not smaller than Sicily or Iceland' with our puny human telescopes, surely the clever Martians, who were certain to be 'far superior to us' due to their planet's older age, would be able to spot our own geometrical signals? The idea of talking to our alien 'cousins in the sky', Flammarion declared in 1892, was 'not at all absurd, and is, perhaps, less bold than those of the telephone or the phonograph'.[7]

One of Flammarion's good friends was Charles Cros (1846–88), a poet and inventor. In an 1869 booklet, Cros laid out a plan for focusing powerful electric lights via a series of parabolic mirrors until such time as a giant beam of light could be broadcast out towards Mars or Venus, literally bathing it in illumination. Presumably for best effect this feat would be performed during the Martian night, so as to attract the most alien attention. Hopefully the ETs would not be too annoyed at being woken up, and realise that the glowing signal emanated from Earth. Then, they could keep a look-out for further communications, and some means of interplanetary telegraph involving the flashing of lights could be devised; a number of ideas for just such a 'light-language' were laid out by Cros in his pamphlet, which was reprinted by Flammarion in one of his popular books, to lend the idea publicity. Just in case Flammarion's efforts did not work, Cros also began writing fiction intended to promote his ideas himself, notably 1872's *Une Drame Interastral*, which told the tale of two young lovers, one on Earth and one on Venus, who romantically broadcast giant mirror-images of flowers to one another across the vast depths of inky space.[8] The Finnish mathematician Edward Engelbert Neovius (1823–88), another Flammarionophile, had similar ideas, with his 1875 book *The Greatest Mission of Our Time* providing very detailed calculations as to what precise strength of light would be needed to communicate with our future Martian friends.[9] Or, rather than shining bright lights up to Mars, why not simply lower some already existing ones here on Earth instead? In 1892, a letter writen to London's *Pall Mall Gazette* suggested simultaneously dimming every light in London to let the Martians know we were here –an idea which won support from the prominent astrophysicist, J. Norman Lockyer (1836–1920).[10]

## From Ants to Antennae

The most significant contribution to this debate in Britain came from the noted Victorian man of science Sir Francis Galton (1822–1911), a cousin of Charles Darwin. Whilst Galton was undoubtedly a genius of sorts, being a pioneer in such fields as statistics and fingerprinting, he was also a bizarre nutter of the highest order. Some of his published papers, like 1865's *On Spectacles for Divers*, 1885's *The Measure of Fidget* and 1906's *Cutting a Round Cake on Scientific Principles*, testify to the vast range of his sometimes comical quest for obscure knowledge. Probably his most famously eccentric paper was 1872's *Statistical Inquiries into the Efficacy of Prayer*, a numerical study into whether or not prayer worked. Galton concluded that it did not, on the grounds that priests had shorter average life spans than either lawyers or doctors, and that ships carrying missionaries abroad were equally as likely to sink as any others.[11] Another bizarre piece of work was his 1896 essay *Intelligible Signals Between Neighbouring Stars*, in which Galton laid out his own unique ideas for contacting other worlds. The article had its origins in Press speculation that Martians might already be trying to signal us themselves; bright spots had been observed on the planet by several reputable astronomers. These silly season stories happened to coincide with a period of ill health for Galton, as a remedy for which he was compelled to spend a 'somewhat dreamy' holiday in the spa town of Wildbad. It was here, relaxing in the curative waters, that Galton devised his paper.[12]

Galton's basic premise was that, one fine day, some 'mad millionaire on Mars, or rather … a mad billionaire' should start up a programme of signalling messages down to Earth. The first sign of this, said Galton, would be 'the sight of minute scintillations of light proceeding from a single well-defined spot on the surface of Mars', produced by an 'immense assemblage' of some alien version of Gauss' heliotrope mirrors. These communications, Galton arbitrarily decides, will be a visual variant of Morse Code, with three distinct elements, a dot, a dash and a line, each of which will last for 1.25, 2.5 and five seconds respectively. He thinks that the Martians will have letters, words, sentences and even paragraphs, which can be laid out by these means. They will also possess numbers, which he imagines will be founded upon a base of eight, rather than the number ten, as with us. The reason for this is that Galton whimsically supposes the Martians will be a breed of giant super-ants 'who count up to eight by their six limbs and two antennae, as our forefathers counted up to ten on their fingers.' Mathematical messages will then become the basis for a method of 'picture-writing', in which various indicated geometrical points are stitched together to make up images of things like Saturn and its rings, in a manner analogous to a join-the-dots puzzle. Once you knew that a dot-picture of the word 'Saturn' had been broadcast to us by the super-ants, you could then watch out for another signal indicating a short word which matched up with this image; dot-dot-dash-dash-line, say. In this way, a giant picture dictionary could be laboriously built up, allowing basic interplanetary communication to occur (alternative arrangements would presumably have to be devised for abstract nouns, verbs, tenses, connectives, etc). Such methods might not allow complex examples of modernist poetry to be meaningfully exchanged between the planets, but at least Mars' giant billionaire ants could warn us if a killer meteorite was headed our way tomorrow.[13]

Across the pond in America, interest in interplanetary communication really began to take off following the 1908 proposal of the astronomer W. H. Pickering (1858–1938) that, at cost of some $10 million, a cluster of 5,000 10-foot parabolic mirrors be constructed across the US so as to reflect light from powerful lamps up onto the Martian surface. Pickering's idea led to a lively debate in the pages of *Scientific American* magazine, in which other ideas for signalling to Mars were discussed. For example, one R. W. Wood (1868–1955) of Johns Hopkins University suggested that, to save money on expensive mirrors, someone should just buy a giant roll of black cloth and periodically roll it out and then back in again in the middle of a desert. Some of these ideas were so impractical that they quickly drew their parodists; an article in *Science* for July 1909 suggested drilling a giant hole straight through the Earth, several miles across, and then using it to flash Morse Code messages through to our planetary neighbours.[14] Another silly idea, put forward in 1894, was that several square miles of useless snow in one of the Earth's frozen zones should be illuminated with electric lamps laid out in the shape of a human being, waving upwards into the air, so as to attract the attention of any passing space farers.[15] It could sometimes be hard to tell which ideas were satire and which were genuine. For example, in 1899 the obscure amateur French astronomer A. Mercier published his booklet *Communications with Mars*, in which he bemoaned the 'loneliness' which must be felt by those on other planets, unable to communicate with any species other than their own. Noticing certain unexplained flashes had been observed on Mars during 1899, Mercier speculated these might have been intended as a response to the Paris Expo of that year, during which many electric lights had been set up to impress tourists. The logical next step was to transform the new Eiffel Tower into a giant Martian signalling device, filling it with mirrors angled to receive masses of sunbeams at sunset, reflecting them up to Mars. Occasionally, a giant screen could be placed over the Tower to obscure the illumination, making it even more obvious that this was no natural source of light, like a giant wildfire. Alternatively, we could cover an entire mountain with mirrors, and perform the same basic trick that way instead. Reckoning all this would cost around 50,000 francs, Mercier set up a subscription fund, towards which various people did actually donate.[16]

Mercier should have made an application to the Estate of Anne Guzman (d.1891), a rich widow who died in 1891, establishing a prize of 100,000 gold francs to be given in the name of her dead son, Pierre, to the first person successfully to establish two-way communication with the inhabitants of another planet. Both Madame and Pierre Guzman had been huge fans of Camille Flammarion's popular books on astronomy, and once he found out about the gesture, Flammarion helped promote the 'Guzman Prize', as it became known, which was to be administered by the French *Académie des Sciences*. Or, then again, perhaps Mercier would not have qualified for the cash prize after all; according to some accounts (namely, that of Flammarion himself), successful communication with Mars was specifically exempt from being rewarded, as the task was considered to be too 'easy' at the time to even be worth celebrating! However, according to other accounts (namely, that of Flammarion himself ... he was not the most consistent of men), Madame Guzman had specifically stated that Mars was the *best* place for prize-seekers to try and contact, seeing as this planet seemed to afford the most plausible prospects of success.[17]

## Under the Volcano

One man who sought to win the Guzman Prize was the world-famous Serbian-American inventor, electrical engineer and noted madman, Nikola Tesla (1856–1943). As befits the archetypal mad genius type, Tesla is famed for a number of things, both positive and negative in nature, from developing the A/C method of power generation used to power our homes today, to falling in love with a female pigeon which he claimed had the ability to fire laser beams from its eyes.[18] Because of his incredible strangeness, Tesla has become something of a hero to the lunatic fringe over the years, particularly to those who inhabit the less reputable outposts of ufology, with a Space Age religion having been founded in his name during the 1950s in America;

a persistent rumour even emerged that Tesla wasn't actually a human being at all, but a Venusian who had agreed to be born on the wrong planet to save humanity from itself.[19] In particular, Tesla was supposed to have faked his death in 1943 and then retired to some secret location in the Venezuelan jungle, where he used his skills to build a fleet of real-life flying saucers.

In this endeavour, Tesla was supposed to have been joined by his arch-rival Guglielmo Marconi (1874–1937), another great scientist who is generally acclaimed as being the inventor of radio – although many insist that the credit should instead have gone to Tesla. According to legend, the Irish-Italian Marconi is supposed to have invented some kind of death-ray device which he demonstrated to the Italian dictator Benito Mussolini (1883–1945) in 1937. Mussolini is said to have been 'quite pleased' by the death-ray, as possession of it would mean Italy would be invincible. However, when Pope Pius XI (1857–1939) heard about this, he supposedly stepped in and ordered Mussolini to make Marconi stop his research, an assertion so laughable in its lack of knowledge about which man was really the more powerful in Italy at this time as to need no criticism here. Allegedly, an irritated Marconi then boarded his 'floating super-laboratory', which was cunningly disguised as a yacht, and set sail for South America after faking his own death. Marconi used his riches to lure some ninety-eight of the world's leading scientists with him, it is said, and, like some Bond villain, set up a top-secret underground laboratory inside the crater of an extinct Venezuelan volcano. Known as '*La Ciudad Subterranean de los Andes*' – 'The Underground City of the Andes' – this secret lair was where Marconi apparently developed the advanced anti-gravity technology, based upon Tesla's own design, which he used to create his own working flying saucer with. Then, in 1943, Marconi swooped down on Tesla in his spaceship and forced him to fake his own death too, before flying away with him back to his hidden base. We owe most of this 'knowledge' to one Nacisso Genovese (1911–82), a physics teacher in a Mexican high school, who published a book called *My Trip to Mars* sometime during the late 1950s. According to Genovese, he was a former student of Marconi, and had been lucky enough to have been flown into outer space by his mentor several times; it only took a few hours to get to the moon, but trips to Mars lasted several days.[20]

## Electric Dreams

Absurd and false though these tales are, the question does have to be asked as to why it is that both Tesla and Marconi ended up being associated so closely with the idea of UFOs? The most probable answer is that during their lives both men had made some very public claims to have contacted aliens using their newly minted invention of the radio set. Tesla's boasts came first, in a February 1901 edition of *Collier's Magazine*, where his article 'Talking with the Planets' gave fascinating details about experiments

Tesla had been making with high-voltage wireless equipment in his isolated desert laboratory in Colorado Springs during 1899. Optimistically declaring interplanetary communications to be 'the dominating idea of the century that has just begun', Tesla threw naïve-sounding opinions out like confetti, opining that it was not implausible the moon might be hollow and so have 'intelligent beings' living inside it. Considering signalling to such creatures with lamps or mirrors to be impractical, Tesla proposed instead using a kind of wireless telegraphy. Even more sensationally, Tesla claimed to have received some wireless messages himself in 1899, which he judged likely to have originated from space. His first observations of these signals 'positively terrified' him, he said, 'as there was present in them something mysterious, not to say supernatural, and I was alone in my laboratory at night'. The electro-magnetic disturbances he was receiving, he said, did not emanate from natural sources, as they occurred 'periodically, and with such a clear suggestion of number and order' that they had to be artificial. 'The feeling is growing on me,' Tesla concluded, 'that I had been the first to hear the greeting of one planet to another.'[21]

This was quite a claim, and not one Tesla was willing to relinquish – although, had he lived into the 1950s, he may have been disappointed to see his signals explained away as being due to Jupiter's moon Io periodically emitting regular pulses when it passes through a certain belt of space plasma during its orbit.[22] In 1919, Tesla was back in the newspapers again, stating his opinion quite definitely that the signals must have come from Martians, and now claiming to have signalled straight back to them, replies he was sure must have been received and noted. Proposing we try transmitting pictures to Mars, Tesla recommended we 'flash' an image of the human face to them, and see what they send us in return. Explaining that he didn't believe in the usual Hertzian theory of radio waves, Tesla then set out to explain his own personal method for broadcasting to Mars, something which asked us to imagine that the Earth was 'a bag of rubber filled with water', which could be manipulated via an air pump in order to produce waves and oscillations that would eventually reach Mars. Basically, Tesla seemed to be proposing that we make the whole Earth vibrate somehow, and then hope that Martians picked up the energy waves this would send out – or, then again, maybe not, his whole thinking here is rather obscure.[23] In years to come, Tesla even began picking up unknown voices over the radio, a 'mystery' he chose to explain by recourse to aliens, but which must just have been down to the simple interception of stray transmissions, especially seeing as some of the invading voices were talking in English, French and German.[24]

Whilst Tesla's ideas were dismissed as overexcited nonsense by most scientists at the time, and marked arguably the beginning of the decline in his reputation from genius to crank,[25] evidently Marconi himself was not one of these sceptics, as in January 1920 a long article appeared on the front page of the magazine section of the *New York Sun*. Headlined 'Marconi

Credits Mystery Flash to Far Planet', it brought together interviews with the three main electrical geniuses of the day, Marconi, Tesla and Thomas Edison (1847–1931), all of whom seemed happy to provide confirmatory opinion that talking with Martians via wireless was not only a possibility, but highly likely to have already occurred. Edison even claimed that maybe the recent Spanish Flu epidemic had originated in outer space, and proposed that anomalous fluctuations in magnetic compasses could be due to them picking up alien messages. Marconi himself had been hearing strange signals which he had cautiously attributed to possible alien beings since 1919, telling a reporter that:

> For all we know, for years past, possibly for centuries uncounted, many of the stars [re: planets] that we see in the firmament may have been in [radio] communication one with the other ... We upon the Earth have worked alone and stumblingly. If those dwelling in other worlds have had the help of other beings, so that each finds itself in possession of a progression [in science] representative of the sum of the wisdom existing among the communicating spheres, they may have far surpassed us ... During my experiments ... I have received signals which quite conceivably might have arisen somewhere in interplanetary space! The marvels of the cosmos are unlimited.[26]

However, as is generally the case, such sensational claims received way more coverage than did Marconi's later more sober statement of 1924 that the idea of receiving radio waves from Mars was 'a fantastic absurdity', which he had 'never ... attempted', nor even 'given ... serious thought' towards. The reason for the shameless *volte-face* was that, in the intervening years, it had become increasingly clear that the weird signals picked up by Tesla and Marconi were simply the result of interference from the ever-increasing numbers of radio masts then being constructed.[27] Far from being close to winning the Guzman Prize for his efforts, Tesla was as far away as anyone had ever been. In 1937, the then eighty-one-year-old Tesla gave an interview to the *New York Times* in which he was still talking wistfully about claiming his 100,000 francs:

> The money, of course, is a trifling consideration, but for the great historical honour of being the first to achieve this miracle I would be willing to give my life. I am just as sure that prize will be awarded to me as if I already had it in my pocket. They have got to do it ... This discovery of mine will be remembered when everything else I have done is covered with dust.[28]

In fact, Madame Guzman's fortune never was successfully claimed, by anybody. Eventually, in 1969, some medals made in her name were minted by the French and awarded to the *Apollo 11* astronauts after they had just proven the exact opposite of what she had wanted them to find, at least as far as the moon went[29] – a development I'm sure Tesla would have regretted immensely.

## Radio Days

As this suggests, the fad for signalling other planets moved with the times, with ideas about how to perform such a task increasingly shifting from signalling with mirrors and lamps to communication via radio waves, in line with the march of new technology. The size of this alien radio craze can be seen in the extent to which the idea infiltrated popular culture. You can find echoes of it in advertising, for example, as in a 1901 poster for soap, in which a Martian beams down some words to Earth, namely 'The First Message from Mars: Send Us Up Some Pears' Soap!'[30] As shown by the American academic Robert Crossley in his study of such subjects, *Imagining Mars*, the literature of this period, too, contained its fair share of stories in which signals between men and Martians were exchanged. One particularly interesting such narrative was Lois Pope Gratacap's (1851–1917) emetic 1903 fable *The Certainty of a Future Life in Mars*, which tells the story of an astronomer named Randolph Dodd who promises to send his son Bradford radio wave messages from Mars following his death. He thinks he will be able to do this because of his belief that each human soul is reincarnated and evolves from planet to planet, a belief which, as we shall eventually see, was once shared by a surprising number of persons. A belief only possessed by Gratacap, however, was that Mars was filled with undead scientists like Galileo and Newton, whose reincarnated bodies were so subtle that, whenever they had a shit, it instantly evaporated into nothing in the Martian atmosphere.[31] Meanwhile, pulp magazines serialised tales like 1922's *The Moon Terror* by A. G. Birch, in which mysterious signals from our satellite are picked up across the world, bringing panic to mankind.[32] As radio became ever more popular, so did fears about its ability to do evil – fears which, due to its usefulness to men like the Nazis in terms of spreading inflammatory propaganda, were eventually to be justified. The academic Jeffrey Sconce, in his own study of this subject, has suggested that the 'Message from Mars' genre of horror story might be some kind of coded reflection of this fear.[33]

In the early days of radio, the medium was more of a personal hobby than the gigantic commercial enterprise it was to become as the 1920s progressed, and for many amateur radio hams, particularly young boys still naïve enough not to know any better, the idea that they might get the scoop of being the one to make first contact with Mars whilst fiddling about with the dials one night was a major part of the devices' initial appeal. Indeed, throughout the mid-twenties, several large cities across the US forced commercial stations to stop broadcasting for one night a week, so that such amateurs could continue to scan the wavelengths in search of spies, Reds, gangsters and aliens trying to talk to one another in secret. The idea of receiving transmissions from space seemed quite plausible to many, given the new media landscape of the time. Radio was presented to a still astonished public as the new wonder invention which was rapidly shrinking the world, allowing messages to be shared across the Atlantic or Pacific almost instantaneously, so why should the distance between actual worlds not be shrunk down to more manageable proportions, too? If a ham in Washington could pick up stray broadcasts from London or Tokyo, then why not Venus?[34]

## Watchers of the Skies

Perhaps it was the all-pervasive nature of the idea of interplanetary radio during this period which led to even members of the nascent US military-industrial complex getting involved in the search for alien life in the years following the First World War. In 1920, inspired by Marconi's claims, the chief engineer of General Electric, Charles Steinmetz (1865–1923), opined that, should America 'go into the effort to send messages to Mars with the same degree of intensity and thoroughness with which she went into the War, it is not at all improbable that the plan would succeed', though he did warn the 'cost of the attempt might be a billion dollars' and would probably involve the nation's entire electricity supply being channelled into a single sending station. Canny makers of war-time searchlights also got in on the game, with a Brooklyn inventor named Elmer Sperry (1860–1930) trying to keep his order books full during peace-time by proposing the sale of 200 of his products to the government for Mars-signalling purposes, whilst several newly unemployed military cryptographers happily offered their services to decode any alien messages which might arrive on Earth.[35]

The most significant attempt made by the US military to facilitate contact with aliens came in August 1924, when Mars passed as close to Earth in its orbit as it had done since as far back as 1804. Given this fact, might not friendly Martians try and contact us when the distance their radio waves had to travel was at its lowest? An American astronomer named Dr David Peck Todd (1855–1939) thought that they might. Sometime prior to 1909, Todd had proposed travelling as far up into the atmosphere in a balloon as he could, armed with a wireless telegraph receiver, in search of Martian signals, but the plan came to naught.[36] By 1924, Todd thought the time had come to try again, and contacted the US Army and Navy, as well as various national radio broadcasters, in an attempt to persuade them to shut down all broadcasts, whether military or commercial, for the day of Mars' closest approach, so as to let him listen out for alien signals properly. The military proved happy to help their former colleague (Todd had worked at the US Naval Observatory from 1875 to 1878), but the broadcasters slightly less so. The best the Radio Corporation of America was willing to do was to suspend all broadcasts for five short minutes at the top of each hour for the space of one day only, whilst US Army and Navy Signals Corps men scanned the skies for a three-day period around the time of the closest approach of Mars on 22 August.[37] A telegram from Edward W. Eberle (1864–1929), Chief of US Naval Operations, instructed all relevant units as follows:

> NAVY DESIRES CO-OPERATE ASTRONOMERS WHO BELIEVE POSSIBLE THAT MARS MAY ATTEMPT COMMUNICATION BY RADIO WAVES WITH THIS PLANET WHILE THEY ARE NEAR TOGETHER THIS END ALL SHORE RADIO STATIONS WILL ESPECIALLY NOTE AND REPORT ANY ELECTRICAL PHENOMENON UNUSUAL CHARACTER AND WILL COVER AS WIDE BAND FREQUENCIES AS POSSIBLE FROM 2400 AUGUST 21 TO 2400 AUGUST 24 WITHOUT INTERFERING WITH TRAFFIC.[38]

Regrettably, nothing was picked up by the Navy but meaningless static – though this did not prevent a yarn arising in later years stating the exact opposite. According to the tale as it has been told in various paranormal books down the years, amateur radio hams were meant to have detected 'freak signals of unidentifiable origin' on the night of 24 August 1924, whilst Dr Todd himself, monitoring events at a Naval observatory using a special Jenkins Radio-Camera, was surprised to find a long strip of photographic tape being printed out from it. Upon examination, this tape supposedly showed 'a fairly regular arrangement of dots and dashes along one side', whilst on the other 'at almost evenly spaced intervals' were 'jumbled groups' which, when considered together, each took the form of 'a crudely drawn human face'.[39] Maybe Sir Francis Galton was correct about there being a race of super-intelligent millionaire ants living on Mars after all!

## Dishing It Out

Despite the sorry history of the field, there have been a few serious, well-funded and professional attempts made by mankind to communicate with alien civilisations via the use of radio telescopes. Most famous is the SETI (Search for Extra-Terrestrial Intelligence) programme, which is generally said to have begun on 8 April 1960, when the American astronomer Frank Drake (b. 1930), of West Virginia's Green Bank Observatory, began his so-called 'Project Ozma', named after the princess of the faraway land of Oz in the *Wizard of Oz* books. Drake and his colleagues surveyed two nearby stars named Tau Ceti and Epsilon Eridani, deemed promising in terms of their solar systems' ability to harbour life, on the

21 centimetre wavelength of 1420 Mhz, which the American physicists Giuseppe Cocconi (1914–2008) and Philip Morrison (1915–2005) had already calculated in a 1959 paper would be the most likely to carry extraterrestrial communications. (The numbers are associated with hydrogen, the most abundant element in the universe, which emits radiation at a spectral wavelength of 1420 Mhz, the same value as that of a 21 centimetre wavelength; thus, 21 and 1420 were guessed to be numbers with universal 'cosmic significance'.) Project Ozma took place in complete secrecy, because Drake feared media ridicule for his efforts. Perhaps he was wise not to draw any attention to the experiment, as he found nothing useful in any case. With subsequent experiments proving equally unsuccessful, in the 1970s many SETI operations were transferred to a new 300-metre-wide radio telescope, then the world's largest, located at the Arecibo Observatory in the US semi-colony of Puerto Rico. Here, on 16 November 1974, a different tack was used, with a three-minute message to the stars from humanity being beamed out for listening alien civilisations to pick up. Directed towards a galaxy known as M13, and expected to be intercepted by any waiting ETs in about 21,000 years' time, it consists of a stream of binary digits which, if assembled correctly by the aliens, will give them crude 8-bit NES-style pixel-block illustrations of a human being, a DNA double-helix, and our own solar system to examine.[40]

The SETI programme is not limited to any one single nation or observatory, of course, especially now that it is largely funded in the West by private individuals rather than NASA, and the latest contribution to the field comes from China, in the shape of the 500-metre-wide Aperture Spherical Telescope located in Guizhou Province, which now overtakes Arecibo as the world's largest such device. Beginning operations in September 2016, it is specifically billed by Chinese officials as being aimed at discovering signs of alien life in the cosmos. So serious are the Chinese in this aim that over 9,000 local residents were forced to move away from the area to make way for the dish, whilst those few who were allowed to remain have been banned from owning mobile phones in case their signals interfere with the quest for alien life – let's hope they haven't been allowed to keep hold of their old microwave ovens, either.[41] The SETI programme undoubtedly represents a noble cause, but it seems an indisputable fact that, thus far, it has produced very little of note in terms of evidence for life beyond our planet. Given this, funding has often been in question for the programme,[42] and so new angles have recently been tried out, such as the SETI@home project set up in 1999 by the University of California, Berkeley, in which home PC users are invited to connect to the Internet and 'donate' their own PC's processing power to the task of assessing the vast reams of data produced by the radio dish out at Arecibo.[43] Whilst the analysis is automated, to avoid overexcitable UFO spotters from phoning Puerto Rico every five minutes claiming to have discovered a new message from Alpha Centauri, home users' amateur interpretations of this data could hardly have been more misguided than some of the blunders made by professional astronomers involved in the SETI programme down the years.

## Getting Your Signals Crossed

Most famously, in 1965 three Soviet scientists, Gennady B. Sholomitsky, Nikolai Kardashev (b. 1937) and I. S. Shklovskii (1916–85), rather incautiously held a Press conference in which they more or less announced to the world that they had received a message from an alien civilisation; or, at least, that was how the media reported their findings. The Russians had studied a far-off astronomical object named CTA-102, and found that it was emitting a suspiciously regular pattern of radio waves, which made it seem like an artificial construct. Sadly, it subsequently transpired that the object was what we now term a 'quasar' – a phenomenon then unknown, and even now imperfectly understood. Thought to be luminous super-massive black holes, quasars are now proven to emit a regular radio signal of the exact kind observed by the Soviets, although at the time there was no way in which the men could have known this. Still, their efforts were not entirely in vain. In 1967, the American band The Byrds released their album *Younger Than Yesterday* where, amongst all the Bob Dylan covers, lurked a song entitled 'CTA-102', billed as an attempt at something called 'space-rock', in which lyrics about the Russians' supposed communications with ETs were mixed in with simulated alien voices babbling away happily. The keen lover of astronomy Jim McGuinn (b. 1942), who co-wrote the song, was later delighted when real-life scientists made a sly reference to his band in an article in *The Astrophysical Journal*, thus making the whole debacle worthwhile – for one man, at least.[44]

In 1967, a group of British scientists from Cambridge's Cavendish Laboratory led by Professor Tony Hewish (b. 1924) and Jocelyn Bell (b. 1943), his student, could have fallen into exactly the same trap, after detecting another unknown radio source which seemed to be switching itself on and off with such amazing precision that they initially thought it had to be artificial – and thus alien – in origin. One suggestion, reluctantly made, was that it might have been some kind of space-based navigation beacon, intended to aid travel for spaceships. As such, the scientists involved privately dubbed it LGM1 – Little Green Men 1. However, they held off from holding any Press conference, preferring to investigate further. It was a good job they did, as additional study led them to conclude the source of the radio waves was a new kind of star, hitherto unknown to science, termed a pulsar. As the name implies, pulsars are dying neutron-stars which send out regular pulses of radiation across the universe during their final death throes. Instead of becoming infamous for having made a terrible mistake, like the foolish

Russians, the LGM-1 team thus became famous for having discovered an entirely new kind of phenomenon to be recorded in the text books; Hewish received a Nobel Prize in 1974. He didn't get a Byrds song written about him, though.[45]

As things stand, the SETI programme has enjoyed only the one possible direct hit, and even that now seems in danger of falling. The so-called 'WOW! signal' is the most celebrated event in SETI's entire history. It occurred on 15 August 1977, when an astronomer from Ohio State University named Jerry Ehman examined a print-out of the data from his radio telescope – nicknamed Big Ear – and its latest sweep of the night sky before noticing something very strange; a 72-second blast of radiation which was like nothing Ehman had ever seen before. So impressed was Ehman that he took out a red pen, circled the relevant piece of data, and wrote beside it, simply, 'WOW!', hence the signal's popular name (which is certainly more catchy than its actual print-out signifier of '6EQUJ5'). Seeing as the whole operation was a SETI one, Ehman's tentative conclusion was that the blast may have come from an alien world ... especially seeing as it travelled on a frequency close to the fabled 1420 Mhz. Consulting a star map, Ehman found that it emanated from a single point in space located in the constellation of Sagittarius, known as 'The Teapot'. Just to the east of The Teapot's handle, near a cluster of stars known as M55, was where the radio signal appeared to originate. The problem was that there was nothing there. Unless it had come from a moving spaceship, there was no good explanation as to what the signal had been. It certainly looked artificial ... but was it? The cases of CTA-102 and LGM-1 should have taught astronomers not to jump to conclusions and, to be fair, Ehman did not. He has always maintained he doesn't really know *what* the signal was.[46] The problem, however, might be about to be solved. In April 2016, Professor Antonio Paris, of Florida's St Petersburg College, examined astronomical records and discovered that, at the exact moment the WOW! signal was received, two comets were passing by the precise patch of sky scanned by Big Ear on their regular path through the heavens. However, neither of these two comets was discovered until 2006, meaning that Ehman could not have known about them at the time. Furthermore, the comets would have been carrying with them a large cloud of hydrogen – from which a short signal at 1420 Mhz could well have been emitted down towards Earth. The comets are due to fly by again in 2017, when Professor Paris hopes to be able to test out his idea – although, as he himself admits, he rather hopes his theory fails. 'There's still a little bit inside of me that hopes it was aliens,' he told the media in 2016.[47]

It seems that mankind's efforts to contact other planets have so far been in vain. Perhaps the futility of the whole exercise was best summed up in a satirical short story penned by the French humorist Tristan Bernard (1866–1947), entitled *What Exactly Can They Tell Us?*. In a direct parody of the Mars-signalling fads of his youth, Bernard has Earth's astronomers all aflutter with excitement after observing light signals on Mars. Bernard's eager astronomers then lay out a huge sheet of paper across the Sahara, onto which are written, in gigantic letters, the question 'I beg your pardon?' The Martians quickly signal back. 'Nothing', they say. 'Why are you making signs, then?' ask Earth's puzzled scientists. 'We are not talking to you, we are talking to the Saturnians', comes back Mars' rather deflating reply, and the interplanetary conversation ends … forever.[48] What makes us Earthlings so sure that alien beings would be so eager to start a conversation up with us anyway? By galactic standards, both our planet and our race might well be extremely boring. Perhaps this is why Earth's astronomers have spent the last 150 years talking only to themselves.

# 15

# Girl From Mars: Dr Robinson's Interplanetary Telegraph of Love

My own favourite attempt to contact aliens involved an eccentric former lawyer and town clerk of Shoreditch in the East End of London who was habitually described in Press reports of the time as being called *Doctor* Hugh Mansfield Robinson. This is misleading, as it implies he was either a medic or a scientist, when in fact he was only a doctor in the sense he had the right to put the letters 'LLD' after his name; that is to say, he was a 'Doctor of Laws', a type of legal qualification. A 1912 oil painting of him dressed in full legal regalia seems to show a self-possessed, calm and serious-looking man, with perhaps just a hint of pompousness about him, but no clue at all of what he was later to become – but then, official portraits of distinguished public servants tend not to present them as being raving lunatics who claim to have established intimate relations with telepathic big-eared beauties from Mars via the use of a home radio set.[1]

Robinson's term as town clerk came to an end in 1911, and subsequent media photographs show an elderly balding man, so I would presume that his eccentric alien exploits took place during a long period of at least semi-retirement. Either way, he first got onto Mars' wavelength in 1921, when a London department store named Gamages took delivery of some unusually high-powered radio receivers. Together with a local radio engineer named Ernest B. Rogers, Robinson approached the store and made the unusual request that these new receivers be opened up for some extraterrestrial messages which the two men were soon expecting to receive. The store agreed; presumably it was good advertising. Robinson said that at a certain specific time the Martians would provide him with a Morse Code broadcast of certain letters of the alphabet already known to him, and which he had written out and placed within a sealed envelope, now in the possession of Rogers.

According to the two men, the message from Mars came in at the correct time, and read thus: 'UM GA WA NA'. At first this seemed disappointing, but it transpired that the letters contained within Rogers' sealed envelope were the very same. Robinson then translated the baby-talk into the following trite communiqué: 'God is All in All'. Delighted, Robinson explained that he knew the message would come through safely as it had been his birthday that day, and he was sure his friends the Martians didn't want to disappoint him.[2]

How did this all begin? It transpired that Robinson, a dabbler in Spiritualism, had long had both an esoteric bent and an unquenchable love for free publicity. A photograph of his son Leonard, then a fourteen-year-old naval cadet described as being 'England's youngest aviator', appeared in British newspapers in January 1911. Sitting behind the controls of a monoplane, he was described by the Press Association as having 'already made some successful flights'.[3] One evening during 1918, however, Leonard had accidentally been responsible for another significant feat of aeronautics, when a chance remark he made had the effect of causing his father to fly away up to Mars – *sans* any kind of mechanical transport at all. Father and son were at the cinema, where presumably there was a sci-fi feature of some sort showing. Intrigued, Leonard asked his dad how it was possible for Martians to send signals to Earth. Robinson had to admit he did not know – but, fortunately, someone else in the cinema that night *did* know how the trick was performed. Feeling a pain in his left temple, and sensing an invisible female presence right there in the stalls with him, Dr Robinson heard a voice inside his head saying 'Come with me. I will show you.' The voice was emanating from a lovely Martian lady named Oomaruru – the name meant 'Loved One' in her native tongue – who was soon to become both Robinson's spirit guide and his interplanetary sweetheart. In an interview given to the Hungarian parapsychologist Nandor Fodor (1895–1964) in 1928, Robinson explained what happened next. Apparently, his astral body became converted into psychic radio waves somehow and was then broadcast out across the stars:

> I caught hold of my son's hand to remain earthbound. With my other hand – which was now a hand of my phantom body – I clasped Oomaruru. With the speed of light we flew and flew. Halfway I felt jerked back. That was the point where the radio waves from Mars and Earth clashed and created a chaos. The grace of God helped me through. I saw a giant red globe in front of me: Mars. We got nearer and nearer and alighted inside a radio station. Oomaruru called out: "That man is a medium, jump into his body!" I did so and looked at things through his eyes. I saw many radio towers, with sparks flashing, and antennas for reception. But the revolutions of Mars imposed a terrific strain on the etheric band [ghostly umbilical cord] that tied me to my body. I felt I was in deadly danger. As if drawn back by snapped

elastic, I felt I was rushing back to Earth. In four minutes I was back, and told my son of my experience.[4]

So what was young Leonard doing whilst all this was going on? Didn't he panic, thinking his dad was having a sudden fit or seizure? Inexplicably, Robinson neglected to say.

## London Calling

Robinson failed to gain any major Press attention until 1926, when Mars once again came into opposition with Earth (as indeed it had done in 1918, when Leonard had asked his dad his fateful question at the cinema). 'Opposition' between Earth and Mars occurs about every twenty-six months or so, when the orbital path of the two planets brings them closer together than they would ordinarily be – giving excellent conditions, you might have thought, for interplanetary signalling sessions.[5] During early October 1926, during the run-up to the much anticipated opposition of 4 November, Robinson claimed to have received a psychic message from Mars, informing him that its people would beam a three-word Martian phrase – *opesti nipitia secombra* – to a specific radio-receiving station at a specific time soon. When word leaked to the Press, the radio station in question refused to listen out for the alien communication, fearing it would be made to look silly, although it was legally obliged to send a message consisting of three 'M's to 'the Good Ship Mars' from Robinson after he had paid it to do so. Fortunately, London's Fleetway House radio station stepped in to receive the message at the last minute, intercepting a different communication instead – two Morse Code 'M's, presumably standing for 'Mars'. These signals apparently were genuinely received, in the presence of several 'expert telegraphists' and remained unexplained; presumably someone was in on the trick of sending them out, but who?[6]

Later that month, Robinson – by now habitually referred to as 'Dr Robinson' – hit upon the wheeze of asking the General Post Office to send some radio telegrams out to Mars on his behalf, something which led confused foreign newspapers to report that this was an official attempt by the British Government to talk to aliens.[7] His first attempt failed when he handed in a written telegram form addressed simply to 'Mars' at London's Central Post Office, which was promptly handed back with the words 'address unknown' stamped on it,[8] but our hero quickly tried again and this time stood his ground. Again prefixing his message with three 'M's – the 'international call-sign for Mars', so he said – the message was once more *opesti nipitia secombra*. The Post Office clerk said the best he could do was to send it on to Britain's most powerful GPO transmitter of the day, located in Hilmorton, and get them to beam it out into the ether. Seeing as there was no specific price guide for sending messages to other planets, the GPO man charged Robinson at the standard long-distance rate of 18p per word. As a GPO official later commented, 'If people wish to send messages to the moon and the man thereon, and are prepared to pay for them, there does not seem to be any valid reason why the Post Office should refuse the revenue.' The man behind the counter, fulfilling his ethical duty, did warn his customer that the GPO 'could not guarantee delivery' of the message, seeing as Martians may not actually exist, but Robinson decided to sacrifice his small change and paid for the telegram to be sent anyway. It was to be the best 54p he would ever spend.[9]

Newspapers finding Robinson's jape amusing, he was soon giving long interviews in which he outlined in great detail the peculiar details of life on Mars, apparently not realising the laughing stock he was busily making of himself. In Robinson's own, incredibly bizarre, words:

> I know the Martians well through telepathy. They are from seven to eight feet tall and have huge shocks of hair which stand up straight. They have Chinese features and they smoke pipes and drink tea from the spouts of kettles.

Allegedly, this tea-related habit was in itself 'not unlike [that of] the Chinese, many of whom refuse to use teacups'. Furthermore, it transpired that these Martians had elephant-like ears which were described as being 'a bit over-prominent', and which 'might flap if a hurricane were to strike' the planet. Drawings of the Martians produced under Robinson's expert guidance showed a race of weird cone-headed beings whose skin had the distinct texture of walnuts, which hardly lent his claims much credibility. It was also during these interviews that the figure of Oomaruru first began being mentioned in print, together with descriptions of her as being Robinson's 'girlfriend'.[10] The public would be reading a lot more about this touching romance during the years to come ...

## My Favourite Martian

In 1928, Dr Hugh Mansfield Robinson really hit the big time. The planets would once more enter opposition on 21 December, and Robinson began his next campaign of attention-seeking a full two months early, in late October. He started off by performing basically the same stunt as he had done in 1926, contacting the GPO once more and asking them to transmit some 'secret messages' (later revealed to read 'Love to Mars from Earth' – or '*Mar la oi de Earth*', in Mars-speak – and '*Com Ga Mar*', or 'God is Love') over their best new transmitter, located at Rugby. To ensure the Martians didn't miss it, the Morse Code would be sent out twice, once at 2.15 a.m. and then again at 2.30 a.m. on Wednesday 24 October, whilst some GPO engineers waited up long past their bedtime at a specially configured receiving station in St Albans to see if Mars would respond, this extra service being provided at Robinson's own expense. Predictably, Mars stayed silent. Robinson blamed poor Post Office equipment. This disappointed him, but he wasn't the only one. According to Robinson, 'The Martians were very annoyed that the signals could not come up to them. They were sitting up hours to receive them!'[11]

Newspaper men themselves, having not really expected Mars to reply anyway, were less disappointed; after all, they had uncovered a grade-A loon with a taste for the limelight who was only too willing to provide them with a litany of absurd details about his numerous successful trips to another world using only his astral body. Robinson was even capable of calculating the precise speed at which his soul had flown up to Mars during these visits – some 35,000,000 miles per hour. ('At this rate, even an afternoon excursion would be possible', commented the *Glasgow Herald*.) Once there, Robinson quickly found that Mars was a very curious place. There were a few different races of Martian on the planet, some of whom 'look like rats', whilst others had 'heads shaped like that of a walrus'. Some Martians lived underground, in caves, and many thought atheism to be a form of insanity. The dominant and most cultured Martians were Oomaruru's kin, the tall Chinese people who drank tea straight from the kettle, and who also had a curious approach to getting their five portions of fruit and veg per day. According to Robinson, these humanoid Martians 'treat and electrify their fruit trees in a peculiar way' so that 'a fruit resembling an apple' would end up containing 'all the constituents necessary for the human body'. Eating a mere three of these per day would be enough to keep the Martian doctor away forever, apparently.

Predictably, most Press interest focused upon the figure of Oomaruru, Robinson's ET girlfriend. Describing her as 'a woman in green', Robinson happily detailed her attractive appearance. 'She has a sweet face and big ears', he said, which Dumbo-like appendages 'did not especially detract from her beauty'. She was also six feet tall, and in constant telepathic communication with Robinson down here on Earth ('If I want to tell her something about, say, a blue book, I think "a blue book" – visualise it – and use no words.'). He even provided reporters with some drawings of Oomaruru, created by a spirit medium, which were described as showing 'a round-faced woman with long hair – not bobbed – a whimsical, half-smiling mouth; dark, penetrating eyes; curious nose, and very large ears.' Robinson also implied that Oomaruru was quite scantily clad: 'She has a very artistic flowing green dress, and her clothes indicate that the country is not very cold.' Robinson's overall assessment? Oomaruru was 'really very sweet'. He didn't specifically say she was his lover, merely his 'woman guide and collaborator', but the Press clearly thought it would be funny to imply that she was, and Robinson appeared in no great rush to dispel such illusions. For his wife, it was a different matter entirely. Now described as living in Royden, Hertfordshire, the Robinsons were evidently going through a sticky patch. Reporters who traipsed out there to question Mrs Robinson found the house in darkness, and a forbidding sign reading 'OUT'. Paying no attention, a *Daily Express* man stood there and repeatedly 'pounded' on the front door until eventually the angry human housewife emerged to give her side of the interplanetary love triangle. 'I don't know anything about this Mars affair,' she said, sounding every inch the woman scorned. 'I have refused to have the experiments conducted in this house while I remain in it. I don't know whether anyone encouraged my husband, but there will be no more of that foolishness in *this* house!' She added that her husband had gone off to London again then slammed the door, leaving the Press pack to trudge its weary way back to the capital. There, they were to find Robinson in ebullient mood.[12]

## Radio Ga-Ga

Back in London, Robinson was to be found lurking within the Chiswick laboratory of a man named Professor A. M. Low (1888–1956), described as being 'one of Britain's best-known younger scientists and writers'. Low was also happy to court Press publicity, proudly telling interviewers that:

> I have a friend who says that he has been to Mars, and who tells me that all Martian women have two thumbs on each hand. He also says they have x-ray telescopic eyes and already know all about the Earth.[13]

As this quote suggests, Professor Low was another eccentric who, like 'Dr' Robinson, assumed a title to which he was not entitled – he was no qualified professor at all. Unlike Robinson, however, he *was* a genuine *bona fide* scientific genius, albeit a deeply individualistic one. An enthusiastic experimenter from an early age – neighbours were always complaining about him causing explosions – as a young man Low joined his uncle's engineering firm where he produced several inventions, the most profitable being a whistling egg timer called 'The Chanticleer'. He also gave the first public demonstration of what was later to become television, although typically he never saw the project through to fruition, continually flitting from ambitious scheme to ambitious scheme without completing most of them, something which, together with his love for publicity, led many of his less flamboyant peers to dislike him. A skilled radio engineer, during the First World War Low had successfully led research into the invention of electrically steered wire-guided rockets. Because of such achievements, Low claimed that German secret agents made two unsuccessful attempts to assassinate him with guns and poison cigarettes, but he survived and went on to become both the Chairman of the RAC and a founder member of the British Interplanetary Society. He was also the author of several popular books on science and invention, and of various terrible-sounding sci-fi novels and short stories for children, such as his now probably unprintable 1934 racial fantasy *The Time-Traveller*, which not only depicts a world threatened by evil black pirates who wish to conquer the globe in the name of their dark-skinned dictator, but also successfully predicted the eventual coming of decimal coinage.[14]

Low's role was to provide Robinson with access to the advanced radio equipment sitting in his Chiswick lab. Refusing to simply accept the GPO's word for it, whilst expert telegraphists were monitoring the wavelengths at St Albans on 24 October, Robinson and Low were huddled around a confusion of hi-tech aerials and radio sets, listening out. Apparently, their wait was not in vain; eight minutes after the GPO had signalled to Mars, 'two mysterious messages, in unknown code' came down to Chiswick, over the ether. Low heard them too, not just Robinson, but was cautious about their nature, saying he 'could not recognise them as intelligent messages'. Apparently, they sounded like 'a long series of undecipherable dots and dashes' such as 'eight or ten dots, as if disturbances were confusing them' – or, sceptics may suggest, as if radio disturbance was *all* the sounds really were. Low was expecting nothing less: 'I regard the experiment as being an amusing one', he explained. Moreover, at 18p per word, he thought the GPO's rates for contacting other planets were an absolute bargain. However, Low specifically said that he did not believe any message could reach Mars, no matter how reasonably priced, doubting that radio waves 'would be able to traverse space beyond the Earth's atmosphere'. In his view, a better approach to the issue would be to have 'a fleet of aeroplanes lay down a smoke screen 14 miles long and 7 miles wide', providing a form of sky-writing large enough to be seen from space 'if the Martians have telescopes'. So, whilst describing the whole charade as 'foolish', the clearly amused Low nonetheless adopted the generous attitude that it is always 'best to be helpful if one can' and that anything was worth trying once, just to see what would happen. Robinson himself was rather less guarded in his words. Presuming that the meaningless late-night radio interference really was a message from Mars, he cracked open a bottle of champagne and 'drank the health of all Martians indiscriminately'. Possibly Robinson was still feeling the effects of this celebratory tipple when he emerged from Low's lab to announce the following words to the journalists who had managed to track him down:

This is the greatest event in the history of the human race! It is the big thing foretold in the message from the pyramids. I have a theory that the whirlwind experienced in London [on] Monday night was in the nature of a hint from Mars, and I am investigating this.

The word 'hubris' springs to mind. There had indeed been a freak whirlwind in London's West End on Monday 22 October, which was strong enough to cause some £15,000 of damage within a mere 30 seconds, but it had been nothing to do with Mars.[15] The doctor's claims were getting out of hand.

Soon, even an embarrassed Oomaruru was back in psychic contact with Robinson and ordering him, like a naughty child, to 'go to bed'.[16]

## Keep Your Ears Open

Like many a failed experimenter, Robinson's next step was to blame his equipment. The Martians, he said, now 'laugh' at the British Post Office. 'Mars received neither message. Do not attempt to use the Rugby station again, but make the next attempt in America,' Oomaruru advised. Robinson even took time out to blast the BBC for refusing to cancel their radio programmes and let him commandeer the airwaves for the night, beaming out a constant stream of gibberish Mars speak, 'for fear that children who might be listening would be afraid.' 'Is that science?' he fumed contemptuously, to which the only possible answer is surely 'no'. The French weren't much better, either. According to Mars, unknown Frenchmen had been sending broadcasts up to their planet in some mysterious code. Robinson's 'Martian collaborators' had duly sent these back down to him to investigate, which led to Robinson approaching the French Consulate in London with a piece of paper bearing various random letters and demanding to know what they meant. The Consul simply 'shrugged his shoulders' and told Robinson to go away, conduct the Englishman found outrageous. The basic problem, besides the incurable rudeness of the French, was that none of the European stations were able to transmit at a high enough wavelength to get through to Mars. 'I only hope some American millionaire will put up a few shillings' and get a 30,000-metre wavelength radio set operating, Robinson sighed, but no such generous donor ever appeared.[17]

By mid-November Robinson was back, announcing that he had managed to arrange yet another attempt to contact Mars, this time in Rio de Janeiro, where a far superior radio transmitter was eagerly awaiting his attention. This attempt was to take place on 15 December, a week before Mars entered into its next opposition. To travel to Brazil in 1928 would have been an expensive proposition, so Robinson simply provided the Brazilians with a message to send off to Mars and awaited developments in what was now described as either 'his London office' or 'his little surgery in Spital Square, Whitechapel', listening in to a 'special apparatus which he confided sadly cost him $50' (presumably the fabled 'psycho-telepathic motor-meter' he later boasted of inventing[18]). The Companhia Radiotelegraphica Brasiliera sent out his chosen message ('God is Love – Earth to Mars') as requested at 11.30 p.m., but even though a group of ten engineers and journalists donned 'special earphones' and listened out for a reply until 12.15 a.m., no message from Mars came flying down to Rio. Visiting him that night, a United Press reporter found a depressed Robinson sitting with 'his bald head buried despondently in his hands', wondering whether Oomaruru was leading him on, or even cruelly 'toying with his spiritual affections'. However, suddenly his mood was lifted as a telepathic signal sent out from Old Big Ears herself arrived, informing him that the radio message had been received loud and clear on Mars after all. The Martians had quickly replied, but the 'atmospherics were very strong' that night, so they had not got through properly. 'Oomaruru has told me,' explained Robinson, 'that the Martians are agog. They are doing their damndest to get through, she told me – although not in those words, of course. She is a *lady* in every sense of the word.'

Robinson made a second attempt to contact Mars from Rio on 21 December, the actual day of the opposition, but this too came to naught, which was surprising seeing as Robinson had by now wisely enlisted a psychic greyhound named Nell to assist him in the operation. According to the dog's owner, a Mr Thomas John Littlejohn of Falmouth, Nell would prove useful to Robinson as she had the power to 'transmit her ideals' into people's brains from a distance. However, the media seem to have taken Nell's failure to produce the goods as their cue to abandon Robinson back to his former obscurity – for a couple of years, anyway.[19] His influence, however, lived on, with the occasional imitator emerging to keep his flame alive. One woman who directly piggybacked off his long-distance love affair with Oomaruru was one Irene Briares, a Frenchwoman described as 'a young authoress', who during Robinson's 1928 Brazil-based escapade took the opportunity to send her own telegram to Mars via the Poste Telegraph in Paris, asking Martian males 'not to neglect Parisian girls'. Bemoaning the fact that all Frenchmen were unworthy of her love, Mademoiselle Briares prevailed upon her suspicious local telegrapher to send out her message to Mars in a special code of her own devising, 'because it was not for Earthly eyes to see it'. Pining for 'a Martian intellectual', Briares said she was certain that some clever 'Prince Charming of Mars' would be able to decode her signal, despite the fact that 'I have written it in the most lyrical code any daughter of Eve could have conceived' (her novels must have been of the Mills & Boon type). Briares admitted she was unsure how she would be able to meet her extraterrestrial suitor face to face once he had received her message, but seeing as this was clearly a PR stunt being pulled with the aim of advancing sales of her books, I doubt she was really expecting the problem to be one she would ever end up having to actually deal with.[20]

## Sensitive Flowers

By 1930 Robinson was back hitting the headlines. In March, the discovery of the outermost planet (since downgraded) Pluto had been announced to the world. Within a few days, Robinson was making a Press announcement of his own, declaring to downbeat reporters that radio communication with Pluto would prove impossible due to its great distance from Earth. In any case, he added, any aliens living there were bound to be 'so different from Earthly life' that it would be 'silly' to think any meaningful sort of conversation could be conducted between them and us.[21] Much more sensationally, at the end of May Robinson revealed that he had secretly founded a new educational institute, the College of Telepathy, whose entire staff was described in one newspaper headline as consisting of 'Six Teachers and a Telepathic Dog'. Of course, the hound in question was none other than Robinson's old pal Nell, and a closer reading of the article revealed the disappointing information that she wasn't actually a teacher at the College at all – that would be ridiculous. Instead, the animal merely 'sat on the College advisory committee'. Robinson's initial student body consisted of a mere seven students, all of whom were female. The Student Union of the institute was made up of a single young lady, described as being 'very pretty' with 'blue eyes and golden hair'. When United Press' Henry T. Russell asked her name, he was brushed off with the reply 'Neophyte No. 1', but eventually managed to squeeze it out of her that she was really called Claire. Claire's curriculum at the College (which in reality would presumably have been Robinson's own house – where had Mrs R gone by this stage?) appeared to consist largely of sitting around in the garden sunbathing and eating flowers; pleasant, fairy-like activities which the establishment's Head claimed could facilitate contact with Martians. According to Dr Robinson:

> Purifying body and soul, sun-bathing, dancing and the eating of flower-petals are conducive to telepathic receptiveness. Lilac and roses, also violets, are exquisite food. Raw salad, carrots, cabbage and other vegetables which have not been deprived of their vitamins through boiling, make for that health necessary for beauty of body and soul. Moral and physical health are conducive to telepathy ... Few people seem to realise it, but telepathy is the missing link in progress. It is being scandalously neglected by our scientists in spite of its obviously tremendous value. Were it practiced to a greater extent, then it would solve our telephone difficulties. The telephone is not much use anyway, one in three calls do not materialise. You either get a wrong number, a busy signal or a "they don't answer" from the operator. All this will be done away with when the world learns to telepathise.

The trouble was, explained Robinson, that in our hurried modern age, most men were far too busy with their important jobs to be bothered making the effort to become psychic; several millennia ago, the Biblical Prophets ('who were the telepathists of that time') had followed a primitive version of Robinson's regime so they could speak to God, but those days were long gone, and no men of such quality and self-discipline could be found walking the Earth today. As such, it would be more practical to train up women – but most modern girls were far too frivolous to be bothered with it all, either. The solution, said Robinson, was to actively appeal to women's inherent stupidity by telling them that becoming telepathic would also make them beautiful. Fortunately, dancing around in his garden would make women fit and slim, sunbathing would cause them to develop nice healthy tans, and eating flowers and raw vegetables would improve their skin and radiance. By urging women to 'think beautiful thoughts' whilst doing all this, Robinson hoped to 'tackle the women' by making them beautiful in both body and soul, thus facilitating the development of their latent telepathic tendencies. 'Complete chastity and abstinence from flesh, alcohol and tobacco' could work wonders for a woman's ability to transmit her thoughts to Martians, he said. Eventually, Robinson hoped to raise enough money to open a new campus on the South Coast, where 'ugly women – if any can be found who will admit it – may flock and be made beautiful, then telepathic'.[22]

## Carry On Cleo

The final Press appearance of Robinson I could find occurred in April 1933, with a series of ever-more desperate claims being made by him about the twin topics of agriculture on Mars and the surprising revelation that Oomaruru was in fact the alien reincarnation of Cleopatra, Queen of the Nile – or 'Marc Anthony's erstwhile honey', as American newspapers insisted upon describing her. Apparently picking up on the disasters then facing US farmers in the era of the Dust Bowl and the Great Depression, Robinson decided to begin peddling the line that, on Mars, a farmer's lot was a happy one. Indeed, so well respected were Martian farmers that Cleopatra (AKA: Oomaruru) had specifically chosen to be reincarnated in the role of a farmer's wife, and now lived happily within a glass house amidst the alien corn. Robinson explained farmers' immense popularity on Mars by pointing out that Martians, like the students at his College, were all vegetarians. Another titbit revealed by Robinson was that, through some unaccountable oversight, Martian inventors had never managed to work out how to create a phonograph. As soon as he discovered this fact, Robinson quickly communicated the relevant blueprints to the aliens via psychic means, and consequently is now known on Mars as a great inventor. He also claimed to have used his own Earth phonograph to record three separate Martian voices (including that of Oomaruru) on wax cylinder, emanating from within the mouth of a psychic clergyman's wife named Mrs St John James. Journalists described these recordings as 'unintelligible', but Robinson said he would set to work on a translation presently, to pass on Mars' message to mankind. This was a vital task, as during the last five years Oomaruru had correctly 'forecast five or six world-shattering events'.[23]

If queenly Oomaruru really had ever managed to predict such coming tragedies as Pearl Harbour or the Holocaust then she certainly kept quiet about it, though, and tales of Dr Robinson and his Martian girlfriend disappeared from the pages of the Press as newspapermen worldwide began finding rather more important issues to spill their ink over. Perhaps this was because, as Robinson had told reporters back in 1929, Oomaruru was in fact 'not a good seer', which directly contradicts his claims in the 1933 report. Actually, Robinson had declared some four years earlier, the best 'prophet and seer' on Mars was a man named Ookonga, who was Oomaruru's boyfriend at the time – this must have been the noble veggie-farmer she later married after realising she was really Cleopatra. According to Robinson, Ookonga had recently predicted a litany of imminent doom for planet Earth, consisting of 'a series of earthquakes, tidal waves, famines, floods, sudden death, destruction and six-day bicycle races'. Supposedly, Ookonga had predicted the Wall Street Crash to Robinson six months before it had actually happened, although curiously the normally publicity-hungry Londoner had failed to reveal this message to the media in good time. Presumably this was because the 'Great Ones' of Mars had also passed on a message to Robinson to the effect that Americans 'are too fond of gold' and have to 'cut out this worship of Mammon' if they didn't want worse things than a stock market crash to befall them some day.[24]

What are we to make of all this? Surely nobody really took Dr Robinson seriously in his claims? Most newspaper reports certainly didn't, being scattered with silly cartoons of Oomaruru and jokey headlines like 'Ears Stretch for Reply to Mars Message' or 'Small Ears and Long Antennae in England Strain in Vain'.[25] Occasionally, though, Robinson's less sensational escapades (i.e. those not involving psychic dogs and telepathic flower-eating) were described using terms like 'a serious and semi-official effort', so some small minority at least did seem to fall for it.[26] After all, sometimes Robinson's claims were reported directly alongside similarly bizarre claims about life on Mars from other, more reputable scientists and astronomers, such as the following statement released to the Press by Professor Philip Fox (1878–1944), of the prestigious Dearborn Observatory, during the very same week in 1926 that Robinson was making his first attempts to enlist the GPO to his cause. According to Fox:

> It seems certain that vegetable life exists on Mars, and where vegetable life exists animal life is certain to be – all conditions being equal. The Martian animal is probably a fur-bearing one, equipped by nature to live in the waters around the polar snow-caps. It must of necessity be quite small to be able to migrate rapidly with the changing seasons. It probably would be something like our seal, enabling it to swim along with streams of icy water melted from the snow-caps.[27]

Not even Oomaruru ever claimed there were any seals on Mars. What about Robinson himself though? Was he mentally ill, or just a publicity seeker? It seems to have been a little bit of both. Undoubtedly he loved the glare of publicity, and many of his sillier statements seem likely to have been simply designed to gain column inches. But, then again, some of his statements appear profoundly unbalanced. According to Nandor Fodor, Robinson – a 'chubby-faced, blue-eyed, yellow-browed gentleman, bald in front, grey-haired in back, stout and vital, with a faraway look that never seemed to concentrate on this Earth'[28] – was 'as slippery as a human eel' and constitutionally incapable of refraining from boasting about his many bizarre experiences to anyone who would listen. Apparently, 'It was nothing' for him to 'pour out statements' of the following kind:

> From this chair in which you see me sitting, without loss of consciousness I can make myself visible at a distance of thousands of miles and move objects. I am the disciple of an Indian Mahatma who could project his mental body in nine minutes into the sun … With humility we prepare for the Second Coming … I tell you, the Lord Jesus is already active in London … See this black line that goes through my head? This is the positive tube through which my mental energy is discharged. This white tube is the negative one. Here is a photograph … of an insane boy. Do you see the mess of criss-crossed lines? The telepathic mechanism has broken down. The boy was hearing voices … This sphere here is the *boule d'or*, the vehicle of the soul … Here are some thought-photographs … This is an elemental … And here is the picture of Ka-Ka-Hotep, the Egyptian mummy at the British Museum that brings disaster to everybody who meddles with it. This is the spirit that inhabits the mummy. The odd thing is that … [it is identical to] my old Indian nurse.[29]

## God Save Our Martian Queen

There is also the suspicious fact that the nomenclature of Robinson's Mars appears to have been modelled upon a series of barely altered place names from Namibia. The capital city of Mars, for instance, was apparently Ookalonga[30] – and there is both a village and a tribe called Okalongo in Namibia. Even Oomaruru herself possesses a name curiously similar to that of the Namibian city of Omaruru. But why? Possibly, during the days of Empire, Robinson had once had some connections with Namibia. Alternatively, perhaps he simply opened up an atlas and, in a most non-PC way, chose to make use of some exotic African language to lend him inspiration. According to Nandor Fodor, the Martian phrase *opesti nipitia secombra* was 'allegedly a greeting in some African tongue', too, whilst Robinson himself spoke teasingly of how he had since discovered there was a place in New Zealand called Umaruru.[31] Or maybe he simply made up the daft names from scratch; it is impossible to know. Even more suggestive was a letter which appeared in the London *Evening Standard* during Robinson's 1928 attempts to contact Mars. When reader John C. Budden heard about one of Robinson's earlier alleged trials, in which the words

'UM GA WA NA' had been received, he was outraged. Did people not know, asked Budden, that this was a common mantra chanted by Tibetan Buddhist monks?[32]

On the other hand, Robinson does appear to have been a declared Spiritualist, with connections to other people from within the faith. For example, there was Mrs St John James, the medium whom he used to record going into trances and talking in Martian – or 'emitting strange noises', as one hard-nosed journalist described it. According to Nandor Fodor, Robinson claimed that the Martians would 'speak like giants' through this woman's mouth, with their voices so loud that they 'roll and rock the room'. Fodor disagreed, saying that she simply spoke 'loud gibberish', but she also had a nice sideline in singing Martian songs which were then recorded by Robinson onto wax, her repertoire including both a 'recessional Swan-Song' of regret, intended to be sung whenever Mars went out of opposition, and, most impressively of all, the Martian National Anthem.[33] These valuable recordings were handed over for safe-keeping to the National Laboratory for Psychical Research in Kensington, run by England's most famous ghost hunter Harry Price (1881–1948), but they appear to have since gone missing. Price himself met and investigated Mrs St John James in 1926, and in a later 1942 account of this meeting included the intriguing information that she also sang Martian love songs. However, he was unimpressed by them. 'If the "melodies" did not originate in Mars,' wrote Price sarcastically, 'at least they *sounded* unearthly.'[34] Robinson also sometimes made use of an American medium named Suzanna Harris (1854–1932), who upon one occasion 'emitted an ectoplasm' which then took on the shape of a pair of trumpets which were subsequently blown upon by Martian spirits, but the less said about this particular fantasy, the better.[35]

The overall impression I get of Dr Hugh Mansfield Robinson is that of a bored, lonely old man who attempted to spice up his life by engaging in a very public fantasy which he half-knew was entirely fictional, and half-came to believe. Fodor himself reckoned that Oomaruru was an idealised version of Robinson's dead mother, with whom he wished to be reunited in some distant Other World.[36] Perhaps the most peculiar aspect of the whole affair for modern readers, though, is the way in which Mars and Spiritualism mix together into what would appear today to be a woefully incompatible melange. Newspaper reports of the time about Oomaruru & Co just seem to have accepted that these areas were connected somehow, whereas nowadays we would consider outer space to be the realm of science and astronomy, and inner-space to be the realm of fantasy and psychology. This was not always the case. Contrary to popular belief, space travel did not begin in the mid-twentieth century. Instead, human beings have been travelling to and exploring other planets and moons in search of alien life for many centuries now; doing so not with rockets, but with nothing more technologically advanced than the powers of the human mind …

# 16

# You Look Familiar: The Alien in the Mirror

One man who managed successfully to leave our Earth behind without learning how to pilot a spaceship was our old friend Johannes Kepler, whose tract *Somnium* (*The Dream*) is an extended fantasy about travelling up to the moon in the company of a flying demon. The book only appeared in print in 1634, four years after Kepler's death, but is set in 1608 when, Kepler wrote, he had developed a taste for researches in the field of magic. Falling asleep one night whilst gazing at the stars, he has a dream about a man named Duracotus, whose mother, Fiolxhilda, is a 'wise-woman' – one of those marginalised village folk who once acted as healers and fortune tellers, making a living by selling herbs and amulets, but who were often mislabelled by critics as being witches. Annoyed by various aspects of her young son's misbehaviour, Fiolxhilda gives him away to a sea captain one day, and he begins to travel the seven seas. At one point Duracotus sails to Denmark, where he falls in with the eminent astronomer (and possessor of a fake brass nose) Tycho Brahe (1546–1601), who teaches him the science of the starry heavens.

So far, anyone who knew Kepler would have easily realised that his text was a kind of romanticised autobiography; he was Duracotus, and his mother Katherine was Fiolxhilda. He had not been given away to a sea captain, but was forced to travel far from home to further his learning, and had indeed ended up in Denmark, acting as Brahe's assistant. Unfortunately, in *Somnium* the fictional counterpart of Kepler eventually returns back to his mother, who has long regretted her impulsive act of abandonment. Reconciled, the two begin to talk about the night sky and Duracotus is puzzled to find that this entirely uneducated woman knows as much about the topic as he, the pupil of the great Tycho, does. Asked to explain, Fiolxhilda admits that, as a wise woman, she is contact with the 'daemons of Levania', a race of moon spirits whom she can summon down to Earth and order to fly her up towards the lunar surface. This is supposed to be just a narrative device for framing Kepler's subsequent speculative essay about what life might be like on the moon, but unfortunately certain people at the time did not take it in this way. In 1615, after the manuscript version of *Somnium* had been circulating for some time, generating local gossip, Kepler's mother Katherine (1546–1621) was arrested and charged with witchcraft. Katherine's aunt had apparently been burnt alive as a witch some years beforehand, and Katherine herself – by her own son's admission a 'small, thin, swarthy, gossiping and quarrelsome' woman, 'of a bad disposition' – was known for collecting herbs and concocting potions, so the text of *Somnium* seemed almost tantamount to a public accusation against her to some. In particular, Frau Kepler was accused of having cursed a little tin jug from which several persons who had sipped had either died or fallen ill, and of making attempts to steal the skull of her own father from his grave in order to refashion it as a macabre drinking vessel. For five long years, on and off, Katherine's guilt-ridden son sought to clear her name, but during much of this time she was kept in prison as a sorceress, where conditions evidently took their toll. Six months after she had been found innocent and released, Mrs Kepler keeled over and died.[1]

The second half of Kepler's book is more scientific in nature than the first, containing conjecture about how human beings might be enabled to breathe in outer space (by sticking wet sponges up their nostrils!) and upon the likely effects of space travel upon a person's body. Flight to the moon is demon-powered, but seems to take account of the force of gravity, here dubbed 'magnetic force'. Once on the moon, Duracotus describes no intentional fantasy world, but an accidental one instead. Based upon the limited telescopic knowledge of the day, the moon is described as astronomers back then might reasonably have thought it would be, not as we now know it actually is. Kepler correctly states that the moon orbits so as to always show the same face toward us, this side being known by the satellite's inhabitants as 'Prevolva'. The Dark Side of the moon, meanwhile, which faces permanently away from us, is dubbed 'Subvolva'. Nights on Kepler's moon are much longer than our own and freezing cold and icy, whilst the equally long days are 'fifteen times hotter than our Africa'. Just as we see a man in the moon, the Prevolvans see Africa as a severed human head. Subvolvans live in circular cities – their walls being the moon's craters – but most life on the moon takes the form of gigantic serpents, some with legs or wings, which can grow to maturity and die within the space of a single lunar day, being kept constantly on the move by a need to escape from the boiling sun by seeking refuge in caves or bodies of water.[2] It's certainly an interesting mental exercise, but I wonder if, had he known the results of writing it, Kepler would still have found his book worth the trouble? Not only was his mother nearly burned as a witch, *Somnium* was later to become the inspiration for one of the stanzas mocking overimaginative astronomers in Samuel Butler's *The Elephant in the Moon*.[3] Dr Hugh Mansfield Robinson got away lightly with his own fantasies; had he lived a few hundred years earlier, he may not have found himself quite so indulged!

## A Font of Little Wisdom

The subsequent generation of men of science who wrote down their visions of life on other worlds were careful not to present them as having taken place anywhere other than within their own heads. By 1656, when the Jesuit scholar-scientist Athanasius Kircher (1602–80) published his *Ecstatic Journey*, in which a figure called Theodidactus is escorted across the cosmos by an educating angel named Cosmiel, there was no suggestion whatsoever from the author that it was anything but speculative fiction, with not a hint of autobiography being anywhere present within.[4]

The most widely read book of this kind was doubtless 1686's *Conversations on the Plurality of Worlds* by Bernard de Fontenelle (1657–1757), the resident historian of the l'Académie des Sciences. Fontenelle's book caused a sensation immediately upon publication and retained its popularity right into the nineteenth century, going through at least a hundred editions, translations and reprints by 1800.[5] Probably this popularity was less because of its scientific merits and more on account of its easy style and wit. As Fontenelle himself wrote, 'Truth and fiction are in some measure blended' in his book, with 'the union of philosophy and amusement' being what he was chiefly aiming at.[6] Structured as a series of nocturnal dialogues between a gentlemanly philosopher (i.e. Fontenelle) and a beautiful young French Marchioness, the book is really little more than an elegantly composed primer for those persons, like the noblewoman in question, who wish to know more about astronomy. It explains, for example, the Copernican model of the universe, the rotation of our Earth in space, and how eclipses occur. During her education the Marchioness expresses a desire to know whether other beings exist in outer space, however, something which gives her tutor the chance to engage in a few amusing flights of fancy to break up the text with. However, all these descriptions of aliens are really very vague, and seem designed to illustrate the natural conditions then thought to be prevailing upon each planet in an entertaining way more than anything else.

For example, in aiming to demonstrate to his pupil that, from the perspective of any putative dweller on the moon, it would be the Earth which would appear to cover the sun during an eclipse, Fontenelle depicts lunarians as cowering in fear whilst such an event occurs, just like superstitious people on Earth still did when they saw the moon cover the sun from our own planet's perspective. As he asks: 'What right have they to frighten us, unless we can frighten them?'[7] Aiming also to prove that the atmosphere on the moon would be as different to our own as the sea is to the land on Earth, he describes lunarians sailing across the surface of our atmosphere and letting down nets to pull us up into their space boats with, whereupon we would begin gasping for our own type of air just as much as a fish does when pulled onto dry land.[8] Explaining that Mercury and Venus are nearer to the sun than we are, meanwhile, leads to the following memorably non-PC description of Venusians and Mercurians being made:

> [The Venusians] are much like the Moors of Granada; a little dark, sun-burnt people, scorched with the sun; full of wit and animation, always in love, always making verses, listening to music, having galas, dances and tournaments ... But what must the inhabitants of Mercury be? ... They must be almost mad with vivacity. Like most of the negroes [of Africa], they are without memory; never reflecting; acting by starts and at random; in short, Mercury is the bedlam of the universe.[9]

There is no way that Fontenelle could have known any of this for sure, even it were really true; he is just drawing logical(ish) conclusions, in a playful fashion, about what type of beings would be most likely to inhabit what type of planetary climate. Saturnians, being distant from the sun, would be 'very wise' and 'very phlegmatic', having cool intellects and never smiling, for example, with Earthlings, occupying the centre of the then-known solar system, having 'no determined character', being merely 'a compound' species due to our middling position.[10] Fontenelle also used contemporary ideas of planetary geography to draw dubious conclusions about the likely kinds of civilisations which might flourish on them; Jupiter is huge, so its nations must be remote from one another, he says, whereas Mercury is tiny, so the aliens there must constitute a single, unified nation, where 'all [are] neighbours, living familiarly together, and hardly considering the tour of their world more than a pleasant walk.'[11] This book represents the true beginnings of a science now known as 'exobiology' or 'astrobiology', a discipline devoted to speculating about the possible origin, nature and distribution of alien life forms throughout the universe, and how we might go about detecting them. When you hear NASA scientists guessing that maybe on icy moons there could be small creatures with fur, or airy little things resembling flying jellyfish floating in the atmosphere of gas giants like Jupiter, then this is where that tradition really begins.

One noticeable thing about Fontenelle's adventures in exobiology is that he at no point attempts to speculate in any great detail about what his aliens may look like – as he says, any such descriptions would be 'entirely chimerical'. The descriptions of their character which he gives, though, tend to lend the casual reader the impression they are basically humanoid in form. However, eager to avoid the 'theological difficulties' relating to the idea of human people being found on other, non-Christian, worlds, Fontenelle does explicitly say that he has 'asserted no such thing: I say there are inhabitants, and I likewise say they may not at all resemble us.'[12] Another dubious racial comparison is made by Fontenelle to ram this point home:

> I don't believe there are men in the moon. We see how much nature is changed even when we have travelled from here to China; different faces; different figures; different manners; and almost a different sort of understandings; from here to the moon the alteration must be considerably greater. When adventurers explore unknown countries, the inhabitants they find are scarcely human; they are animals in the shape of men ... almost devoid of human reason; could any of these travellers reach the moon, they surely would not find it inhabited by men.[13]

By this logic, such 'animals in the shape of men' as African pygmies or Australian aborigines would also be little more than aliens – but even alleged non-humans like these were still built upon an obviously human*oid* plan, and Fontenelle cannot seem to resist thinking of his extraterrestrials in a similar fashion. Beings on Mars, Mercury, Venus, Saturn and Jupiter, he says, are all likely to be variations on a theme, as with the faces of human beings, which all follow the same basic plan of having two eyes, one nose and a mouth, and yet manage to look unique nonetheless: 'In the universe we are but as a little family whose faces resemble each other; the next planet contains another family who have a different style of countenance.'[14] The reason why most of the aliens in sci-fi shows like *Star Trek* look like humans with pointy ears or weird foreheads is not only down to limited budgets, but also because that is what most people naturally seem to *expect* aliens to look like. As the Marchioness admits in the book, each planet seems the centre of the universe to its own inhabitants, who 'always attribute to others what belongs to ourselves',[15] and there can be few readers of the *Conversations* who imagine the aliens spoken of therein to resemble, say, giant silver-bearded frogs or flying purple bubbles with eyes made from cheese. Fontenelle looked into the mirror of the universe, and saw a variation of his own species looking back down at him – which is not altogether surprising, when you consider that this is what mirrors are for.

## The Mighty Micro

Other significant contributors towards this kind of literature included the German mathematician Christian Wolff (1674–1754), whose 1735 *Elementa Matheseos Universae* notoriously contained absurdly specific calculations about the likely eye size and extreme height of Jupiterians (1,440 feet tall, apparently[16]), and the great Prussian philosopher Immanuel Kant (1724–1804), whose 1755 *Universal Natural History and Theory of the Heavens* featured an early version of what was later to become known as the 'nebular hypothesis', which we have already mentioned briefly in relation to ideas of life on Mars. Basically, Kant's idea held that, at the beginning of time, matter floating free within the infinite universe began gradually to condense around other, especially dense, masses of matter, which became planets. These also began to be attracted to other, even larger masses, which we now call suns, thus accounting for the eventual formation of solar systems like our own. However, Kant guessed that any mass closer to the centre of a solar system would be denser and cruder than that at the periphery, where the masses involved would be lighter and more airy; seeing as this mass then goes on to make up the living inhabitants of each planet, it therefore followed that, the further away from its parent sun a planet was, the more clever and less dense-minded its aliens would be. In addition, within this model the outer planets were the older ones, thus allowing evolutionary processes to occur over a greater time span on worlds at the edge of a solar system. Therefore, Kant's Venusians, as occupants of the nearest planet to the sun, were pure morons, whereas more distant men on Jupiter and Saturn were likely to be geniuses by comparison. Earth being in the exact middle of the incompletely known solar system of 1755, Kant assessed us Earthlings as occupying 'exactly the middle rung' on the ladder of interplanetary intelligence. There could even be moral implications to all this, Kant hypothesised. Maybe Saturnians were less bound to matter, more ghost-like than us Earth men, thus freeing them from the irrational desire to sin, or even from the necessity of death itself? In later life Kant recanted most of these ideas as the mere rashness of youth, excising such wild speculations from subsequent editions of his book, but as we saw earlier on, his idea that the further out from the sun a planet was, the more evolved its inhabitants were, was to have some influence upon Communist star-gazers in later centuries.[17]

So well known did such literature become that, in 1752, the French *philosophe* and arch-rationalist Voltaire (1694–1778) published a satire of them in the form of a short story called *Micromégas*. This tells the tale of the titular Micromégas, an alien from the star-system of Sirius, who stands some 120,000 feet tall. Exiled from his home world for 800 years, he sets out on a tour of the universe, landing on Saturn where he encounters a puny dwarf, only 6,000 feet tall, who is really a caricature of Fontenelle himself – 'a spirited man who had not invented anything, to tell the truth, but who understood the inventions of others very well'.[18] Wolff too appears to be directly mocked in the book, when a series of computations manage to work out that Micromégas' nose is '6,333 feet plus a fraction', a measurement as absurdly precise as that attributed by Wolff to Jupiterians' eyes.[19] The fact that all the story's aliens are humanoid is referred to by Micromégas with a comment to the effect that 'the Author of Nature' has 'scattered across this universe a profusion of varieties [of intelligent alien] with a kind of admirable uniformity' in which 'all are different, and all resemble one another'.[20] But is it really possible for such rational beings, no matter how intelligent, to be able to predict what life on other worlds will be like? Will they even recognise it when they see it? Voltaire implies not. When Micromégas and his new friend the 'dwarf' visit Earth, both initially think it is uninhabited until Micromégas accidentally drops some diamonds which act as a magnifying glass beneath which a whale then happens to pass. The whale was a mere 'small atom' to the two alien giants, however, and they initially had 'no reason to believe that a soul was in it'.[21]

Again, this seems like a mockery of Fontenelle's ideas. For Fontenelle, one of the best guarantees that there was life to be seen on other planets if only we had powerful enough telescopes was that life had recently been found lurking within tiny droplets of water using the newly created microscopes of the Dutch scientist Antoni van Leeuwenhoek (1632–1723). As Fontenelle says:

> There have been found even in very hard stones an endless number of worms [bacteria, etc] lodged in every interstice, feeding on parts of the stone. Consider the countless numbers of these little beings, and how many years they could subsist on a quantity of food as big as a grain of sand; and then though the moon should be but a mass of rock, we may let it be eaten by its inhabitants rather than not assign any [inhabitants] to it. In short everything is animated; everything is full of life.[22]

But if so, as Voltaire asked, then how would you recognise it when you saw it? You would imagine it would be an easy thing to do, but not necessarily. If it was possible for Micromégas to mistake even a giant whale for a lifeless atom beneath a microscope, then surely there was no way that the idle conjectures of men like Fontenelle about life on other planets millions of miles away could be trusted? Microscopes and telescopes might alike have been getting better with each passing year, but the actual minds that looked through them were not necessarily keeping pace with technological advancement. That is what Voltaire seemed to be suggesting – and, had he lived some two centuries later, the great satirist might have found that things were not about to get any better ...

## Invasion of the Humanoids

In 1969 Britain's leading UFO magazine, *Flying Saucer Review*, put out a special paperback called *The Humanoids*. A collection of witness accounts of alleged saucer landings, together with descriptions of these craft's occupants, the book's editor, Charles Bowen (1918–87), admitted that the term 'humanoid' wasn't to be found in any dictionary (it is now), but said nonetheless that it summed up exceedingly well the kinds of entities then being described by witnesses across the globe.[23] From a study of UFO landing reports in Italy in 1954 alone, for example, we can read of 'three dwarfs dressed in metallic diving suits' who were caught red-handed in the act of bothering rabbits,[24] 'a strange being about 1.3 metres in height covered with a luminous suit' firing a paralysing ray from his torch at a witness,[25] and some 'figures dressed in light colours and wearing transparent helmets', one of whom 'had a dark face and a sort of trunk or hose' who communicated to one another with 'guttural sounds' after landing inside a sports stadium.[26] Many of the alleged humanoid encounters in the book are so ridiculous as to be amusing, such as an American case from 1957 in which 'a little man' approached a New Jersey resident named John Trasco and asked if he could borrow his dog. 'We are a peaceful people,' said the alien. 'We don't want no trouble. We just want your dog.'[27] Trasco said no, on account of the entity's strange appearance:

[He] was dressed in a green suit with shiny buttons, with a green tam-o-shanter-like cap, and gloves with a tiny object at the tip of each. His face was putty-coloured, [he] had a nose and chin and large, protuberant frog-like eyes.[28]

In other words, he was a humanoid frog dressed as a leprechaun! I also like the 1947 account of a Brazilian surveyor named José C. Higgins, who professed one day to have seen:

… three tall entities in transparent suits covering head and body, and inflated like rubber bags, and with metal boxes on their backs. Their clothing, visible through the suits, resembled brightly coloured paper. The entities, all identical, had huge round eyes, huge round bald heads, no eyebrows, no beards, and legs longer in proportion to our own. Higgins would not tell whether they were male or female, but found them strangely beautiful.[29]

He must have been the only one. Some of the aliens described in Bowen's book were so human-like as to be wearing Earth clothes, meanwhile, such as the tiny 65-cm-tall Chinese-looking men clad in 'very smart' blue suits, 'like the musicians at a fair', who supposedly jumped out of a space balloon hovering over the Spanish countryside one afternoon in 1953 purely in order to slap an illiterate teenage cowherd named Máximo Hernáiz in the face.[30] Some humanoids tried to vary things up by having different numbers of eyes than us Earth men do, such as a three-eyed albino melon head witnessed in Argentina in 1962,[31] or a bald and sphere-headed Cyclops-like creature with 'vivid' red skin but no eyebrows, ears or nose seen in Brazil in 1963,[32] but genuinely 'alien'-sounding aliens are much rarer in Bowen's book. There were a few robots reported,[33] but stories like that from Peru in 1965 of a 'strange being' which 'resembled a shrub', about 80 cm high, with one gold-coloured eye nestling in its head and several 'other smaller eyes' running 'up and down' its blackish-coloured body,[34] or a tale from England in 1963 of 'a dark figure ... completely black, human-sized, but with no head' and a pair of bat-like wings and webbed feet[35] were highly atypical to say the least. As you may have expected, most of the humanoids reported in *The Humanoids* looked pretty much like ... well, like humanoids.

In a field as vast as ufology, of course, you can always find exceptions to this general rule – I think of a 1954 case from France in which four young siblings, alerted by their dog's barking, witnessed some bizarre being 'like a sugar-lump' beckoning towards them with apparent friendly intent. Knowing it wise never to talk to strangers, however, one of the children shot a toy arrow at the entity, which they labelled a 'ghost', leading it to fling him to the ground with some kind of invisible force. It then waddled away towards a nearby meadow, never to be seen again – or so the children said.[36] Possibly the weirdest encounter with aliens of non-humanoid form is supposed to have occurred on the evening of 22 July 1975, when a teenage boy named only as 'Trevor' was climbing Wylfa Hill near Machynlleth in Wales, whilst his parents were examining a nearby cottage. At the top of the hill, he saw something incredibly odd and, horrified, hid behind a boulder. There sat a 'domed disc' which was flashing lights in colours he had never seen before. Inside the craft sat two 'pulsating lumps of jelly' formed up from numerous 'small, white corpuscular discs'. A flap opened in the side of the craft and one of the aliens began moving towards the exit. Trevor fled, and shouted for his dad to come and see the monsters. Trevor's dad ignored him, so the lad rushed back up for a second look only to see the disc and its occupants disappear, changing colours to blend into the grass. Trevor quickly developed negative changes in his personality and, the very next day, lost his voice. A few days later, he became hysterically blind in one eye. When this eye recovered its sight, the other one went blind instead. He stayed partially so for several years afterwards, almost as if what he had seen was so strange that he was trying to block out the possibility of ever setting eyes on such a thing again.[37] Was this the kind of extreme psychological reaction you may expect from someone who had seen something genuinely alien, or was the sighting itself simply the initial expression of an already disturbed personality in the witness? Whichever way, such overtly aberrant cases are rare.

## Grey Eminences

In 1977, the researcher Alvin Lawson (1929–2010) conducted a series of tests at California's Anaheim Memorial Hospital in which various subjects were hypnotised and asked to describe an imaginary UFO 'abduction' which was happening to them right there and then during their trance. Strikingly, Lawson found that all of the imaginary aliens described by his subjects tallied with descriptions of aliens supposedly seen by people in real life, and that all were clearly humanoid in form. In the end, Lawson created a classification system for these aliens, splitting them up into six different types. First were the 'humans', who resembled our own species almost precisely, but dressed appropriately for travel across the galaxies, or with minor differences like pointy ears. Then came the 'humanoids', who were basically human but with certain more obvious deformities like giant domed heads, waxy hairless bodies, or abnormally large eyes, sometimes with vertical pupils like cats. Then were 'animal entities', which resembled Bigfoot or the Creature from the Black Lagoon – large, human-sized apes or reptiles who went about on two legs and used their arms and hands much as people do. Then were 'robot entities', metallic man-shaped machines like the Cybermen. 'Exotic entities' combined features from two or more of the above categories, for example humanoids with robotic hands and arms, or else had overtly grotesque facial features like trunks. Finally came the 'apparitional entities' – ghost-like humanoids able to float, shapeshift, hover, glow or pass through walls like spirits.[38]

At least during Lawson's day the humanoids had a bit of variety to them, though. Today, it seems as if almost all alien encounters involve a now familiar species of extraterrestrial called the 'Greys'. With their big bald heads, smooth and waxy whitish-grey skin and huge, wraparound black eyes, everyone knows about these sinister, dwarfish entities from the star-system of Zeta Reticuli and their alleged love for abducting hapless citizens and taking them back to their spaceships to perform nefarious medical experiments upon them. Fewer people, though, seem to realise that these now ubiquitous aliens were once rare in ufology; flick through *The Humanoids*, and you'll find nary a single one. Prior to the 1980s, such figures made up hardly one in ten alien encounters in the Western world; during the 1950s, it is estimated that fewer than 3 per cent of ETs witnessed were Greys; now it is at least two thirds. The key event in spreading the meme was the 1987 publication of the American sci-fi writer Whitley Strieber's (b. 1945) book *Communion*, which purported to be an accurate record of the author's own abduction experiences at the hands of the Greys. The book was marketed heavily, sold well, and featured an evocative painting of such a creature on its front cover. *Communion* was then turned into a film which also did good business, and there you have it – the Grey was embedded in the minds of the public worldwide, so move over Orthon.[39] But why did Strieber's Greys strike such a chord?

## Greys' Anatomy

The reason Strieber gave his book the title he did was that his wife began talking in her sleep one night in an alien-like voice, saying it should be called *Communion*, 'because that's what it's about'. If so, then it was a pretty strange kind of communion; according to Strieber, when he underwent his first alien encounter in December 1985 the Greys took a foot-long triangular instrument and forcibly inserted it into his anus, apparently in search of stool samples. Strieber himself interpreted the experience as one of anal rape. His tentative conclusion was that such entities – whether literally real, or simply parts of our own psyche – had always been visiting us, formerly in the guise of gods, angels or spirits. Now, however, for the first time in history, we were simply ignoring them and refusing to acknowledge their presence, obsessed as our civilisation now is with the current quasi-religion of materialist science. If so, then it seemed as if modern-day mankind had become alienated from a central part of himself, with the fairies, angels and Nordics of old now coming back disguised as Greys and taking their revenge upon our anuses, such extreme measures being required in order to force mankind into once more acknowledging them.[40] Many other writers have also viewed the Greys as representing sadistic parodies of the contemporary scientific mindset and the cold, clinical and detached way our species treats the natural world around us. John Rimmer, former editor of the British UFO magazine *Magonia*, once put it like this:

> The present form of the UFO abduction may be the result of people's fears of science and technology. The way abductees are treated by their captors sounds rather like the way scientists treat laboratory rats; they are pushed, poked, prodded, examined clinically, treated much like laboratory specimens, then dumped when the experiment is over. Many people would see this as a way in which modern society is dehumanising the individual.[41]

The standard bodily characteristics of the Greys can be interpreted as being in themselves symbolic of their inherently anti-humanistic status. For example, in a 1997 book the writer Peter Brookesmith analysed the Greys' anatomy as follows. The huge heads of the aliens stand in for their 'superhuman brains and intellects'. Their 'attenuated bodies' suggest that the Greys have 'no physical warmth or earthly sympathy', with implications of 'frigid intellect and passionless asceticism'. As their white-hued complexions hint, they would seem to be literally bloodless – reports of their veins being filled with chlorophyll give them 'as much visceral feeling as an aspidistra'. Their nakedness too is revealing, as they tend to 'go straight round like my teddy bear'; Greys are seldom described as possessing either bums or genitals, indicating their 'passionless nature', free from the 'earthy distractions' of digestion and defecation. Most alien of all are the Greys' eyes, big black and staring. They seem to have no pupils, and are as far away from being 'windows into the soul' as possible. It is not hard to imagine such creatures as being cruel and devoid of sympathy.[42]

And yet, the Greys are still humanoids. During his studies, Alvin Lawson noticed that the big-headed, big-eyed Greys strongly resembled the undeveloped human foetus when still in the womb. He wondered if, in undergoing their abduction experiences, abductees were in fact reliving their own birth. Victims would often pass through tubes on the alien spacecraft, corresponding to them being snatched from the safety and comfort of the womb and passing through the birth tunnel. When the abductees then find themselves suddenly within a brightly lit room being subjected to medical examination by strange beings, this is like the moment of sudden exposure of the new-born baby to the delivery room, surrounded by midwives and doctors, just waiting to weigh and examine you, or snip your umbilical cord off. The circular placenta was identified by Lawson as being the flying saucer itself. Interestingly, an examination of eight abductees who had been delivered by caesarean section showed that seven made no mention of being taken down tunnels or tubes aboard the alien spaceship, and the one who did had been delivered by an emergency caesarean only after becoming stuck inside her mother's birth canal.[43] It seems obvious that the Greys are in some sense disguised versions of humanity itself. The closer to the basic human form the alien is, the more obviously alien its differences will seem; black, wrap-around eyes on a giant extraterrestrial insect would not seem as uncanny and disturbing as on a humanoid. In meeting what seems like a cruel parody of our modern selves, we are forced to confront what is most alien within us.

In reality, the shape which aliens might take could well be far more strange and unexpected than the humanoid Greys ever were. Possibly, extraterrestrial life may come in such an unexpected form that we are unable even to recognise it as *being* life in the first place. Sci-fi writers have often exploited such a trope. The novelist Olaf Stapledon (1886–1950) used his 1930 novel *Last and First Men* to depict a massive invasion of Earth from Mars which wasn't even noticed by humanity due to the fact that the Martians took on the form of living clouds with electromagnetic group minds. The Martians, however, correspondingly failed to recognise that there was any intelligent life on Earth either, being disappointed to find that Earth clouds were mute and inert. Stapledon's Martian clouds communicate with one another via a means which is half-radio broadcast, half-telepathy. As such, whilst they dismiss mankind as a species of unintelligent animal due to his inability to broadcast his thoughts to them, when they encounter our radio stations, they *do* think that these structures are intelligent beings – albeit of a primitive, non-gaseous kind. The humans who tend these structures, the Martians presume to be unintelligent beast slaves under the radio masts' control. Stapledon's conceit may be a satire upon the 1920s craze for trying to send signals up to Mars. The very idea that aliens would have radio sets compatible with our own – or indeed minds compatible with our own to facilitate the picking up of psychic brain waves – is the height of folly.[44]

## The Mirror of Man

Genuine alien life forms could easily be as unrecognisable as Stapledon's Martian clouds; perhaps more so. To the American cosmologist and UFO sceptic Carl Sagan (1934–97), the way that most of the aliens people claim to encounter today strongly resemble humans in their basic morphology is simply down to 'a failure of the imagination' and 'a preoccupation with human concerns' upon behalf of abductees. As he put it: 'Not a single being presented in these accounts is as astonishing as a cockatoo would be if you had never before beheld a bird.'[45] So, is there any good reason to expect that authentic intelligent aliens would, in fact, resemble human beings? One line of argument which could be cited to back the idea of humanoid aliens up is developed in a book published in 1698 by the Dutch scientist Christiaan Huygens (1629–95) called *Cosmotheoros*, known in its English translation of that same year as *The Celestial Worlds Discovered*.

Huygens was no hopeless amateur; he is credited with discovering the true nature of the rings of Saturn. Nonetheless, his *Cosmotheoros* contains some very dubious pieces of reasoning, all of which, taken together, add up to the conclusion that there must be creatures very strongly resembling human beings present within our solar system. Whilst doffing his cap to Fontenelle, 'the ingenious French author',[46] Huygens gives no narrative frame to his speculations, presenting them as simply the logical conclusions of a thoughtful mind. His argument is that astronomical observations have proved the planets to be very similar to Earth in the sense that they have moons, are spherical, orbit the sun, etc, so why should they not also be similar to our own world in that they contain life too?[47] If they do contain animals and plants, continues Huygens, then by rights they should be similar to ours down on Earth, seeing as they have 'the same sun to warm and enliven them as we have'. He argues that our own plants and animals in the Old World of Europe are basically the same as those in the New World of the Americas. Yes, they may look a bit different on the surface, but both sets of animals have lungs, feet, wings, hearts, guts and genitals. If it had pleased God to make the two sets of animals completely different from one another then He surely could have done, argues Huygens; but the fact that He did not do so down here on Earth means that there is no reason to expect that He should have done so on other planets either.[48] Furthermore, says Huygens, alien animals must, like animals here on Earth, inhabit environments which are solid, liquid or gas in nature. How would they be able to move in these environments, if they did not develop feet, fins or wings?[49]

So far, if you only substitute the word 'evolution' for 'God', this is standard exobiology. It is a common position today to say that, no matter where you are within the universe, Nature will always develop similar solutions. Maybe the humanoid shape is just the best possible solution Nature could ever devise for intelligent life to inhabit, a basic template towards which all animals aspire. Huygens seems to have thought something similar. Belief in God led him to conclude that the whole point of God creating planets filled with flora and fauna in the first place was to then stock them with 'some creatures or other endued with reason' like humans, who could 'enjoy the fruits' of these worlds and thence 'adore the wise Creator of them'.[50] Seeing as reason is inherently reasonable, however, Huygens then concludes that any intelligent alien beings must by definition enjoy the same kind of logic as humans do; how, for example, could their mathematics or geometry be any different? Two plus two must equal four throughout the whole universe, not only here on Earth.[51] Without our senses, though, how would we be able to reason such things out? Huygens concludes we would not. Therefore, he says, rational aliens must have eyes through which to see. As he can imagine no better way that eyes could work than the way our own do, Huygens attributes such ETs human eyes – two of them, otherwise they would be unable to judge depth properly. Because speech is necessary to spreading knowledge, meanwhile, Huygens says aliens must have a pair of ears too, because 'it will scarce seem credible that two such useful, such excellent things were designed only for us'.[52]

After carrying on in this vein for the other human senses, Huygens then bizarrely goes on to claim that all intelligent ETs must also inevitably possess astronomy, a conclusion which appears to be based upon the simple fact that he himself was obsessed with the subject, and so couldn't imagine a world where such an art was not pursued. The actual argument he presents is that seeing as men on other worlds must have had experience of eclipses, as we have, primitive aliens must have been scared of them and so studied them in order to understand their nature better, thus giving rise to the science of astronomy.[53] Thinking this argument foolproof, Huygens then uses the 'fact' that aliens have invented astronomy to prove that they must also have invented such crafts as metallurgy, optics, arithmetic and writing, since without any of these there could be no such thing as telescopes or astronomical record-keeping.[54] Furthermore, seeing as instrument-making is very fiddly, Huygens lends his aliens human hands, with fingers and an opposable thumb, for otherwise how would they be able to grip and manipulate tools? Had these aliens hoofs, then this would be like a cruel joke of Nature, says Huygens, for clever hoof-handed men might well have been able to devise telescopes within their heads, but certainly not to build them. He admits that elephants can pick things up with their trunks, but who has ever seen an elephant build a telescope? Such an appendage is 'nothing but a nose', says Huygens, and if aliens did not have human-like hands then we might be led to say that 'Nature has been kinder ... even to squirrels and monkeys than to them'.[55] Furthermore, he says that all intelligent aliens must also be upright bipeds, in order to aid 'the more convenient and easy contemplation and observation of the stars', another illustration of Huygens' apparent self-obsession.[56] He does admit that these aliens may be woolly or covered in feathers, or even crustaceans of some kind, but seems unable to admit of the possibility that they would not be essentially humanoid in form.[57] As he confesses:

> I cannot without horror and impatience suffer any other figure for the habitation of a reasonable soul [than the humanoid]. For when I do but represent to my imagination or eyes a creature like a man in everything else, but that has a neck four times as long, and great round saucer-eyes five or six times as big ... I cannot look upon't without the utmost aversion, although at the same time I can give no account of my dislike.[58]

As we have just seen, Huygens was hardly the only one.

# 17

# Wandering Stars: Astral Travels

None of Fontenelle, Huygens or their ilk ever claimed to have actually travelled up to the starry heavens themselves, but they nonetheless provided distinct imaginative models for their more literally minded imitators to copy, particularly when it came to the idea of aliens being humanoid. Take the obscure 1880 Spiritualist tome *Mars Revealed, Or Seven Days in the Spirit World*, whose American author, Henry Gaston, claimed to have flown up to Mars in astral-form, led by his supposed 'spirit-guide' John of Patmos, who showed him the wonders of the distant planet. Clearly Gaston knew little of actual astronomy, describing a golden, jewel-filled landscape, 'a world in ruby, emerald and silver', dotted with seas, rivers and lakes, and complete with lush, luxuriant vegetation. This all sounded very exotic. Sadly, however, the actual Martians were so human-like as to sound positively banal; the moralising Gaston actually takes the time to write about their commendably regular habits of tooth-brushing and hair-combing, the Martians taking good care of their personal hygiene, 'as do all decent people on Earth'. These well-groomed aliens are clearly nothing but mirror-images of Gaston himself, even down to their chosen haircuts. 'Like all men of sense', writes Gaston, Martian males have side-partings. So, we may reasonably presume, did Henry Gaston …[1]

In the narratives of Spiritualists like Gaston, the fictional magical travels of writers like Kepler and Kircher became mixed up with the humanoid alien traditions promulgated by Fontenelle, Huygens and their school in order to create a new sub-genre of narrative mixing both science and the supernatural, of a kind which reads very peculiarly to modern ears. For contemporary readers familiar with the work of the early exobiologists, though, they would have seemed part of an already pre-existing tradition. For example, Gaston's book contained a prefatory letter explicitly linking his narrative to that of arguably the most influential astral-traveller of all time, Emanuel Swedenborg (1688–1772), a Swedish mystic, philosopher, engineer and scientist whose followers founded a religion devoted to his teachings a few years after his death in London in 1772, called the Church of the New Jerusalem, of which William Blake was the best-known member – and Hugh Mansfield Robinson a much later disciple.[2] During his early life, Swedenborg enjoyed a successful career as Chief Inspector of Sweden's Royal Board of Mines, but around 1744, aged fifty-six or so, he underwent some severe crisis of the mind, ending up having any number of bizarre visions in which he spoke with the likes of Plato, St Augustine, Moses, Abraham … and ghosts from other planets.

Once, Swedenborg had been a stable and practical individual, but this all changed one day whilst sitting in a London tavern following a good meal, when a ghostly mist began to form before his eyes. It became thicker, and he noticed that the floor was suddenly covered with snakes and frogs. Then, a man appeared in the corner of the room, looked over at Swedenborg and said 'Don't eat so much.' The next night, this same man appeared again and announced that He was The Lord God Himself, that He had chosen Swedenborg to communicate the true meaning of the Bible to the world, and that he would soon begin to write holy books under His Divine guidance.[3] These books, just like Swedenborg's visions, contained much that was incredibly bizarre. There is his account of being awoken one night by the ghosts of malicious Quakers deciding to possess his hair and transform it into living snakes,[4] his claim that a bout of toothache had been caused by St Paul,[5] and his so-called 'science of correspondences' which held that everything in the material world had its secret spiritual counterpart in the hidden world.[6] Swedenborg's father claimed to have been given a reading list of recommended authors by an angel during his youth, and swore blind that he had once been physically assaulted by Satan, implying that there was already a history of mental illness in the family[7] – not that this prevented his deluded followers from treating the man's long, rambling works as if they really were the products of Divine intervention, as opposed to the fruit of insanity.

## Fantastic Voyages

Swedenborg's books tend to be fairly repetitious, with reams of pious dullness being enlivened by sudden small gems of glorious nonsense. Such is his 1758 *Concerning the Earths in Our Solar System Which Are Called Planets*, a collection of Swedenborg's personal observations upon the nature of living humanoids – and ghosts – on other worlds. All of these beings were in some sense human and worshipped the same God, taught Swedenborg, and together made up something he called the Grand Man, described as a kind of gigantic disembodied human being, inherent within the structure of the universe. The inhabitants of each planet corresponded magically with either a bodily organ or a quality of this Grand Man – his eyes, for example, or his wisdom – thus linking once more microcosm with macrocosm.[8]

Amongst each planet's spirits, said Swedenborg, were a band of knowledge-seekers who wandered from world to world in search of information about their inhabitants. Being a visionary, Swedenborg claimed to have met several different varieties, each of whom told him of life (and after-life) on their home planets. Firstly, he was visited by ghosts from Mercury, who told him that, in the Grand Man, they stood for the faculty of memory. As soon as they met him, they fed off his own memories to get a good picture of life on Earth. Haughty and high-minded, the Mercurial ghosts did not speak using physical voices, but via psychic means. Whenever they did so, their words sounded 'undulatory', and entered into Swedenborg's mind through his left eye. He knew when they were about to appear because of a 'whitish coloured flame' announcing their arrival; then they would suddenly materialise in the guise of floating 'crystalline globes' which could combine together into one much larger globe when travelling between planets. They didn't think much of Earth men, once sending Swedenborg a present of an abnormally long piece of paper, intended to satirise the fact that most of our knowledge here on Earth is contained within books, not within people's heads. They even visited the ghost of Christian Wolff, and let him know how disappointed they were with the standard of his learning. When Swedenborg tried to broadcast psychic images of fields and woods to them to demonstrate Earth's beauty, they responded by filling them with mental images of snakes and making the woodland rivers run black to show how little they cared for such things. The ghosts of Mercury thought they knew everything, so the angels of the Earth decided to teach them a lesson by giving a long speech informing them of all the things they did not know, and would never even know that they didn't know. Watching the Mercurians receive this knowledge, Swedenborg said he saw them literally begin to deflate in size like dying balloons, as their pride was squeezed out of them. Eventually, Swedenborg received a vision of some living Mercurians, who looked exactly like ordinary humans, only with smaller faces. Their cattle, too, were exactly the same as Earth cows, but tiny. Asked how they avoided burning up, given Mercury's closeness to the sun, the ghosts explained to Swedenborg that a planet's temperature really had less to do with its distance from its nearest star and more with 'the altitude and density of the atmosphere'.[9]

Next to visit were spirits from Jupiter, who told Swedenborg they were living through a golden era with no war, crime, poverty or hunger. The Jovian ghosts glowed 'like lightning' when annoyed, or glittered with different patterns of miniature stars to signify agreement or disagreement. When they visited Swedenborg, they invaded his face and forced him to smile, filling him with 'tranquillity and delight'. They swam through the air and had giant, round heads, on account of the fact that the living people on Jupiter were obsessed with their own faces and devoted their whole lives towards worshipping them. They constantly washed and cleaned their visages, and wore special hats made from blue tree bark in order to avoid any facial sun damage, although the rest of their bodies were almost naked. Nobody had spots or facial blemishes of any kind on Jupiter, so the ghosts were shocked to see the dreadful state of some people's faces down here on Earth, thinking them to be 'deformed'. So perfect were their alien physiognomies that on Jupiter, instead of having to speak, it was obvious what a person wanted to say simply from looking at their face, thus making it impossible for them to dissemble. The evolution of language amongst us Earthlings, concluded the Jovian ghosts, was simply the evolution of lying. So in love with their own faces were they that all people on Jupiter slept on their backs in bed, so they could show their faces off to God above.[10] They also adopted the following bizarre mode of locomotion in order to display their faces better:

> With respect to their manner of walking, they do not walk erect like the inhabitants of this and of several other Earths, nor do they creep on all fours like four-footed beasts, but as they go along they assist themselves with their hands, and alternately half-elevate themselves on their feet, and also at every third step turn their face sideways and behind them, and likewise at the same time bend the body a little, which is done suddenly; for it is thought indecent amongst them to be seen in any other point of view than with the face in front. In walking thus, they always keep the face elevated as with us, so that they may look at the heavens as well as the earth; holding the face downwards so as to see the earth alone, they call an accursed thing; the most vile and abject among them give in to this habit, but if they continue in it they are banished [from] society.[11]

In the Grand Man, Jovians represented 'The Imaginative Principle of Thought', and Swedenborg's depiction of life on Jupiter was nothing if not imaginative. A series of bizarre and deadly spirits also shared the planet with the Jovians, he said, such as a chastising form of demon resembling 'a darkish cloud with movable stars in it' which causes joint pain and belly cramps, or inflicts suffocation upon sinners. Some evil Jupiterian spirits, meanwhile, shaped like 'a flying fire', would crouch beneath a man's bum and start telling lies, encouraging him to sin. Even ghosts on Jupiter see ghosts, it transpires; spirits shaped like old men will materialise there, warning other ghosts not to lie, whilst the sudden appearance of a ghostly face in a window indicates that it is time for any other ghosts present to leave the place in question. Apparitions of the top half of a bald head, meanwhile, are an infallible omen that you will die within a year. Jovians don't necessarily mind being told this, however, and are happy to die at the average age of thirty, presumably because then they won't get old enough for their faces to go all wrinkly; the only trouble with this short life span is that they have to get married whilst still children. A final class of ghost on Jupiter are known as the 'sweepers of chimneys', because they have sooty faces. One of these appeared to Swedenborg himself, and spoke into his elbow. Such spirits were rather like butterflies, said the sweep ghost, as they wished to cast off their sooty garments and coverings, enter into Heaven and thus become bright, shining angels. Because of this desire, the sweep ghosts were the 'Seminal Vessels', or testicles, of the Grand Man, Swedenborg realised; they have an immense desire to escape their sooty sheaths and enter Heaven, much as a man's sperm wishes to escape from the restrictive sheath of his balls and spurt out into the Heaven of his lover's chosen orifice.[12]

These two planets were the strangest places seen by Swedenborg during his visions, but his book does contain other useful nuggets of information, such as that yellow-skinned Martians have 'beards' which are not made of hair but instead consist of discoloured black skin which makes them impossible to shave off,[13] that Saturnian ghosts are so holy that, if ever tempted to sin, a *Macbeth*-like ghost dagger appears in their hands, urging them to stab themselves,[14] and that moonmen have very loud voices like 'the sound of thunder', which issue from within their stomachs in order to frighten away any lurking evil spirits. In terms of the Grand Man, these tummy talkers represented the 'ensiform cartilage', apparently.[15] Concerning those planets outside our solar system, Swedenborg was not granted visions of them by itinerant spirits, but instead flew up to see them for himself, in astral form. On these unnamed worlds, he encountered space sheep,[16] an alien Hell filled with female magicians clad in green,[17] windows made from grass,[18] a type of luminous wood which didn't have to be lit to illuminate fireplaces,[19] woolly cows,[20] God in the form of a humanoid cloud[21] and a race of alien women spinning thread with their toes.[22]

## Spirit in the Sky

The surprising thing about these comical fantasies was that they proved so influential. People actually believed what Swedenborg had to say, woolly cows, luminous wood and all. The Church of the New Jerusalem has never had that many members, but a number of its teachings were adopted and adapted by the late nineteenth-century Spiritualist movement, which captured literally millions of followers throughout Europe and North America around a century after Swedenborg's death. And so it was that, thanks to Swedenborg, the idea of ghosts mingling with aliens in outer space gained a much wider currency than it ever really deserved.

The best example can be found in the life and work of Andrew Jackson Davis (1826–1910), the 'Poughkeepsie Seer', a psychic and author from New York State who became one of the major figures in the American Spiritualist movement. In 1843, aged seventeen, Davis attended some lectures on hypnotism and quickly became interested in the subject. Allowing himself to be put into trances by a local tailor, Davis is said to have showed some aptitude for performing accurate medical diagnoses upon sick persons who were brought to meet him whilst he was mesmerised. In 1844, he underwent a visionary experience, flying in astral form over the Catskill Mountains and across the Hudson River, where he met the ghosts of both Emanuel Swedenborg and the ancient Roman physician Galen (*c.* 129–*c.* 216). Galen handed Davis a caduceus, a kind of magical staff of healing, whilst Swedenborg promised him spiritual guidance. Most likely he had heard mesmerists mention Swedenborg during their lectures and shows, and picked up some knowledge of his creed from there, but it was at this point that Davis began to develop his own personal Spiritualist doctrines, which showed a clear Swedenborgian influence. Another big influence upon Davis, meanwhile, was Charles Fourier! This may seem surprising, but between 1845 and 1847, Davis had been going into regular trance states and then giving public lectures in New York upon a variety of philosophical, social and scientific topics he professed to have known nothing about beforehand. Amongst those who regularly attended the 157 lectures he gave was a man named Arthur Brisbane (1809–90), who had founded his own Fourierist Society in New York in 1839. It was doubtless from conversations with Brisbane that Davis picked up his knowledge of Fourier, some of whose less lunatic teachings soon began popping up in Davis' trance lectures, along with those of Swedenborg.

In 1847, Davis' trance revelations were published in book form as *The Principles of Nature*, an incredibly tedious 800-page compendium of wholly invented 'scientific' knowledge about the character of the universe, together with spiritual, political and philosophical theorising of a sort which blended the doctrines of his two main influences, Swedenborg and Fourier, together into one seamless whole. Immensely popular, the book went through thirty-four editions in under thirty years, and became the foundation-stone of American Spiritualism, its 'Bible', almost. Within could be found not only Davis' own Swedenborg-like descriptions of life on other planets, and pleas for more Fourierist communes to be created across the nation, but also echoes of Fourier's belief that it was possible for a person to be reincarnated on another planet after death – or even, in the most extreme cases, be reincarnated *as* another planet after death! Basically, Davis wished for a spiritual regeneration of mankind to take place hand-in-hand with a Fourier-style social regeneration of the world, leading to a real nexus growing up in America between the two seemingly unrelated topics of socialism and Spiritualism.[23]

Predictably, Davis' psychically derived descriptions of life on other worlds tally fairly well with his hero Swedenborg's own astral adventures, though he seemed unable to resist the temptation to add to these fantasies somewhat. *The Principles of Nature* itself begins with an account of how, at the beginning of time, the entire universe was made up from a kind of liquid fire,[24] before going on to give an account of a 'sun of suns', the Great Eternal Sun, which is to some extent really God Himself, and which is forever 'breathing forth a system of concentric circles of suns and systems of suns'. These other suns then emit light particles into the universe, which slowly begin to accumulate and solidify into planets. We are asked to imagine a number of solar systems emanating out from the central sun, each of which itself forms a kind of ring, with a sun at the centre, orbited by several planets, many of which are then orbited by moons.[25] Each planet thus constitutes a solar system in miniature, and each solar system also constitutes a smaller representation of the way that the concentric rings of solar systems orbit the Great Eternal Sun, meaning that the universe continually repeats itself, in yet another kind of microcosm-macrocosm arrangement. Even atoms replicate this sort of solar system-type arrangement in their structure, says Davis, meaning that all is really one, and that, seeing as everything participates within the nature of the Great Eternal Sun, God thus inhabits every element of the universe, from the biggest planet to the tiniest atom.[26]

The basics of Creation having thus been established, Davis takes us on a Swedenborg-inspired tour of our planetary neighbours. Uranus, he says, is uninhabited, and of little interest; being far away from the baking sun, it has not fully solidified yet, with its density being only 'a little more than that of water'.[27] Seeing as the planet is basically a big ball of cold soup, Davis moves quickly on to describing the face-obsessed inhabitants of Jupiter, the false-bearded, yellow-skinned Martians and the memory-obsessed Mercurians,[28] all of whom are dead-ringers for the same aliens seen by Swedenborg. Davis does add a few new details, such as the observation that Venusians have large breasts,[29] but there is not a great deal here to surprise seasoned readers of the great Swedish mystic. What is more surprising is the way that, towards the end of his book, Davis begins hymning the praises of Fourier, whom he compares favourably to Jesus, and whose theories about 'the unspeakable harmony that pervades the universe' he says should be used to enable that same kind of harmony to 'prevail among, and join inseparably, the inhabitants of the Earth'.[30] Davis, we must recall, will not have been able to read all of the really weird sexual stuff Fourier wrote, which was suppressed until the 1960s, and so reserves much of his praise for Fourier's teachings about people's need to work in jobs which interest or suit them, and his ideas upon agricultural reform.[31] Seeing as Davis claimed to have seen various Swedenborgian alien utopias first-hand whilst travelling through space, he knew from personal experience that a better world was eminently possible.[32] Once Fourier's ideas had been put into action, said Davis:

> Then there will be an order in human society, in which every group may represent a planet. And the groups may be so arranged as that their interests will revolve around the central object of their own

industry [as around the Great Eternal Sun] ... Let the Sun of the race be the centre of all human wisdom, whose enlivening influence may generate industry, abundance and happiness. Let each group, society or state be a planet; and let the whole give to and receive from the Central Sun congenial reciprocations, so that there may not exist any inertia, restriction, poverty or unhappiness ... This, then, should be the order of society. Then mankind would represent the harmony of the solar system, in which no disturbance is discoverable, because the Great Central Sun is both the parent and governor, whose pervading influence sustains an indestructible equilibrium ... [The Fourierist society created] will be as a human body in its arrangements and interior movements, but will correspond externally to the structure of the universe.[33]

Once again microcosm shall be made to merge with macrocosm, leading to harmony for all! For Davis, following Fourier's instructions is the key to restoring concord to our currently discordant land: 'It is impossible to escape the conclusion that [Fourier] revealed many truthful causes and principles of reform that must be in some degree practised before the Kingdom of Heaven can be established on Earth'.[34]

## Paradise Regained

One of the people who attended Davis' trance lectures was a former Universalist Minister, utopian theoriser and Swedenborgian named Thomas Lake Harris (1823–1906). At first, Harris was so impressed by Davis' 1847 *The Principles of Nature* that he spent four or five months travelling around giving lectures on it, but whilst speaking in Cincinnati in 1848, he fell in with some 'Spiritual Brotherhood' present in the city, and drifted away from his earlier ideas. Five years later, he was falling into Davis-like trances himself. During one reverie, which lasted on and off for some fourteen days in November 1853, he received visions of Heaven and the other planets, reciting his experiences in verse form through his physical body down here below on Earth.[35] Sometimes, these trance-delivered verses would actually be sung by Harris, and his disciples claimed reason to believe that parts of it were inspired by none other than Dante Alighieri (1265–1321).[36] Apparently, three years previously, in 1850, Harris had been visited by 'certain spirits, some of whom were members of ... a poet's heaven', who had foretold that he would write a great epic one day.[37] The long poem which eventually resulted, however, was certainly no *Divine Comedy*. Harris' *An Epic of the Starry Heavens* tells of how a gang of Swedenborg-like ghosts from other planets approach him during one of his stupors and take him on a guided tour of the cosmos, pointing out a series of surprising facts such as that space is actually purple,[38] Jesus lives in the sun,[39] and that the Garden of Eden is located upon a small island on the planet Mars:

> *By what other name*
> *Than Eden can I call this floral plain,*
> *Whose azure waters flow like odorous balms,*
> *Tongued with sweet eloquence? These trees are palms.*
> *It is an island in an azure sea*
> *That I am borne to. Overhead I see*
> *A firmament of alabaster hue,*
> *Flecked with red rose-leaves, ever falling through,*
> *And melting on the air in crimson dew.*[40]

Meanwhile, Harris' even more twee version of Jupiter featured such delights as flying horses with fiery wings, star-sheep fed on silver lilies, and trees that could talk:

> *Wingéd Pegasus was not all a fable,*
> *Here are flame-winged horses who are able*
> *With rapid feet, traversing upper space,*
> *To fly through heaven like thoughts …*
> *I see a company of angel-men,*
> *And angel-women 'sociate with them;*
> *White sheep in fields of ether star the meads*
> *Of chrysolite. Each flock a woman feeds*
> *With silver lilies which all radiant grow*
> *In spiral pathways, where her bright feet glow …*
> *There's a tree I behold which in earth begins,*
> *It has pulses of crimson in ivory limbs,*
> *And its lance-like leaves are transformed, and bear*
> *Flowers that are lights, and that shine through the air*
> *Like faces of angels through fields of fragrance;*
> *And the leaves are tongues, and with musical cadence*
> *They utter the secrets of life, and tell*
> *Of inner virtues in flowers that dwell.*[41]

There are also electric birds, a sky full of blooms, and 'electric flowers/ That ope their golden cups for heavenly showers' found on Jupiter,[42] and a glorious Imperial City which features canals filled with white waters, streets paved with gold, and plants with stems made from jasper. Here also live a race of 'Immortal Men, each one the paragon/Of all perfection',[43] which is significant, as Harris' basic message seems to be that, one day, human beings shall evolve into such supermen too, once our primitive planet has reached a state of greater enlightenment. The best example of what such a perfect globe will look like is given to Harris on Mercury, when 'a spiritual voice' speaks of a world of wonders which is soon to come on Earth. Listening to what the voice has to say, you may almost be forgiven for thinking it belongs to the ghost of Charles Fourier:

> *Thy world again shall wonders see –*
> *The New Creation yet to be.*
> *Electric steeds shall paw the air,*
> *Electric chariots angels bear;*
> *Electric ships outsail from heaven,*
> *By an interior will-power driven …*
> *Electric lions, beautiful,*
> *Shall seek on lower Earth their kind,*
> *And magnetise with power of mind*
> *Their Earthly mates, till they fulfil*
> *The ancient prophecy, grow mild,*
> *And dally with the unweaned child.*[44]

These sound like anti-lions indeed; and the following sound rather like anti-snakes:

> *And the coiled serpent, quickened by power [shall]*
> *Become a wingéd globe, a spiral flower,*
> *An animated beauty-form, whose flight*
> *Shall be like some fair meteor through the night;*
> *His hiss be changed to tones like any flute,*
> *And heard through air like an Aeolian lute*
> *Distilling liquid cadence; and his tongue,*
> *Poisoned no more, shall be to children young*
> *A lovely flame-flower. He shall lick their hands,*
> *And dwell with doves conjoined in circling bands.*[45]

Fourier's dream of 37 million poets of the quality of Homer inhabiting Earth shall also come true:

> *Genius shall then pertain to all mankind,*
> *And inspiration fill*
> *Each human heart and will*
> *And Deity pervade the common mind.*[46]

Of course, on places like Jupiter, Mars and Mercury, such wonderful transformations had already taken place, said Harris, who will doubtless have been exposed to much gushing talk about Fourier whilst still a loyal follower of Andrew Jackson Davis. Indeed, in another such work, 1858's

*Arcana of Christianity*, Harris actually specifically speaks of visiting an alien 'phalansterian world' in the following terms: 'Had Charles Fourier lived [there] he would have seen the good, the useful, the beautiful and the true in all his speculations far transcended'.[47]

## The Inner Life of Outer Space

In 1911, meanwhile, the clergyman-turned Theosophist C. W. Leadbeater (1854–1934) published his book *The Inner Life*, in which he laid out details of his own astral trips to Mars. Theosophy was a once-popular spiritual movement founded in 1875 by the Russian-born mystic Madame Helena Petrova Blavatsky (1831–91), a confusing mish-mash of Victorian Spiritualism and more traditional aspects of Hinduism, Buddhism, Neoplatonism, Swedenborgianism and Gnosticism, which involved a series of complex teachings regarding such dubious subjects as Atlantis and reincarnation on other planets. Leadbeater was one of Blavatsky's most talented disciples, honestly believing he had psychic powers so well-developed that he could travel the solar system. However, Leadbeater's idea of contemporary civilisation on Mars was clearly based upon his knowledge of the questionable teachings of Percival Lowell, rather than those of Madame Blavatsky. 'Large areas of it are at present desert,' he wrote in *The Inner Life*, 'covered with a bright orange sand which gives the planet the peculiar hue by which we so readily recognise it.' Thankfully, the Martian soil was 'fertile enough' once the canals had been diverted to water the dry regions, Leadbeater continued, before confirming with his own eyes that it was the wide belts of vegetation growing on either side of the canals themselves that could be seen from Earth's telescopes, thus vindicating Lowell's own opinion.[48] As for the Martians themselves, the 'whole civilised population is one race', he observed, with the majority being blue-eyed blondes who were 'somewhat Norwegian in appearance' like Orthon, although a few evolutionary laggards were still brunettes with purple eyes. Both genders wore 'an almost shapeless garment of some very soft material which falls straight from the shoulders down to the feet' and went around barefoot, although some did possess metal slippers.[49] More original were the Martian houses, which were not built, but poured; special molten material was tipped into building-shaped moulds before setting into a kind of coloured one-way glass which the occupant could see out of quite easily, whilst passers-by could not peer back inside in return.[50] The Martian tongue, meanwhile, was entirely artificial in nature, like an alien Esperanto, with all natural Martian languages having been allowed to die out ages ago due to their inefficiency.[51]

Mars was a lazy, carefree world, where everyone had lashings of free time, as domestic animals had been trained to do most of the housework and gardening for their owners, like in *The Flintstones*.[52] There were no tedious family commitments either, as free love was enjoyed, and all resultant children handed over to the state to be raised anonymously, so that nobody ever knew who their parents were.[53] All symptoms of old age and illness were also abolished on Mars, so that everyone looked permanently young and healthy. However, when each Martian grew to be around 100 years old they suddenly became bored of life and went to the alien equivalent of Dignitas, asking to be put out of their misery quickly and painlessly by means unspecified – 'a request which is always granted'.[54] But what came after death? The Martians had no form of religion other than 'what we should call scientific materialism', explained Leadbeater, a hyper-rational but ultimately false outlook upon life which was combated by a secret society of advanced Theosophical sages who 'believed in superphysical worlds' and conjured ghosts at séances. Crossing this hidden fraternity was unwise, as their hypnotic powers could be turned against their enemies, killing them immediately. Some of these super-psychic Martians could even beam their own astral bodies across space and time just like Leadbeater could, showing up at special space séances held by Theosophists down here on Earth. When they could not quite manage this feat, the aliens settled instead for transmitting ideas for poems and novels into the brains of literary-minded Earthlings who foolishly thought they had written their masterpieces themselves.[55] Leadbeater's ultimate conclusion? That life on Mars today was 'by no means unpleasant'.[56]

## Meat-Head Martians

Leadbeater possessed even stranger ideas about the evolutionary past of human beings on both Earth and Mars. Some of these notions were taken straight from Madame Blavatsky, but others appear to be the product of his own psychic travels back through time, to the days when puddings walked the Earth. Leadbeater's system of interplanetary reincarnation is tediously complex, but to simplify, you should imagine that all people today have progressed from being mineral souls to vegetable souls to animal souls and then on to being human souls, with many of these incarnations having occurred on different planets.[57] Any souls which do not move properly up this chain of development are sent to live on our dead moon, 'a place where only refuse gathers ... a kind of a dust-heap or waste-paper basket ... an astral cesspool into which are thrown decaying fragments' of stagnating souls. The moon had once been a living planet where mankind had progressed from animal form to being primitive 'moonmen', but nowadays had become a terrible stellar purgatory.[58] Some souls, however, were a bit more advanced than their fellow kind, but not yet superior enough to move on to the next full stage of evolutionary incarnation; imagine a wise monkey, who is clearly a much more moral monkey than his chimp-chums, but not yet quite wise enough to have his soul incarnate within a lunar cave-man. That monkey's soul might be sent to live on a nearby planet with other particularly clever monkeys, whilst they wait for their less advanced brethren to achieve readiness to become moonmen. Here, these super-monkeys will live in an environment a bit better than normal monkeys do now, but less good than actual human beings will inhabit in the future. (The PG Tips chimps would be an excellent example.) The present-day Nordic inhabitants of Mars, says Leadbeater, are similar; they represent the very best human souls who are ready to move on to the next stage of evolution, but who are kept back from doing so by the rest of us moral retards down here on Earth. That is why these butterflies on a leash are so much more advanced than we are, with their perfect health, well-trained pets and cheerful life of free love.[59]

But what strange evolutionary forms did man once possess during the distant prehistoric past? Using his powers of second sight, Leadbeater was able to find out. During early life on Mars, men had 'definite astral bodies', but they were very wispy and unclear, and had no physical form whatsoever. On Earth, men then developed into 'shapeless drifting clouds' of psychic mind stuff which, towards the end of this evolutionary round, 'began to aggregate around themselves gaseous matter'.[60] Later, humans 'succeeded in developing a certain amount of solidity' but not much, as they were basically made of soft dessert foods, being 'unpleasantly jelly-like in consistency and indeterminate in shape', and greatly resembling 'pudding-bags' on account of 'the curious shapeless projections they had instead of arms and legs'. Apparently, many of these pudding-people 'were so light and tenuous that they were able to drift about in the heavy atmosphere of the time', whilst others simply 'rolled along' the floor.[61] On Mars, these living puddings gradually developed into 'reptilian monkey' creatures, with the basic head-and-four-limbs shape that humans have now. However, they were still 'somewhat jelly-like', said Leadbeater, and 'if one pressed in his skin by a poke of the finger the hole remained for a long time before it filled out again'. Such jelly-monkeys did have bones, but they were very 'rudimentary' in form, and so the pre-humans 'lay about grovelling' in mud all day near rivers.[62] It wasn't until the next stage of evolution down here on Earth that men became proper, solid creatures who were able to stand tall on their own two feet, when a race of meat-people with sausages for heads began to appear:

[Now] men were more complete and began to try to stand upright, though they were still shaky and uncertain, and always fell back to all-fours when pursued or frightened. They began to have hair and bristles upon the body, but they were still loose and flabby. Their skins were dark and their faces scarcely human; strangely flattened, with eyes small and set curiously far apart, so that they could see sideways as well as in front. They had the lower jaw very heavily developed, and practically no forehead, but just a roll of flesh like a sausage where the forehead should have been, the whole head sloping backwards curiously. The arms were much longer in proportion than ours, and could not be perfectly straightened at the elbows, a difficulty which existed with the knees also. The hands and feet were enormous and misshapen, and the heels projected backwards almost as much as the toes did forwards, so that the man was able to walk backwards as rapidly and certainly as in the other direction. This curious form of progress was facilitated by the possession of a third eye at the back of the head.[63]

After this came the moonmen, the first recognisable primitive human beings, whose initial existence was nonetheless nasty, brutish and short because some sausage-people still lived on the moon, with the larger beasts constantly trying to capture and eat their smaller cousins for dinner. However, the lunarians, 'having much more intellect, were presently able to dominate their congeners, and to keep them in some sort of order'.[64] From this point on, the moon people began to evolve into the *Homo sapiens* we know and love today, thus meaning that evolution had a happy ending to it after all, notwithstanding the occasional appearance of atavistic 'water-men' on Mars, who are 'half-reptile, half-ape … with a horrible tarantula-like appearance about the eyes'.[65] And that, according to C. W. Leadbeater, is the Story of Life.

## Rock Legends

Perhaps the most elaborate psychic trips into outer space were made at the behest of the Yorkshire-born American geologist William Denton (1823–83), whose once-popular 1863 book *The Soul of Things* told of his family's adventures in a realm called 'psychometry'. The term, meaning 'measuring the soul', was coined by the physician Joseph Rhodes Buchanan (1814–99) in 1842. Buchanan was talking with the Episcopalian Bishop Leonidas Polk (1806–64) one day, during which the latter chanced to mention the fact that, whenever he should inadvertently touch something made of brass in the dark, he would immediately get an unpleasant metallic taste in his mouth. Intrigued, Buchanan conducted experiments with a group of medical students, who found that they too were able to identify and taste various substances by touch alone. Even weirder, when random medical substances were wrapped up in brown paper and held by the students, the items had an appropriate medical effect; holding an emetic caused vomiting, for instance. Intrigued by this news, William Denton got his sister, Anne Denton Cridge, to hold sealed letters up to her forehead to see what would happen. Supposedly, she experienced mental visions of their writers, describing their physical appearances correctly and in detail. Being a geologist, Denton wondered whether this new skill could be applied to his own field of research. Maybe, by holding up rocks to your forehead and concentrating on them really hard, you could 'read' their entire geological history, via psychometry? If so, then it would certainly save a lot of time and bother with fiddly fieldwork or lab tests.[66]

Some of Denton's experiments do sound initially impressive. Upon being handed a fragment of solidified lava from the 1840 eruption of Mount Kilauea on Hawaii, Anne spoke of 'an ocean of fire ... pouring over a precipice' which she saw 'flow into the ocean, and the water boil intensely', which certainly sounds pretty accurate. Via such methods, an excited William Denton claimed that we might one day be able to read the life story of volcanoes like Vesuvius 'from the time that he was a screaming baby'.[67] Sometimes Denton wrote up these experiments in such a way as to make it seem as if the rocks themselves were speaking to us, as in the following *Autobiography of a Boulder*:

Mercy! What a whirl things are in! I do not know what to make of it. I feel as if I were being belched out of a volcano; there is water and mud, and everything is in a perfect whirl. There are great pieces of rock beside me, some larger than I feel myself to be, though I am of great size. This is the strangest feeling I ever had. I am sent up whirling in a torrent of water, mud and rocks ... it is puff, puff, whirl, whirl, all of us flying round together ... I am out once more. I lie in a basin, in a large open place ... That which lies over me must be ice, for I can see light through it ... How strange! The ice has broken loose, and I am in motion now, travelling southwest ... The noise the ice makes in moving is awful to me ... a terrible grinding noise ... How insignificant a tree or a house would be in its path! ... The ice is leaving me, I believe ... It melts and melts, and keeps sliding on, faster since it melted so rapidly. I have dropped out of the place where I was, onto the ground [where you found me].[68]

And that's straight from the boulder's mouth. The trouble with these accounts is that, whilst the psychometers holding the rock samples may not have known what they were beforehand, the qualified geologist William Denton certainly did. When talking to his assistants, he may inadvertently have given away verbal or other cues which led them along towards giving him the correct answer about the rocks' life stories. If so, then this would not be deliberate fraud, more an accidental collusion between subject and examiner. Perhaps the best evidence this may have been the case can be found in what happened when psychometers examined items whose true origins and nature Denton himself was unsure of – namely, items not of this Earth. Take a fragment of meteorite which Denton once handed a female subject to hold. She correctly guessed that it came from outer space ('her eyeballs were rolled upwards in opposition to her own will' and she began to see stars), but perhaps this was only to be expected, because Denton already knew it was a meteorite himself. What Denton *didn't* know, however, was precisely what meteorites themselves actually were. Their very existence had been resisted by mainstream science until part-way through the nineteenth century, and their precise nature was still a matter for argument. Amongst the numerous now laughable-sounding theories for their origins put forward by Denton in his book included that they were 'ejected from lunar volcanoes', or else were 'fragments of rings once surrounding the Earth, [like] the rings of Saturn'.[69] We now know these ideas to be wholly incorrect but, predictably, subsequent psychometric experiments carried out under the guidance of Professor Denton provided him with some apparently corroborative evidence suggesting otherwise. For example, in July 1860 there was an eclipse of the moon visible from New York State, where the Dentons happened to be staying at the time. Excited, Denton instructed his wife to sit outside so that 'the lunar light could fall upon [her] forehead'. She then had a vision which entirely confirmed her husband's theory that there were active volcanoes on the moon:

> Near the edge of the moon ... I see an enormous crater miles in diameter, and at the bottom is a lake of lava; it is red, and I see it in slow-motion ... it seems as if it should pour out ... I have a view of the moon's crust, and compared with the Earth's it seems a mere shell, enclosing the liquid lava.[70]

Neil Armstrong didn't see anything like that when he went up there about a century later, but during Denton's day the notion of active lunar volcanoes was considered much more plausible than it is now; as Denton points out, Sir William Herschel thought he had seen 'what he considered to be the flames of an active volcano' lurking away up there through his telescope, for instance.[71] Less mainstream was Denton's peculiar notion that Earth may once have had Saturn-like rings but, sure enough, when handed prehistoric fossils to fondle, psychometers saw things like 'a long circular belt or arch of light' in the sky, 'about as broad as the full moon', which Denton interpreted as being a large orbiting solid ring of sun-reflective space debris, from which some rocks occasionally detached themselves and fell to Earth as meteorites.[72] Denton's other hobby was giving people fossil fragments to hold so they could have psychic visions of what complete, living dinosaurs looked like, but through similar means he hoped also to gain 'fossil' visions of a now long-extinct version of the night sky, in which dead stars that had burned out aeons ago would be seen again in their original positions.[73] Given that William Denton was an honest man, and that many of his early psychometric findings seemed fairly astronomically plausible for the time, it is little wonder that he and his subjects ended up so comprehensively fooling themselves over the truth of these trials. When Denton moved on to experiments in psychometric visions of what actual *life* was like on other planets, however, the results were so ridiculous that their fictional nature really should have been spotted by him straight away ...

## Sherman Tanks

It wasn't until the less well-known Volume III of *The Soul of Things*, published separately from the original first two Volumes in 1874, that Denton revealed to the world the results of his experiments in remote-viewing life on distant planets. He really shouldn't have bothered. Things began one evening in 1866 when Denton was standing outside in his Ohio orchard with his nine-year-old son Sherman. Venus was visible, shining brightly in the sky, and Denton told his son to 'Look at that star and then shut your eyes, and tell me what you see.' What he saw was a strange world full of odd creatures, 'half fish and half muskrat', and a very peculiar species of alien tree:

> I see a tree larger than those round here, just like a toadstool. It is a kind of purple colour. It has a monstrous trunk, larger than any I ever saw before ... They are as thick as the woods down here ... Inside of the trees is jelly-like stuff as sweet as honey. There is something hard inside that I spit out.

As well as tasting jelly from inside of giant toadstool trees, young Sherman also encountered some very strange-sounding water, 'heavier than our water, and it does not wet ... It is clear as crystal; but it is hardly water.' Indeed not, as it seemed to be both liquid and solid at the same time. 'It runs, and, where there is a fall, it comes down hard ... *clump, clump!*'[74] Instead of dismissing all this as a childish fantasy, Sherman's dad carefully compared his son's witterings to certain genuine observations made about Venus by reputable astronomers and, in a totally delusional fashion, concluded he was telling the truth. As a result, many subsequent years of Denton's life were totally wasted in getting various members of his long-suffering family to shuffle outside, gaze up at the stars and spend hours hallucinating wildly.

Besides visiting an unknown planet full of levitating ball-bearings[75] and travelling through the centre of the Earth before emerging on the other side again in Japan,[76] Sherman also flew to the sun via the medically dubious method of baring his forehead to its cancer-causing rays. His insightful conclusion? 'I tell you, it is hot.' Thanks, Sherman. Thanks also for informing us that sunspots are really solid lumps of blackened lava, which a person 'could run on in places', if they didn't mind scorching their feet.[77] William Denton actually managed to dig out the opinion of some obscure astronomer or other who also shared the fringe opinion that sunspots were solid, to back up his son's imaginary testimony – but not even he was able to discover any scientist prepared to back up Sherman's opinion that space itself smelled distinctly of strawberries.[78] He did try to get his sister to travel to the sun to see if she saw the same things there as Sherman did, but all that Anne managed to discover on her sojourn to our nearest star was that the sun was actually a giant magnet which attracted all the planets to orbit around it, thus explaining how gravity worked.[79] Anne found that Jupiter was well worth a visit too, as it was full of highly evolved, blue-eyed, blonde Aryans who enjoyed floating through the air around the planet's equator for recreational purposes.[80] Jupiter also possessed trees with super-stretchy rubber leaves, flocks of hairy birds, square-winged scaly flying creatures with mouths in the middle of their bodies,[81] and giant snail-like animals which had no teeth, therefore obliging each beast to 'suck its food off' at mealtimes, thereby ensuring its victims at least died happy.[82] By far the most popular destination for the Denton family's travels, however, was Mars.

## They Must Be on a Different Planet

On 12 January 1867, Sherman Denton enjoyed a psychic vision of Mars for the first time. He saw an unknown type of apple filled with thorns, and 'a monstrous animal like a turtle', eight or nine feet long, with a 'stumpy tail' and covered all over in 'conical spines'. In some water, he also spied thousands of baby spiked turtles, all of them 'no larger than a hen's egg'.[83] Far stranger were the people of Mars, who wore square hats and dressed like Indians, had darkish skin but yellow hair and blue eyes, possessed three fingers and one thumb to a hand, and had eyes 'like cats' eyes' with long vertical pupils.[84] Sherman's dad thought this all sounded very silly; but then he considered the possibility that Sherman's Martians might have looked so similar but different to men on Earth because the 'human form may be the highest form that matter can assume, and toward which it is everywhere striving', an echo of Christiaan Huygens' thinking.[85] Two years later, Sherman was told to look up towards Mars again, and travel back there. His dad didn't tell him the shining object in the sky was Mars, however, and he had no reason to believe it was the same planet at all. Perhaps this was why his Martians now sounded completely different. For one thing, the men now all had long moustaches and were unshaven, whereas on his first visit Sherman had specifically described them as having 'no beards'[86] and for another they seemed to have developed curious new forms of transport which the child had made no mention of before. Not only did they now have flying bicycles, they had also invented a means of attaching springs to their feet and hands so that they could bounce across the Martian surface with incredible speed.[87] Upon being told by his father that he was meant to be on Mars, however, Sherman suddenly changed his tune, expressing puzzlement at not seeing the three-fingered, cat-eyed folk he had met there last time.[88]

Sherman then went on to describe the alien culture. Martian theatres often staged gastronomic plays in which the actors pretended to eat food on-stage. However, they each had 'a false mouth, connected with a bag', into which the food actually went, thus allowing them to carry on saying their lines whilst they only pretended to feast.[89] Martian clothing manufacturers, meanwhile, made use of peculiar machines shaped like fierce dragons. Thread is laced between their teeth, then the jaws go up and down, weaving the clothes, until eventually the finished article is spat out from the beast's mouth. As the jaws move, the dragon's eyes roll around and its legs move, making it appear alive.[90] Martian fireplaces were equally as odd, being designed to funnel smoke downwards into a body of water. The water then turns black and solidifies, with the resultant sludge carried outside for manure.[91] The Mars-men also hold mock fights in which the Martian Satan is put to the sword. The alien Devil, played by a Godzilla-like man in a suit, has eight horns on his back, six-inch claws on his fingers, lobster eyes, foot-long teeth and green scales. The Mars-Devil stamps his foot and sends sinners through trapdoors to Hell, before the Martian Army, dressed in red hats and white feathers, arrives to kill him. Rather spectacularly, they pepper Satan with mortar fire until his head is blown off and flies across to the other side of the stage, trailing blood.[92] Even fruit-picking on Mars sounds fun. People hop on special foot-powered hover devices, float to the top of trees, and pluck the fruit from their branches. If their crops are threatened by birds, they leave this task, fly after them, and hit the feathered pests with big sticks.[93] Winter sports are much loved on Mars, too. Grown men and children delight to dress up in suits of metal armour and slide down icy hills like speedy humanoid bobsleighs.[94] Martians also have electric generators powered by lightning strikes, ingenious wind-up cars, and special hats made of 'green stuff' which act as wearable umbrellas, but stick to the head 'like wet leather to a stone' during rain, making them impossible to remove.[95]

William Denton's method of testing the truth of his son's claims was to send his other relatives off to Mars too, and then compare their separate accounts. Next to try was Denton's sister Anne, whose visions sounded entirely unlike those of her nephew. The Martians, she said, were 'chunky and short in stature', and the planet was full of huge dinosaurs.[96] The squat Martian humanoids were now mainly redheads with bright pink skin and bodies which circulated not only blood but also electricity, as could be seen in their static-filled hair. They lived in triangular cities and, when they couldn't be bothered walking, emitted electricity down from their feet into the ground, the charge produced moving them along in 'an undulatory motion ... in wave-like curves, from one point to another', slightly above the ground. This 'method of half-flying' was considered 'delightful', and Martians often gathered together in groups to do it *en masse* for fun.[97] An 'inferior race' of humanoid pygmies with tiny heads, who were 'almost black, their bodies nearly covered with dark hair' also lived on Anne's version of Mars, but the less said about them the better, apparently.[98] Far better to dwell upon the lovely buildings of the pink-skinned folk, which looked at first sight as if they were made from vegetation, but in fact were skilfully carved from stone.[99] Just about the only similarity between Anne's Martians and those of Sherman was that they too possessed strange eyes. However, they were not cat-like, but had pupils formed like 'a many-sided lens, enabling them to see more from the side than we can'.[100] On the other hand, the lower part of their face was 'dish-shaped', and they had big heads and abnormally long hands and feet, totally unlike the Martians Sherman encountered. They also spoke Italian.[101]

So, who was right about the Red Planet and who was deluded? William Denton ordered his wife to fly up to Mars as well to settle the matter for good. Unfortunately, she saw yet another entirely different world, with huge, turreted houses and revolving roads upon which you simply had to stand still to reach your destination.[102] In appearance Mrs Denton's Martians were highly evolved Nordics like the floating people of Jupiter, although again there was an inferior race of black men present on the planet too.[103] A trip to a Martian library revealed that the Nordics had special books with multi-directional pages which could be read by four people at once,[104] and they seemed to possess some special variety of horseless carriage, as well as advanced flying machines built in a variety of styles.[105] The most useful Mars machines, however, were the automatic dusting contraptions which could clean entire rooms and sweep whole floors, leaving them literally spotless, with the emancipated Martian housewife's only task being to wind it up beforehand.[106] As William Denton had to admit, there were 'several apparent discrepancies' between these three accounts, to say the least. However, rather than concluding the obvious, he chose to claim that each of his three brave psychometers had travelled to different parts of Mars and seen different races and civilisations at work and play. After all, if three Martian psychics had travelled to London, Tokyo and the Belgian Congo on Earth, then they would each come back with totally different descriptions of what us Earthlings were like. With this in mind, Denton then sought to rank the Martians in order of racial superiority and plot their respective homelands upon a map.[107] Truly, the human capacity for self-deception knows no bounds.

# 18

# Jung Dreamers:
# From Freud to Flammarion

Following the explosion of Sigmund Freud's (1856–1939) ideas across Europe during the late nineteenth and early twentieth centuries, another attempt at explaining such 'psychic' extraterrestrial fantasies came from within the new realms of psychiatry and psychoanalysis. A number of pioneer practitioners of 'the talking cure' encountered patients who felt that they had travelled to other planets via supernatural means; C. G. Jung, that early theoriser about flying saucers, was a case in point. One of his patients was an eighteen-year-old girl who had been admitted to a mental hospital after hearing voices and becoming mute. It turned out that, when aged fifteen, she had been seduced by her own brother. Over the course of several weeks in the asylum, Jung gradually managed to get her to start speaking to him. What she had to say was most surprising; it turned out she was a regular visitor to the moon. She told Jung that the lunar surface was inhabited, but only by males. All female lunarians had to live underground in caves, due to the presence of an evil winged vampire lurking on 'the high mountains of the moon', just waiting to swoop down and kidnap them. Outraged, the abused girl returned to the moon bearing a knife, planning to kill the vampire. Eventually he flew down towards her, beating his wings, which were so large they obscured the true nature of his form. Curious, the girl wished to see what he looked like before she stabbed him, whereupon he opened his wings to reveal a 'man of unearthly beauty' who left her feeling 'spellbound', before picking her up and flying off with her. Annoyingly, however, her talks with Jung had put an end to the girl's visions at this point, which depressed her unutterably; life on the moon was 'beautiful' and 'rich with meaning' by comparison with her dismal lot here on Earth. Jung explained to her that, even though this world was indeed pretty awful, in the end its imperfections had to be faced and she could not go on running away into fantasy land forever. Ultimately the girl accepted this diagnosis, stopped flying to the moon, and went on to live an ordinary, sane life. This was one of Jung's earliest cases, and gave him an important insight into the lives of the mentally ill; that, whilst outside observers might see only the person's 'tragic destruction', their inner life may well have a 'richness and importance' which was wholly unexpected.[1]

A similar case involved one of Jung's own relatives, a fifteen-year-old girl named only as 'S. W.', whose sufferings provided the material for his doctoral thesis. S. W. claimed to be a spirit medium who could channel the voices of dead family members through her body during séances. Jung did not believe that she was genuinely speaking to the dead, but what he discovered during his investigation of her was most interesting nonetheless. S. W. had developed the habit of falling into somnambulistic trances, during which she would see spirits. Some of these were frightening and demonic, but most were the friendly shades of her departed ancestors who appeared from out of a 'foggy brightness' in the form of 'shining white figures ... all wrapped in white veil-like robes', whilst the female spooks 'had things resembling turbans on their heads and wore girdles'.[2] These spirits would encourage S. W.'s astral body (which was Jewish for some reason) to leave her sleeping physical one and float away with them on curious adventures, many of them in outer space. Her grandfather was her main guide, and he would show her the correct path to take through 'that space between the stars which people think is empty but which really contains countless spirit worlds.' S. W. talked of some system of interplanetary reincarnation being inherent within the universe, and of the existence of non-human alien races she termed 'star-dwellers'. The Martians were just such a race, with S. W. claiming to have visited the Red Planet and seen sophisticated flying machines which the ETs used to fly across their famous canals in, thus negating the need for any bridges to be built. Apparently, these aliens were not humanoid, but possessed 'the most laughable [shape] imaginable, such as no one could possibly conceive'. During her astral travels, however, S. W. was unable under intergalactic law to venture down any lower than seventy-five feet above the surface of the planets she visited; it was the same with any aliens visiting Earth. Violation of these laws led to star dwellers being penalised by being forced to take on human form and waste their lives away down here, an experience which made them cold and cruel. Napoleon was one such alien miscreant, as was an old man S. W. had once seen on a train. You could recognise such exiles due to their 'hairless, eyebrowless, sharply cut faces' and 'peculiar expression', the girl said.[3]

Jung's ultimate conclusion was that these were all mental fantasies, created by the girl's subconscious mind. Of limited intelligence, poorly educated, with no real interests or spark, S. W.'s father had died during her adolescence and her mother was less than affectionate towards the dullard child. As Jung says, 'In these circumstances it is no wonder that she felt shut in and unhappy.'[4] Therefore, through her trances and astral travels, S. W. developed a kind of 'secondary personality' which travelled through space having all kinds of interesting adventures in order to compensate for the unbearable dullness of her real life and boring character. As an adolescent, S. W. had a new personality just waiting to break through and overthrow her old childhood one, but in her oppressive home this was an awkward feat to achieve, leading to her bouts of space-based trance as a means of temporary escape. Eventually, though, the girl grew up and things improved. S. W. abandoned Spiritualism, got a job and became noticeably 'pleasanter and more stable' in character – before dying, aged twenty-six, from tuberculosis.[5]

## Ghosts of Mars

The curious connection between outer space and the dead in such cases seems to reflect the apparent overlap between psychoanalysis and parapsychology in the worldview of many of those (like Jung) who investigated them. Psychology, hypnotism and Spiritualism had several links at the time, with practitioners of each discipline often maintaining a quasi-professional interest in the others, too. For example, in 1895 one Colonel Albert de Rochas (1837–1914), a Frenchman with a liking for both psychical research and hypnotism, was called out to help a family friend, given the pseudonym 'Mireille', who was suffering some ailment. Knowing of his skills in hypnotism, Mireille hoped de Rochas could help alleviate her sufferings. He did, but during one trance session Mireille claimed to be rising up through outer space, which she deemed to be luminous and full of 'phantoms', one of whom was a dead childhood friend of hers named Victor. Mireille spoke of visiting Mars and its canals, before one day it seemed as if her body had been suddenly taken over by the spirit of Victor. He was initially puzzled as to why he was wearing women's clothing, but once he had calmed down Victor explained that, following his death, he lived on amongst the planets – a form of existence about which he could give little information beyond the curious detail that all dead persons had arms which also functioned as genitalia, or 'organs of affection'. According to Victor, these arms grew larger and larger over time and were often mistaken for angel wings, when they were really giant celestial penises. Perhaps unsurprisingly, de Rochas refused to believe Victor's story, choosing instead to deem him a mere secondary personality lurking within the mind of Mireille herself.[6]

Another such instance came to the attention of an eminent American psychical researcher named Professor James Hyslop (1854–1920) in 1901, when he was contacted by a clergyman he dubbed 'Mr Smead' and asked to examine his wife, who claimed to be in mental contact with Mars. Mrs Smead had been familiar with the idea of automatic writing since childhood, and in 1895 had begun experimenting once more within this field. Communications quickly came through her 'possessed' pen, purporting to be from three dead children of the Smeads. At one point, the Smeads asked where it was these souls lived after death, to which the words 'some spirits are on the Earth and others are on other worlds' was written in reply. It seems the dead Smeads were currently living – if that's the right word – on Mars. Mars itself was inhabited by living people too, though; they resembled American Indians and were experts in canal-making. The ghosts kindly drew a map of Mars through Mrs Smead's agency, labelling it with several distinct climatic zones, ranging from Zentin (Cold Zone) to Dirnstze (North Temperate Zone) to Emerincenren (Equatorial Zone), and criss-crossed with canals and other bodies of water. The Smeads abandoned their séances for the next five years, but in 1900 took them up again. Evidently the long hiatus had given Mrs Smead's subconscious mind the chance to think up more details about life on Mars, with one of the first new communications being a labelled diagram of a 'dog-house temple', containing statues of canines intended for religious worship. Snippets of a Martian language complete with basic grammar also began to appear, with 'Mar*e*' meaning 'Man' and 'Mar*en*' meaning 'Men', for example. However, some attempts at conjugating Martian verbs proved unsuccessful; asked what the Martian was for 'The boy runs', the communicating ghost replied primly that 'People do not run on Mars, but only walk'. Amusingly inept drawings of Martians resembling melted witches and big-hatted folk wearing giant nappies also appeared, together with appropriately childish descriptions of their appearance:

The women wear bag-like skirts and funny hats. Their hair is hanging down their long shoulders. The men put theirs up and keep long hair under their hats. We went all around. The people are different in different places.

Very good – give that dead child a gold star. We can give further credit to the ghost children for providing us with useful information about the strange political system on Mars, in which aristocrats were denied the vote and their children forced to work the fields before marriage, to give them a taste of what ordinary life was like. Elaborate diagrams of Martian clocks and flying machines also emerged, as did detail regarding the disappointingly small size of Martian chickens. Pleasingly, the Martian diet centred around cake and bread, and they liked to drink 'something like water', said the spirits. Regrettably, however, no further data about Mars was henceforth received as a new ghost named Harrison Clarke suddenly appeared on the scene and put a stop to all communication with the Smeads' departed infants. Clarke was a dab hand at mirror-writing, and spun an elaborate back-story involving his supposed exploits during the US Civil War. However, through diligent research Professor Hyslop was able to prove that no such person as Harrison Clarke had participated in the specific battles the ghost claimed to have done, facts which embarrassed the 'spirit' when first put to him. Clarke then quickly rallied by claiming to have enlisted under a false name – which he refused to give, thus making it conveniently impossible for Hyslop to prove or disprove the spook's former existence either way. Hyslop's ultimate conclusion was obvious; all of the 'ghosts' Mrs Smead had been contacting were merely secondary personalities lurking away somewhere within the recesses of her mind, and all claims made about Mars just unconscious fictions.[7]

## Miss Smith Goes to Mars

The two cases given above differ from the ones involving Jung's patients because none of the persons involved were mentally disturbed. The secondary personalities involved arose either inadvertently, following medical hypnosis, or after a deliberate effort to contact spirits via the séance room. They were not, as with Jung's patients, spontaneous attempts by a damaged psyche to escape from unpleasant home lives or personal tragedies. At other times, secondary personalities can emerge simply as a response to boredom, as with the celebrated case of Hélène Smith (1861–1929), a Swiss medium whose supposed trips through space (to Mars) and through time (to a bygone India) were expertly catalogued by the eminent psychologist Théodore Flournoy (1854–1920) in his classic study *From India to the Planet Mars*. Smith – whose real name, protected in the book, was Catherine Elise Müller – was a non-professional medium with a responsible day job in a commercial firm, who was not in any way troubled when Flournoy met her. Indeed, he described her as being 'a beautiful woman about thirty years of age, tall, vigorous, of a fresh, healthy complexion, with hair and eyes almost black, of an open and intelligent countenance, which at once evoked sympathy'.[8]

However, her childhood had not been quite so happy. A quiet and reserved girl, Hélène had spent hours by herself, 'passed motionless in an easy chair', drifting into states of semi-trance in which she had seen various wonders, such as 'highly coloured landscapes' and 'a lion of stone with a mutilated head'. These hallucinations, she said, 'bore a close resemblance' to her later visions of India and Mars.[9] More bizarrely, when embarrassed, she used to see 'a visual hallucination of a man clothed in a long, brown robe, with a white cross on his breast, like a monk', who would stand beside her in supportive silence until the feeling had passed.[10] Bored by her hum-drum circumstances, says Flournoy, Hélène was primed from an early age to have elaborate mental fantasies of a kind intended to compensate for this feeling of 'instinctive inward revolt against the modest environment in which it was her lot to be born', feelings which were so strong that she once 'seriously asked her parents if it was absolutely certain she was their daughter', and not some stranger who had been brought into their home accidentally.[11] Seeing as her mother was also an occasional visionary, once awakening in the night to see 'an angel, of dazzling brightness' standing beside the bed of her dying three-year-old child, Hélène clearly *was* her mother's daughter.[12] Reputedly, several of her relatives had psychic powers, so it was no especial surprise, after being introduced to Spiritualism by family friends in 1891, that Hélène began going into trances and demonstrating apparent qualities of telepathy and clairvoyance, whilst tables began tipping and pianos playing by themselves in her presence. When introduced to her in 1894, Flournoy, who was more than willing to accept the reality of paranormal phenomena, was impressed by the way she seemed to know certain obscure facts about his family history which, by rights, she really shouldn't have done.[13]

However, one thing Flournoy was less willing to believe in the reality of were Hélène's accounts of her trips to ancient India and present-day Mars whilst utilising the powers of her astral body. These, Flournoy realised, were but yet more fantasy products of her subconscious mind, which was at work all the time in the background spinning such tales. Even whilst Hélène was fully awake and occupied, some residual sense of her childhood boredom seemed to linger on somewhere, leading her to concoct strange and exotic fairytale worlds into which she could escape either in her dreams, during séance trances, or even spontaneously, at random waking moments. The astral jaunts to India came first, but were followed in 1894 by journeys to Mars as well. There was much talk within Hélène's circle at the time of matters astronomical, including the work of Camille Flammarion, the French-speaking world's most famous astronomer, whom we met earlier on encouraging others to try and make contact with Mars, a planet he felt certain was inhabited.[14] In 1894, Hélène had made the acquaintance of one Auguste Lamaitre, a friend of Flournoy's, who let slip in her presence the comment that 'It would be very interesting to know what is happening upon other planets.' The first time that she ever went into a trance and flew to Mars, on 25 November of that year, her favourite haunted table had rapped out the words 'Lemaitre, that which you have so much wished for!'[15] Clearly, Hélène Smith's unconscious mind was quite obliging, wishing to make the dreams of those around her come true, as well as her own. This can also be seen in the fact that, during her first astral trip to Mars, Hélène claimed to have encountered the reincarnation of a young man named Alexis Mirbel, whose bereaved mother had attended Hélène's séance with the specific hope of hearing from her dead son.[16]

## The Red-Light Zone

It was not until February 1896 that Miss Smith returned to Mars, a rather lovely place to which she floated away during trances, up through 'a dense fog' of many colours before encountering a star which grew larger and larger, 'as large as our house', even, before an appropriately coloured light, 'very brilliant and red', suffused all that she saw, letting her know that her soul was now safely on the Red Planet.[17] She was then taken down to solid ground in 'a curious little car without wheels',[18] whereupon she was obliged to greet her hosts with 'a complicated pantomime expressing the manners of Martian politeness', involving 'uncouth gestures' such as slapping her hands, tapping her nose, chin and lips, bowing in a twisted fashion, and much 'rotation [of her body] upon the floor'.[19] Hélène also demonstrated the ability to speak Martian, initially by spouting pure gibberish but, as time went on, this random chitter-chatter was actually developed by Smith's mind into a fully functioning alien tongue with its own alphabet and rules of grammar – albeit one which, as Flournoy was comprehensively able to show, was based rather closely upon the ordinary Earth language of French.[20] As usual, 'Mars' was really fairy-land, as descriptions of 'a beautiful blue-pink lake with a bridge the sides of which were transparent and formed of yellow tubes like the pipes of an organ' demonstrates clearly.[21] Surrounded by supernatural red light, meanwhile, Hélène would enjoy glorious visions of Martian flowers resembling bizarre pineapples in her bedroom, which she would feel compelled to draw, her pencil being guided by 'an invisible force'.[22]

These escapades had no coherent continuous plotline to them,[23] consisting instead of a series of fragmented scenes, such as the sight of a little Martian girl using a magic wand to make 'a number of grotesque little figures dance in a white tub'.[24] Really, these Martian visions were just the same as Hélène's earlier Indian ones, with Martian architecture looking suspiciously oriental in nature,[25] and with some recurring characters tagging along too. For example, a dark-skinned Martian gentleman named Astané, who owned a hand-held flying machine of curious design which spat out flames,[26] was really the alien reincarnation of a Hindu fakir and sorcerer named Kanga who had previously appeared in Hélène's Indian visions. A man of many talents, he would appear to Hélène when she was in the bath and beam her up to the Red Planet.[27] The main way you could tell that Astané was an alien was not through his entirely human-like appearance, but on account of the bizarre animal he kept as a pet, a sort of living draught excluder possessing 'the head of a cabbage' with 'a big green eye in the middle' and five or six appendages which could have been either ears or paws hanging off it at either side. Apparently this animal 'united the intelligence of the dog with the stupidity of the parrot' as it was able both to fetch Astané objects (like his favourite telescope) under instruction and to copy its owner's handwriting![28]

Smith herself thought these 'findings' of hers would 'cause all the discoveries of Monsieur Flammarion to sink into insignificance',[29] and this opinion is not entirely to be despised. As with Charles Fourier, the mad ideas of Hélène Smith were later taken up and celebrated by André Breton and his fellow jesters in the Surrealist movement as absolutely excellent examples of the creative powers of the unconscious mind let loose to play. Flournoy himself was somewhat less impressed, however. Bored of the Martian tales, he confronted Hélène one day about the utter implausibility of her stories. Why did all the Martians look like humans, why was their language based on French, why was their society basically just the same as our own, only lit by red light? Take a dinner party she had once described, for example, which was exactly the same as an ordinary European dinner party in every single respect bar the fact that the guests had been eating what appeared to be cats.[30] How could she account for all that? In response, Hélène's subconscious began to concoct what Flournoy called an 'Ultra-Martian' cycle of visions, set on a different planet full of much stranger aliens:

> The men, with arms and bodies bare, had for all clothing only a sort of skirt reaching to the waist and supported by a kind of suspenders thrown over the shoulders ... Their heads were very short, being about three inches high by six inches broad, and were close-shaven. They had very small eyes, immense mouths, noses like beans. Everything was so different from what we are accustomed in our world that I should almost have believed it to be an animal rather than a man I saw there.[31]

Here, perhaps, we can see Hélène's subconscious trying almost *too* hard to be inventive. Personally, I much preferred the Martian house pet with the cabbage for a head.

## Fantastic Flim-Flam from Flammarion

The influence of Camille Flammarion upon not only the fantasies of Hélène Smith, but upon interplanetary 'psychics' in a wider sense, appears to have been key; he was speaking optimistically about the prospect of meeting and talking with 'our brothers in space' long before men like George Adamski were claiming to have encountered just such Space-Siblings for real.[32] Historically, astronomy has been seen as a field which is welcoming to the enthusiastic amateur, and Flammarion perhaps deserves the title of having been the most enthusiastic amateur of all time. Obsessed with the heavens, by age sixteen Flammarion had written a 500-page manuscript (never fully published) explaining the supposed origins of the world, named *Cosmogonie Universelle*, something which brought him to the attention of the famous director of the Paris Observatory, Urbain Leverrier (1811–77), who offered him a job as an apprentice astronomer. Already an eager if wholly unqualified observer and recorder of the skies, Flammarion jumped at the chance to work under the great Leverrier, whose calculations had in 1846 helped in the discovery of Neptune.

Working under the strait-laced, methodical and data-driven Leverrier was not all Flammarion had imagined it might be however, and, filled with romantic fervour, the young apprentice continued writing in his spare time. In 1862, aged twenty, Flammarion completed his first published book, *La Pluralité des Mondes Habités* (*The Plurality of Inhabited Worlds* – apparently never translated into English, although he claimed it had been), arguing upon dubious but entertaining grounds that life was present throughout the universe. The book went through numerous editions and, for better or worse, made Flammarion's name. He was also a writer of science fiction, and some unkind critics might well classify some of his ostensibly non-fiction works under this same heading. Nonetheless, he almost single-handedly transformed astronomy from fringe-pursuit to popular pastime, and in 1882, in recognition of his labours a rich admirer unexpectedly donated Flammarion a small estate near Paris, where he established his own private observatory and happily spent the remaining forty or so years of his life.[33] Whatever the astronomical controversy of the day might be, Flammarion was happy to comment upon it – often dubiously, but always entertainingly. Following Giovanni Schiaparelli's cautious initial 1877 findings about Mars' *canali*, for example, he was happy to talk about how it was 'almost certain' that Martians could fly, and that the planet's moons were populated with 'reasoning microscopic mites'.[34] His reckoning behind these amazing conclusions? Simply that he was Camille Flammarion, the most famous astronomer in the world, and he could say whatever he damn well liked!

However, Flammarion was interested not only in astronomy, but also Spiritualism, in which field he later became another leading figure; titles like *Haunted Houses*, *Mysterious Psychic Forces* and *Death and Its Mystery* proved equally as interesting to the public as his books on astronomy. That the two topics became somewhat conflated by him can be seen by the fact that he considered one of the greatest influences upon his career to have been the 1854 book *Terre et Ciel* by Jean Reynaud (1806–63), which argued that, after death, our souls transmigrated from planet to planet, with each new planetary rebirth representing an advance upon the karmic scale.[35] By the later expanded editions of his *Pluralité*, Flammarion was directly imitating his 'master and friend' Reynaud, so much so that he added a new section to his book titled 'Humanity in the Universe', in which he advocated a very similar system of interplanetary reincarnation, calling humans 'citizens of the sky' and saying we were all part of the same celestial family as Martians and Jovians, these being alien races into which we would one day be reborn.[36] It has to be admitted that Flammarion was an extremely well-known writer, as famous then as Stephen Hawking (b. 1942) is today, whose 1880 *Popular Astronomy*, translated into English in 1894, is often said to have done more to help create worldwide interest in the science of the stars than any other book in history.[37] Given this, it seems reasonable to presume that some ordinary readers – if not necessarily any actual *professionals* – would have taken Flammarion's weird fantasies about interplanetary reincarnation as being established fact, or at least a reasonable leading scientific opinion of the day, not mere fanciful speculation. With his endless newspaper and magazine appearances and unshakable celebrity status as the 'face of astronomy', Flammarion's ideas became part of common currency in the late nineteenth and early twentieth centuries. The patients of Jung, Hyslop and Flournoy would be overwhelmingly likely to have heard of them even if they had not actually have read his books, in much the same way the general public consume many popular science ideas today.

## The Reign of Urania

So committed was Flammarion to spreading his teachings that he sometimes tried his hand at a bizarre form of 'educational fiction', in which his theories were embedded within an often rather inept storyline framework. Most relevant was his 1889 book *Urania*. The book is split into three distinct parts, the first of which seems like a semi-autobiographical account of the young Flammarion's apprenticeship in Leverrier's Paris Observatory, with the significant difference that, whilst toiling away there, the narrator (known only as Camille) falls in love with a statue of Urania, the Muse of astronomy. In a nod to the Greek myth of Pygmalion, as well as the likes of Fontenelle and Swedenborg, this goddess suddenly comes to life in young Camille's dreams one night and takes the grateful youth on a flying astral tour of the cosmos – that is to say, the cosmos according to Flammarion. Travelling outside the solar system itself, Camille is taken to an unknown planet which has three different-coloured suns, one blue, one green, and one orangey-yellow, all of which bathe the planet in gorgeous, multi-coloured light. The atmosphere is further enlivened by the sound of a 'soft, delicious music' emanating from the wings of a race of humanoid dragonflies, which seem 'woven out of light'. These flying people live in 'vegetable cities' of semi-sentient flower trees, whose branches lower or raise themselves to receive them as they return home. The whole place seems less like an alien world, and much more like Disneyland. This is because 'the law of progress' prevails upon every planet in existence, says Urania, implying that one day we gross humans, too, may become more like the radiant and peaceful dragonfly-people, living in their fairy-like paradise.[38] On other planets Camille visits, life takes all kinds of forms, from humanoids with glowing eyes, to people with not only eyelids but also earlids, designed so they can close their minds off from whatever they do not wish to hear, to aliens whose very thoughts appear written publicly upon their foreheads. Strangest of all were the race of men who hung upside-down from trees in chrysalises made from their own hair, from which they eventually emerged as butterflies. On none of these planets is life as hard or imperfect as it is on Earth.[39]

The second part of the book is an exceedingly mawkish tale of romance, in which one of Camille's young astronomer friends, George Spero, falls in love with a Scandinavian beauty named Icléa. This Icléa would seem to be a representation of Flammarion's own ideal woman, since most of the young lovers' time is spent with her hanging on his every word as he lays out his syrupy ideas about the true nature of the universe, and the central place of Spiritualism within it. In particular, George is obsessed with the question of whether or not souls are, as he puts it, 'the seeds of planetary humanities';

that is, whether or not they can 'transport themselves from one world to another' for the purposes of reincarnation, as Flammarion's hero Jean Reynaud thought they did. According to George, the idea is a true one, with the process of souls travelling between planets being compared to the way that 'the electric telegraph instantly sends human thoughts across continents and seas'. 'We are citizens of the sky,' is George's trite conclusion. 'Whether we know it or not, we are really living in the stars', and the most evolved amongst us will end up flying between them one day, too, in astral form, up towards the next planet of our fleshly incarnation.[40] There is to be no happy ending for George and Icléa upon this imperfect sphere of Earth, however, seeing as this section of the book ends with them both dying tragically in a freak hot-air ballooning accident. Upset, Camille attends a séance, where he has a vision of his dead friends living on Mars. Having seen the reincarnated form of George (who is now female, whilst Icléa is male – their souls wished to experience one another's love from all angles, it seems), Camille comes to realise that his friend's Spiritualist theories were correct, that the soul is indeed immortal, and that the shades of the most advanced humans (i.e. astronomers and Spiritualists) 'seed' life on other planets following death. 'The religion of the future', Camille concludes, will be revealed first of all to 'the apostles of astronomy'. Only 'the Uranian soul' – that of the star-gazer – can ever hope to be truly saved![41]

## Phantasms of the Planets

Section three is the oddest part of the book from a literary perspective, seeing as a good chunk of it is made up simply of Camille going through the volumes of his vast library and presenting the reader with verbatim real-life accounts of stories involving ghosts, uncanny coincidences and psychic powers, taken from a number of genuine sources, including the archives of Britain's Society for Psychical Research, of which Flammarion served as President in 1923.[42] By interspersing this data with recent information regarding telescopic observations of Mars, the narrator aims to illustrate that, as he has it, 'astronomy, or the knowledge of the world, and [para]psychology, or knowledge of being' are indissolubly connected.[43] In particular, Flammarion focuses upon reproducing data from a famous 1886 book called *Phantasms of the Living*, about so-called 'crisis apparitions', in which people's spirits appear before their friends or relatives at the moment of their death, with the book's authors concluding that these 'spirits' may in fact be a kind of telepathic projection from the dying person's brain. If so, asks Camille, then why should it not also be possible for a dying person to 'broadcast' their soul onto another planet at the moment of death, to await rebirth there?

The distance from here to Mars is equal to zero for the transmission of [gravitational] attraction; it is almost insignificant for that of light, since a few minutes are enough for a luminous undulation to travel millions of leagues. I thought of the telegraph, the telephone and the phonograph; of the influence a hypnotiser's will has on his subject kilometres distant; and I wondered if some marvellous advance in science might not suddenly throw a celestial bridge between our world and others of its kind in infinity.[44]

The apparent distance between the planets, Camille concludes, is but an illusion, caused by 'the insufficiency of our [human] perceptions'. Much as Edgar Mitchell was later to conclude, 'Instead of being a void separating the worlds from one another, space is rather a connecting link'. If so, then psychic communication between souls on different planets should be simple: 'Cannot the emotion which starts from a brain reach [another] brain vibrating at no matter what distance, just as a sound crosses a room, making the strings of a piano or violin vibrate?'[45] To further hammer home this message, Flammarion has Camille fall asleep under a tree one day before travelling up to Mars in astral form, like one of the Dentons. Later, whilst still awake, he is then visited by the now-Martian soul of George, whose corporeal body snoozes upon the Red Planet; proof positive that two-way interplanetary soul transmission is indeed possible. On Mars, the living people are ethereal like disembodied souls in any case, a sure sign that they are more evolved than we are. The Martian atmosphere itself is nutritive, meaning there is no need for Martians to eat any solid food, merely to keep on breathing. As such, they are light and weightless beings, freed from 'the coarseness of Earthy needs'. Sex in the human sense is unknown, just a plant-like form of almost asexual reproduction; George tells Camille that Martians are really little more than 'winged, sentient, flying flowers' which, despite their giant size, have evolved to be so very airy due to the lesser gravity up there. Martian science is much more advanced than ours, too, with 'a kind of telephotographic apparatus' being used to record every second of Earth's development as it unfolds. George tells Camille that the Martians have already sent us humans numerous messages by way of Gauss-style geometrical signals on their plains, but complains that nobody has ever had enough imagination to realise what they were. The problem says George, is that 'terrestrial humanity ... is a child, and still in primitive ignorance.' Humanity hasn't yet reached the point where the only religion believed in is that peculiar mixture of astronomy and Spiritualism which Flammarion was to make his own.[46]

The book ends with a piece of poltergeist activity, as a picture jumps off Camille's wall revealing a sheet of paper filled with George's final testimony of his beliefs to the world, which is reproduced in full, and which we may presume that Flammarion himself shared. Among George's main tenets are the ideas that souls can travel between worlds, that all planets are in constant invisible contact with one another, and that as such 'the universe forms a single unity'. Seeing as all spirit is contained within matter, in terms of souls appearing within living creatures, and astronomy is the study of matter taken to its most massive extremes, George concludes that this must mean that astronomers are actually the most spiritually developed of all men, even priests; 'Hence, astronomy must be[come] the basis of all philosophical and religious belief.' The human soul's destiny is to 'free itself more and more from the material world, and to belong to the lofty Uranian life, whence it can look down upon matter and suffer no more.' In short, we must all become astronomers – or see humanity's soul perish amidst the dismally diminished perspective of those who are doomed to see the universe from the vantage point only of their own puny planet.[47] Such were the bizarre beliefs of the most famed astronomer of the age – beliefs he was not exactly shy about publicising.

## Fallen Star

Flammarion's speculations inspired a number of imitators to pen their own books advocating a doctrine of interplanetary metempsychosis, thus spreading his doctrines ever further. One of these men, André Pezzani (1818–77), called Flammarion 'the master of a school', whilst another of his pupils was Luis Figuier (1819–94), who sought consolation for his young son's death by imagining that he lived on elsewhere in the solar system, claiming that comets were but 'pleasure trains' laid on for the benefit of departed souls so that they could take a nice long tour of the universe before returning to their new home inside the sun.[48] Another Flammarion fan was Allan Kardec (1804–69), the chief figure in the French Spiritualist movement. Kardec was an early reviewer of Flammarion's famous *Pluralité*, who was so impressed by the work of the young author that he actually suggested a ghost must have helped him write it without his knowledge![49] Kardec's initial ideas on the topic preceded the publication of Flammarion's book, but he too wrote texts claiming that reincarnated souls flitted from planet to planet; according to one disciple, Kardec's basic belief was that 'the globes of each solar system form a series of temporary residences – of progressive training grounds, of places of reward or punishment – for the spirits who are being educated in them'. Flammarion became close to Kardec prior to the older man's death, and in his *Discourse Pronounced Over the Tomb of Allan Kardec*, acclaimed him as being 'the first' to have shown 'a sympathetic view' of his ideas, later 'taking in hand' the *Pluralité* and placing it 'at the base of the doctrinal edifice' he created.[50] There can be no doubt that the two men influenced one another greatly, as they did many other people across the world. It seems that the only person not impressed by the publication of Flammarion's breakthrough 1862 book was his po-faced boss Leverrier, who promptly sacked him following its appearance. It is unlikely the arch-mathematician was jealous of it – just embarrassed.[51]

A man who thought nothing of making hyperbolic public pronouncements along the lines of 'May there not exist between the planetary humanities psychic lives that we do not know of yet? We stand but at the vestibule of knowledge of the universe!'[52] was hardly going to win much favour with a man like Leverrier.

The subconscious mind of a person today, wishing to construct a plausible secondary personality to come through at a séance, would not construct some silly narrative about reincarnation on distant planets, as the idea is no longer part of our modern-day culture, and would be immediately laughed at. During the heyday of Camille Flammarion, however, the situation was clearly quite different. He also seems to have helped add some plausibility to the strange confusion which arose at the beginning of the twentieth century over the twin topics of radio waves and psychic powers. In 1907, Flammarion told a reporter from the *New York Times* of his opinion that 'I dare say the Martians tried to communicate with us hundreds of thousands of years ago, when mammoths were roaming around our comparatively youthful planet'. However, with no radio reply from the mammoths being forthcoming, Flammarion guessed that the Martians would have concluded Earth was uninhabited and given up on us.[53] This complete fantasy was reported as if it were fact, not mere speculation, reflecting the regard in which the Great Man was still held at the time. By 1923, however, when the elderly Flammarion had but two years left to live, the very same newspaper was reporting very critically upon his equally unverified assertion that, even though radio communication had failed with the Red Planet, it would surely be possible to use 'psychic waves' to perform the task instead. 'Telepathy will overcome space,' he proclaimed, a statement which led the *NYT* to run an editorial labelling the man 'pathetic' and implying that he had become totally senile.[54] Flammarion had always had his critics – in 1906, a cruel wit from within the *NYT* itself had dubbed him 'Camel Flim-Flammarion', and published an insulting cartoon of him smoking an opium pipe and hallucinating the appearance of swimmers in the canals of Mars[55] – but as time went on he began to be regarded more and more as a kind of gigantic crank. If crank he was, though, then as this book has conclusively shown, Flammarion was far from alone.

# Conclusion: Owning Your Own Space

Has each astronomer, then, in physical as in moral matters,
a "manner of seeing"?

Camille Flammarion

There are simply no limits to the strange ideas some people are willing to hold about outer space. Take the case of Theodore B. Dufur, an American gentleman who in 1955 made the inventive proposal that, in order to solve the problem of food storage taking up precious cargo weight during future trips to the moon, spaceships themselves should be constructed from edible matter – namely, blocks of frozen margarine. Then, during flight, astronauts could take periodic nibbles from their craft, making it get lighter the further it went, and speeding it up. After lunar touchdown had been achieved, further ships made from things like cheese could then be shot up onto the moon's surface, to keep the explorers well fed during their mission.[1]

Odd though Dufur's idea was, however, perhaps we should not laugh too loudly at it. Whilst self-evidently flawed beyond redemption, it did at least demonstrate a commendable level of original thought upon behalf of its proponent – and without original thought we would get nowhere. Even today, there are some very odd-sounding notions currently being advanced by valiant space scientists who have the guts to face the possibility of being horribly wrong. Take Lisa Randall (b. 1962), a Harvard physicist who has recently proposed, in her 2016 book *Dark Matter and the Dinosaurs*, that there is a gigantic, hitherto unknown 'disc of death', made from a hypothetical dense form of dark matter, running through the Milky Way galaxy. Our solar system orbits around the centre of the Milky Way once every 225 million years or so, but this orbit does not take place upon a wholly level fashion; every 35 million or so years, we oscillate above and below the mid-point of the galactic plane. As our solar system bobs up and down like this, proposes Randall, it crosses the disc of death, altering the dark matter's gravitational pull and causing it to knock objects out from a far-off cluster of thousands of icy rocks circling around beyond Pluto called the Oort Cloud. These icy rocks, released from their usual home, become comets – a particularly large one of which famously crashed into Earth 65 million years ago, leading to the death of the dinosaurs. Randall's theory, then, is that every 35 million years, as we oscillate past the disc of death, there is a greater chance of huge comets crashing into our planet and causing extinction-level events through their impact.[2] Randall is honest enough to admit she is not sure that this *definitely* happened, but is willing nonetheless to put the theory forward and risk looking stupid, an attitude which I find highly commendable.

Another pair of scientists who share this admirably adventurous mindset are Mike Brown (b. 1965) and Konstantin Batygin (b. 1986) of CalTech, who in January 2016 proposed that there may be a secret ninth planet in the solar system, dubbed simply 'Planet Nine'. Of course, there used to be nine planets in the standard model of the solar system anyway, but in 2006 the outermost one, Pluto, was downgraded to the status of a mere 'dwarf-planet' thanks largely to the work of none other than ... the very same Mike Brown of CalTech! Maybe with this latest proposal Brown wishes to make up for his earlier act of planet-murder, but whatever his motives, his idea is a striking one. Up to ten times the size of Earth, Planet Nine is said to operate upon an orbit so huge that it would take 10,000 years to circle the sun, as opposed to Earth's 365 days, and is guessed to be a kind of 'ice-giant' which is permanently frozen due to its distance from our central star. Brown says he has the data to prove Planet Nine's existence, but has yet to actually observe it through a telescope.[3] Maybe one day he will see it, and go down in the annals of science as a truly great man. Or, on the other hand, maybe he won't. It is a similarly challenging calculus of risk v reward being run by Lisa Randall with her daring idea about the dinosaurs.

## Journey to the Centre of the Self

Brown, Batygin and Randall are all professional scientists who, whatever their own levels of personal ambition, must presumably have as their main motivation the admirable goal of expanding the treasury of human knowledge. However, the idea I have been pushing throughout this book is that a surprisingly large portion of mankind cannot stand the idea of living within the dead universe which science has revealed to us and have attempted to rebel, somehow, against the death of the old geocentric model of the cosmos brought about by the likes of Kepler and Galileo. By whatever means, people left cold by such an idea have tried, again and again, to restore man's central place in the order of things – to make Creation all about them once more by resurrecting Plato's long-dead world animal. Take, for example, the bizarre 2010 story of Radijove Lajic, a fifty-year-old Bosnian man who made Press headlines after claiming to have been targeted for bombardment by aliens after (so he said) five separate meteorites had struck the roof of his home during 2007 and 2008, leading him to have to pay for his roof to be reinforced. Then, in 2010, came the final straw, as another rock from above hit home yet again. The alien attacks, Lajic said, only occurred when it was raining, and his reasoning behind blaming ET was as follows:

> I am obviously being targeted by extraterrestrials. I don't know what I have done to annoy them, but there is no other explanation that makes sense. The chance of being hit by a meteorite is so small that getting hit six times has to be deliberate.[4]

I don't know what the actual explanation for these events was (presumably, the rocks were not really meteorites?) but the personal myth which Lajic then built around it all is one which, yet again, tries to restore mankind to his rightful place at the centre of Creation. Here is an obscure, middle-aged man of no note whatsoever, who has apparently somehow become the target of close, stone-throwing attention for a race of advanced space creatures. Who would ever have thought such amazing beings would have cared a fig for the life of Mr Lajic? As the Biblical Job once discovered, even getting negative attention from a superior celestial being still counts as getting attention, in the end. No matter how ridiculous, in the tale given above a meaningful connection has once more been restored between humanity and the heavens. Romantics, fruitcakes and religious fundamentalists need not have bemoaned the loss of the old Ptolemaic system after all; mankind has still, as a kind of compensatory mechanism, left himself standing tall and proud at the centre of the universe. Probably he always will.

One of the most peculiar attempts to restore mankind to his true cosmic position was made by the Russian astrophysicist Nikolai Kozyrev (1908–83). In his youth Kozyrev had made many useful discoveries, having a crater on the moon named after him, but in 1936 he was denounced for harbouring counter-revolutionary tendencies and sentenced to a decade in the Soviet gulag. His imprisonment turned out to be either the making of him, or an ordeal of such horror that it caused Kozyrev to lose his mind, depending on your viewpoint. Several of Kozyrev's fellow inmates were Siberian shamans, imprisoned for failure to absorb the Communist creed, and as he mixed with them, the great scientist slowly began to adopt aspects of their mindset. Observing the night sky from the camp, he came to feel that the entire universe was alive and that the stars and planets were engaged in a constant bout of telepathic communication with one another – and with him. He resurrected the old concept of 'ether', a mysterious 'fluid' (for want of a better term) permeating the cosmos and connecting all things, thus allowing for telepathic contact to occur between, for instance, the moon and himself. Once upon a time, all mankind had such abilities, but they had since died out, leaving only shamans and gifted mediums able to talk to the stars, said Kozyrev. In the early 1990s an organisation called ISRICA (Institute for Scientific Research in Cosmic Anthropoecology) was established in Siberia by modern-day disciples of Kozyrev, to develop means of talking to the celestial bodies more easily. As part of their researches, ISRICA claim to have discovered that, by shielding a person from the surrounding natural magnetic field of the Earth by placing them within a specially lined cylinder they term a 'Kozyrev Mirror', it is possible to restore direct radio contact between macrocosm and microcosm. Here, a subject is kept in a state of sensory deprivation and told to meditate, until hallucinations occur. These hallucinations, say ISRICA, are probably messages from distant living planets and galaxies. According to one subject who agreed to lie within the device, the experience left her 'not knowing where my body finished and the world outside began'.[5] No doubt Charles Fourier or Edgar Mitchell would have loved to have spent some time within such a contraption. But, if they had done so, then what would it have been that they had actually experienced? The answer, I would suggest, is some inner aspect of themselves, being reflected back to them. Maybe that's why the machine is called a Kozyrev *Mirror*?

## The Laws of the Universe

There are many ways in which we can claim our own ownership over the cosmos these days. In 1967, disturbed by the prospect of colonisation of the moon by the Superpowers during the Space Race, a piece of legislation called the Outer Space Treaty was signed into global law by the UN. Article II of this document makes clear that it is not legal to claim ownership over any celestial body – but clearly this does not apply to the concept of mental ownership of such things, which is up for grabs to anybody. In any case, even the concept of banning claims of physical ownership over bodies in space is now coming under increasing dispute. Since 2010, an English GP named Philip Davies has been spending his nights standing somewhere in a field in Hampshire, pointing a telescope at Mars and using an attached laser to beam 'a few quadrillion' light particles at its surface, something which he believes now entitles him to ownership of the Red Planet. Davies' aim is not actually to colonise Mars, but to force changes to be made to the 1967 Treaty, which he views as increasingly out of date. Amongst other problems, the US never fully ratified the clause stating that private individuals, as opposed to nations, were banned from owning property on other planets or moons. At the same time, international law allows individuals to claim ownership of non-titled barren land without even setting foot on it, provided they can demonstrate they will make improvements. By firing quadrillions of photons at Mars from his laser, Davies reckons that by now he will have caused a few molecules of $CO_2$ to have been liberated from the Martian soil, thus making the planet's atmosphere marginally more habitable for humans, and so qualifying for ownership of the planet. He has no expectation his claim will succeed, but hopes to take his case to an international tribunal, prompting legislators into action; Silicon Valley entrepreneurs like Elon Musk are already making plans to land machines on asteroids to begin mining them for valuable minerals, and by not acting now, says Davies, we risk an interplanetary free-for-all.[6]

With such potential legislative flaws in mind, there have been a number of recent cases of crafty individuals making attempts to claim ownership of some of the best-known objects of our solar system. In 1997, for example, three men from Yemen filed a lawsuit against NASA, accusing them of trespassing upon their alleged ancestral home-planet of Mars. NASA's unmanned *Pathfinder* craft had landed on Mars on 4 July that year and released a space buggy called *Sojourner* which began roaming the landscape searching out data, 'without our approval', said the three men, who claimed to have 'inherited the planet from our ancestors 3,000 years ago'. When threatened with arrest by Yemeni authorities, the three men quickly withdrew their claim.[7] More persistent in her quest was a fifty-five-year-old Spanish housewife named Maria Angeles Duran Lopez, who in 2010 laid claim to sole ownership of the sun before attempting to sell it off piecemeal on eBay, charging one Euro per square metre. However, the scheme fell through when eBay found Lopez had violated their so-called 'intangible goods' policy. Lopez tried to sue eBay because it had taken commission from her sales without passing on the actual payments from her clients, but when one of Lopez's neighbours threatened to sue her too on the grounds that she was now legally liable for all the damage that sunlight had caused to his skin and eyes down the years, it seemed as if the litany of litigation was about to get out of hand.[8]

Lopez got her idea from the most famous such scheme of all, devised by the US entrepreneur Dennis Hope, of California, who in 1980 claimed sole ownership of the moon and began selling it off in plots at his local bar for $20 an acre ($25 including mineral rights). By 1997, Hope had proclaimed himself 'Master of the Solar System', giving him ownership rights to every object bar Earth which lay within the sun's orbit – even Planet Nine, had he known of it. Hope wrote 'official' letters to the White House, Kremlin and UN staking his claim, and said that the lack of response indicated he had tacit international approval for his move. Land on the moon was still his best-seller, however, so much so that he was now able to earn an income of $4,000 per month by selling it off, so he said, and employed a dedicated three-man real estate sales team, dubbed the 'Lunar Ambassadors', to drum up sales. One unnamed Californian local even went so far as to pay out $42,500 to Hope, buying enough lunar land to set up his own personal 51st State of the USA there, called 'Lunafornia', complete with its own legislative body (recognised only by him, presumably).[9]

## The Mirror Crack'd

According to C. S. Lewis, each age gets the mental picture or model of the cosmos that it deserves. In *The Discarded Image*, after discussing the vanished medieval picture of Creation which we examined earlier on in this book, Lewis asks us not to dismiss it in its entirety simply because it happened to be untrue. Instead, he says, we should cherish it as a valuable insight into the disappeared mind of former men now long-dead, as we shall all one day be in our turn, too:

> We are all, very properly, familiar with the idea that in every age the human mind is deeply influenced by the prevailing model of the universe. But there is a two-way traffic; the model is also influenced by the prevailing temper of mind. We must recognise that what has been called "a taste in universes" is not only pardonable but inevitable. We can no longer dismiss the change of models as a simple progress from error to truth. No model is a catalogue of ultimate realities, and none is a mere fantasy. Each is a serious attempt to get in all the phenomena known at a given period, and each succeeds in getting in a great many. But also, no less surely, each reflects the prevalent psychology of an age as much as it reflects the state of that age's knowledge ... It is not impossible that our own model will die a violent death, ruthlessly smashed by an unprovoked assault of new facts ... But I think it is more likely to change when, and because, far-reaching changes in the mental temper of our descendants demand that it should. The new model will not be set up without evidence, but the evidence will turn up when the inner need for it becomes sufficiently great. It will be true evidence. But Nature gives most of her evidence in answer to the questions we ask her.[10]

Given these conclusions, what does it say about our own contemporary society that the entire solar system is now apparently becoming viewed as little more than an avenue for interminable legal disputes over issues of ownership and an infinitely large vehicle for making gigantic financial profits? The question our own civilisation now seems set to ask of the universe, it would appear, is '*cui bono?*' If Creation really is a gigantic mirror, like that hallucination-inducing tool named after Nikolai Kozyrev, then it reflects all our flaws as well as our qualities to us. If what we see staring back whenever we look into it is some form of Caliban disguised as a Grey, a cosmos filled with universal deadness, universal blackness, universal ice, universal dialectic or the universal opportunity to turn a quick profit, then we really have only ourselves to blame for the fact. As the great Irish poet W. B. Yeats (1865–1939) once put it, the cosmos and man alike are both little more than 'mirror on mirror mirrored'.

*PER ASPERA, AD ASTRA*

# Notes

NOTE: *All websites accessed between March 2016 and December 2016.*

## Introduction: Black Mirror, White Ice

1. Moore, 1972, pp.49–56
2. http://www.universetoday. com/8298/oops-the-universe-is-beige/#; http://www.telegraph. co.uk/news/science/space/6487622/ The-universe-is-beige.html; https://en.wikipedia.org/wiki/ Cosmic_latte; http://mentalfloss. com/uk/space/26659/why-is-it-so-dark-in-outer space; http:// www.universetoday.com/117284/ why-is-space-black-2/#; https:// en.wikipedia.org/wiki/Olbers%27_ paradox
3. Clegg, 2014, pp.41–5; Herrick, 2008, pp.233–7; Gardell, 1996, pp.144–6
4. Gardell, 1996, p.144
5. Clegg, 2014, pp.41–5; Herrick, 2008, pp.233–7; Gardell, 1996, pp.144–6
6. Gardell, 1996, pp.147–51
7. Gardell, 1996, pp.147–51
8. http://thetimetranscripts.blogspot. co.uk/2014/02/part-55.html
9. Buck, 2015, pp.226–32
10. http://www.zamanipress.com/page6. html
11. http://www.zamanipress.com/page2. html
12. Cited in Kossy, 2001, pp.111–12
13. Cited in Kossy, 2001, p.111
14. http://1stmuse.com/maria_orsitsch; http://www.bibliotecapleyades.net/ ciencia/ciencia_flyingobjects55.htm
15. Theo Paijmans, 'The Vril Seekers', *Fortean Times* 303, pp.42–6
16. Goodrick-Clarke, 2002, pp.164–70
17. Goodrick-Clarke, 2009, pp.201–2, 212–25
18. Goodrick-Clarke, 2009, pp.219–20
19. Pauwels & Bergier, 2001, pp.173–4
20. Tucker, 2016, p.56
21. Information about Hörbiger complied inextricably from Willy Ley, 'Pseudoscience in Nazi-Land', *Astounding Science Fiction* May 1947, online at http:// www.alpenfestung.com/ley_

pseudoscience.htm; Gardner, 1957, pp.37–41; Moore, 1972, pp.66–72; Goodrick-Clarke, 2002, pp.131–4; Wilson, 2006, pp.202–6; Pauwels & Bergier, 2001, pp.145–65; Essers, 1976, pp.26–7, 220; http://www.mpiwg-berlin.mpg.de/en/research/projects/deptiii-christinawessely-welteislehre; http://en.wikipedia.org/wiki/Hanns_H%C3%B6rbiger; http://en.wikipedia.org/wiki/Welteislehre

22. Cited in Gardner, 1957, p.39
23. Gardner, 1957, pp.38–40
24. Cited in Pauwels & Bergier, 2001, pp.164–5; Brugg may have been Bellamy, writing under a pseudonym
25. *Daily Mail*, 11 September 2015, p.31

# 1 Sphere of Fear: Sputnik Spooks the World

1. Dickson, 2001, p.249
2. Dickson, 2001, pp.128–9
3. Dickson, 2001, pp.113–14, 164–5
4. Kris Hollington, 'Lost in Space', *Fortean Times* 233, pp.32–8
5. *The Spokesman-Review* 16 October 1957; Dickson, 2001, p.115
6. Dickson, 2001, p.126
7. Dickson, 2001, p.117
8. *Pittsburgh Post-Gazette*, 1 November 1957
9. Sagan, 1996, pp.284-9
10. http://security.blogs.cnn.com/2012/11/28/u-s-had-plans-to-nuke-the-moon/; https://en.wikipedia.org/wiki/Project_A119
11. Moore, 1972, pp.155–6
12. Oliver, 2013, pp.25, 56; Dickson, 2001, p.115
13. http://www.oxfordamerican.org/magazine/item/719-prayers-for-richard; https://www.theguardian.com/music/2010/nov/28/john-waters-met-little-richard
14. Sagan, 1996, p.72
15. Dickson, 2001, p.166
16. Clarke & Roberts, 2007, p.104
17. Bowen, 1977, pp.200–38
18. Clarke & Roberts, 2007, pp.95–105
19. Keel, 1976, pp.210–11
20. McGovern & Rickard, 2007, pp.41–2; Clarke & Roberts, 2007, pp.4–5
21. https://illuminutti.com/ancient-aliens-debunked/ancient-aliens-debunked-part-3/
22. See http://www.crystalinks.com/ufohistory.html for a good gallery of such images, and http://ancientaliensdebunked.com/references-and-transcripts/ufos-in-ancient-art/ for a good debunking.
23. http://www.express.co.uk/news/weird/674711/David-Icke-claims-the-moon-is-HOLLOW-and-built-by-ALIENS-in-crazy-new-theory
24. Oliver, 2013, p.48
25. http://www.jasoncolavito.com/blog/spaceship-moon-and-soviet-scientific-politics; http://www.jasoncolavito.com/blog/the-soviet-search-for-ancient-astronauts

# 2 Flat-Out Lies: Flat Earths and Unvisited Moons

1. *The Dispatch* (Lexington, North Carolina), 21 Oct 1957
2. Garwood, 2007, p.226
3. *Sydney Morning Herald*, 26 August 1962
4. Garwood, 2007, pp.243, 245–6, 266
5. Garwood, 2007, p.231
6. *Eugene Register-Guard*, 7 June 1960

7. *Sydney Morning Herald*, 26 August 1962
8. Garwood, 2007, pp.224–5, 277
9. Moore, 1972, pp.27–30; *Eugene Register-Guard*, 7 June 1960
10. Garwood, 2007, pp.220–2
11. Michell, 1999, p.23
12. Garwood, 2007, p.391
13. Shenton, 2007, p.223
14. Garwood, 2007, pp.255, 257
15. Pamphlet online at http://library. tfes.org/library/Plane%20Truth%20 1%20(web).pdf
16. Garwood, 2007, p.236
17. *The Spokesman-Review*, 17 August 1967
18. Garwood, 2007, p.228
19. Moore, 1972, p.28
20. *Sarasota Journal*, 3 January 1969
21. *Birmingham Evening Echo*, 17 April 1969
22. Garwood, 2007, p.230
23. *The Spokesman-Review*, 17 August 1967; Moore, 1972, p.26
24. Garwood, 2007, pp.254–5
25. *Birmingham Evening Echo*, 17 April 1969
26. Garwood, 2007, pp.342–3; https:// en.wikipedia.org/wiki/Modern_Flat_ Earth_societies
27. David Percy, 'Dark Side of the Moon Landings', *Fortean Times* 94, pp.34–9
28. Garwood, 2007, pp.347–8; http:// library.tfes.org/library/footballer_ arrested_to_hide_fake_moonwalk. html
29. http://news.bbc.co.uk/1/hi/world/ Americas/2272321.stm
30. Moore, 1972, pp.39–42
31. Garwood, 2007, pp.77–8; https:// en.wikipedia.org/wiki/Modern_Flat_ Earth_societies
32. Cited in Kossy, 1994, pp.77–8
33. Garwood, 2007, pp.341–6; Michell, 1999, pp.31–2
34. Kossy, 1994, p.78; Garwood, 2007, p.346
35. Garwood, 2007, p.268
36. Garwood, 2007, p.228
37. Henriet, 1958, pp.2, 4–5, 10–11, 12–13, 15, 19, 37
38. Henriet, 1958, p.24
39. Henriet, pp.25, 33
40. Henriet, 1958, p.28
41. Henriet, 1958, pp.64–5
42. Henriet, 1958, pp.63, 67, 68
43. Koestler, 2014, pp.3–4
44. Crowe, 1999, pp.115–16
45. Henriet, 1958, p.23
46. Henriet, 1958, p.26
47. Henriet, 1958, p.27
48. Henriet, 1958, pp.33, 35, 36
49. Henriet 1958, pp.40–1
50. Henriet, 1958, pp.49, 69
51. Henriet, 1958, pp.56–9
52. Moore, 1972, p.19
53. *Lodi News-Sentinel*, 12 July 1969
54. Garwood, 2007, pp.272–9

## 3 Diana's Dust: The Accidental Murder of a Moon-Goddess

1. Wilson, 2004, p.194
2. Crowe, 1999, p.18
3. Brueton, 1991, p.36
4. Naylor, 2015, pp.20–1
5. Roud, 2006, pp.317–19
6. Brueton, 1991, pp.56–65
7. Graves, 1999, p.10
8. Graves, 1999, p.20
9. Oliver, 2013, p.194
10. Nicolson, 1960, pp.14–15
11. Rabkin, 2005, pp.55–7
12. Nicolson, 1960, pp.71–85
13. Nicolson, 1960, pp.93–4
14. Nicolson, 1960, pp.96–8
15. Nicolson, 1960, 121

16. Nicolson, 1960, pp.210–14
17. https://en.wikipedia.org/wiki/Space_advertising
18. http://www.universetoday.com/35725/company-looks-to-etch-advertising-on-the-moon/
19. http://www.space.com/30333-moon-billboard-pocari-sweat-sports-drink.html; https://en.wikipedia.org/wiki/Pocari_Sweat; https://www.otsuka.co.jp/en/company/release/2014/0515-01.html
20. Lewis, 2010, pp.92–3
21. Lewis, 2010, pp.3–4; Russell, 2006, p.198
22. Woolley, 2002, pp.147–58; Koestler, 2014, pp.259–62
23. Koestler, 2014, p.261
24. Koestler, 2014, p.7
25. Lewis, 2010, pp.96–7
26. Lewis, 2010, pp.111–12
27. Lewis, 2010, p.112
28. Koestler, 2014, p.15
29. Lewis, 2010, p.113
30. Lewis, 2010, pp.114–15
31. Russell, 2006, pp.196–8
32. Russell, 2006, p.69
33. Russell, 2006, pp.205–6
34. Russell, 2006, p.144
35. http://www.iep.utm.edu/platoorg/
36. Russell, 2006, pp.492–5
37. Russell, 2006, pp.484–5
38. Russell, pp.487–9
39. Koestler, 2014, pp.358–63
40. Russell, 2006, pp.491–2
41. Crowe, 1999, p.37
42. Crowe, 1999, p.104
43. Noble, 1999, pp.115–16
44. http://hoaxes.org/archive/permalink/the_holy_foreskin; https://en.wikipedia.org/wiki/Leo_Allatius
45. Smith, 2014, pp.107–8
46. Smith, 2014, pp.194–7; Young, 2012, pp.163–70
47. Rosenthal, 1997, pp.196–8; Young, 2012, pp.152–3; Smith, 2014, pp.164–5, 206; http://tsiolkovsky.org/en/the-cosmic-philosophy/citizens-of-the-universe-1933/; http://tsiolkovsky.org/en/the-cosmic-philosophy/creatures-higher-than-a-man-1939/; http://tsiolkovsky.org/en/the-cosmic-philosophy/creatures-from-different-stages-of-evolution/
48. Rosenthal, 1997, pp.196–7
49. Smith, 2014, pp.222–3
50. Bainbridge, 2015, p.127; Smith, 2014, pp.229–30
51. Rosenthal, 1997, p.198
52. Essers, 1976, p.iv
53. Essers, 1976, pp.28, 32
54. Essers, 1976, pp.43–5
55. Essers, 1976, pp.223–6
56. Smith, 2014, pp.30–1
57. Noble, 1999, pp.124, 126
58. Cited in Bennett, 2008, p.147
59. Cited in Noble, 1999, p.126
60. Noble, 1999, pp.125–6
61. Cited in Pynchon, 2000, p.1
62. Cited in Noble, 1999, p.128
63. Noble, 1999, p.123
64. Smith, 2014, p.351

## 4 Heavenly Bodies: NASA, Alien Sex and Marrying the Universe

1. Oliver, 2013, p.108
2. Oliver, 2013, pp.41; Noble, 1999, p.139
3. Oliver, 2013, pp.11, 18
4. Oliver, 2013, p.113
5. Oliver, 2013, p.127
6. Oliver, 2013, pp.106–7, 120; Noble, 1999, pp.140–1
7. *Daily Mail*, 22 February 2016, p.15
8. Jenny Randles, 'UFO Casebook', *Fortean Times* 345, p.41

9. Oliver, 2013, pp.28–9, 38–9, 98
10. Oliver, 2013, p.138
11. Noble, 1999, pp.134–5; Oliver, 2013, pp.146–53, 158–9
12. Noble, 1999, p.129
13. Noble, 1999, pp.133–4
14. Oliver, 2013, p.29
15. Noble, 1999, p.131
16. Noble, 1999, p.136
17. Oliver, 2013, pp.109–10
18. Charlotte E. Bennardo, 'Thoughts in Space', *FATE* Jan-Feb 2011, pp.60–4
19. Oliver, 2013, pp.104, 117
20. Oliver, 2013, p.110
21. *The Times*, 8 February 2016, p.47
22. Oliver, 2013, pp.116–18
23. Puharich, 1974, pp.39–40
24. Puharich, 1974, p.207
25. Puharich, 1974, p.177
26. *The Times*, 8 February 2016, p.47; *Fortean Times* 339, p.24
27. Cited in Evans, 1987, pp.146–7
28. Redfern, 2010, pp.47–55; Evans, 1987, pp.145–7; Devereux & Brookesmith, 1997, p.30
29. Keel, 1976, pp.203–8; Devereux & Brookesmith, 1997, pp.31–2; Redfern, 2010, pp.113–15; Evans, 1987, pp.142–3
30. Evans, 1987, pp.151–3, Devereux & Brookesmith, 1997, pp.30–1; http://laurenbeukes.bookslive.co.za/blog/2010/02/19/the-woman-who-loved-an-alien/
31. http://laurenbeukes.bookslive.co.za/blog/2010/02/19/the-woman-who-loved-an-alien/
32. Klarer, 2008, Ch5
33. Thurston, 2013, pp.123–33
34. Barker, 1997, pp.177–89
35. Cited in 'Alien Sex Fiends', *Fortean Times* 339, p.10
36. http://hybridchildrencommunity.com/about/
37. http://hybridchildrencommunity.com/reptilian-alien-hybrid-children/
38. http://hybridchildrencommunity.com/my-reptilian-hybrid-child/
39. http://hybridchildrencommunity.com/how-hybrid-children-are-created-from-orgasmic-ecstasy/
40. http://www.crystalinks.com/aliensex.html; http://www.reptilianagenda.com/exp/e100799a.shtml; http://www.greatdreams.com/reptlan/reps.htm
41. McGovern & Rickard, 2007, pp.45–6; Koestler, 2014, pp.4–5
42. Henriet, 1958, pp.45–7
43. Henriet, 1958, p.72
44. Henriet, 1958, pp.70–1
45. Henriet, 1958, p.75
46. Lewis, 2010, p.109
47. Brueton, 1991, p.114
48. Naylor, 2015, pp.44–5
49. Brueton, 1991, pp.170, 172
50. Brueton, 1991, p.69
51. Naylor, 2015, pp.1–6, 60; Brueton, 1991, pp.66–7
52. Naylor, 2015, pp.185–6; Brueton, 1991, p.67
53. Brueton, 1991, pp.66–7
54. Naylor, 2015, pp.47–9

# 5 Planetary Pareidolia, Part I: Seeing Ourselves Through a Telescope, Via the Moon

1. https://blogs.kcl.ac.uk/kingshistory/2015/05/10/Samuel-butlers-the-elephant-in-the-moon-c-1670/
2. Brueton, 1991, p.82
3. Cited at https://en.wikipedia.org/wiki/Man_in_the_Moon
4. https://en.wikipedia.org/wiki/Man_in_the_Moon
5. Harley, 1885, p.64; https://en.wikipedia.org/wiki/Moon_rabbit;

https://en.wikipedia.org/wiki/Lunar_
pareidolia

6. Naylor, 2015, p.10
7. Harley, 1885, pp.21, 32, 54
8. Plutarch, 1911, pp.18–19
9. Ley, 1964, p.224
10. Sagan, 1996, p.45
11. https://en.wikipedia.org/wiki/Lunar_
pareidolia
12. https://www.youtube.com/
watch?v=9j_F4WJ0UCM
13. Moore, 1972, pp.122–3
14. Ivan T. Sanderson, 'Mysterious
Monuments on the Moon', *Argosy*
Vol.371, No.2, August 1970; online
at http://astrosurf.com/lunascan/
argosy_cuspids.htm
15. Sagan, 1996, p.50
16. Naylor, 2015, pp.18–19
17. Leonard, 1977, pp.46, 197, 204–5
18. Leonard, 1977, pp.13, 214–15
19. Leonard, 1977, pp.41–2
20. Leonard, 1977, pp.49–53
21. Leonard, 1977, pp.56–7
22. Leonard, 1977, pp.65–70
23. Leonard, pp.160–7
24. Leonard, 1977, pp.61–2
25. Leonard, 1977, p.169
26. Leonard, 1977, pp.184–5
27. Leonard, 1977, p.101
28. Leonard, 1977, p.181
29. Leonard, 1977, pp.175, 204
30. Leonard, 1977, p.136
31. Leonard, 1977, pp.124–5
32. Leonard, 1977, p.175
33. Leonard, 1977, p.208
34. Leonard, 1977, p.212
35. Crowe, 1999, pp.70–1; Goodman,
2008, p.186
36. Crowe, 1999, pp.63–5
37. Crossley, 2011, p.30
38. Crowe, 1999, p.219
39. Crowe, 1999, p.221
40. All extracts taken from http://
hoaxes.org/text/display/the_great_
moon_hoax_of_1835_text/
41. Crowe, 1999, p.213
42. Crowe, 1999, pp.214–15
43. Crowe, 1999, p.199
44. Crowe, 1999, p.198; Goodman,
2008, p.278
45. Goodman, 2008, pp.276–80
46. Goodman, 2008, pp.224–7

# 6 Planetary Pareidolia, Part II: Seeing Ourselves Through a Telescope, Via Mars

1. http://www.tampabayskeptics.org/
Mars_morefaces.html
2. http://www.ibtimes.co.uk/
bizarre-things-spotted-mars-ball-
traffic-lights-faces-finger-huge-
penis-1467342
3. Sagan, 1996, p.50
4. http://news.bbc.co.uk/1/hi/
sci/tech/8594101.stm; http://
www.bbc.co.uk/news/science-
environment-20510109
5. http://exopolitics.org/european-
commission-president-says-he-
spoke-to-leaders-of-other-planets-
about-brexit/; http://www.mirror.
co.uk/news/uk-news/nigel-farage-
clashes-eu-chief-8299792
6. http://www.badastronomy.com/bad/
misc/hoagland/bunny.html
7. http://www.badastronomy.com/bad/
misc/Hoagland/glassworm.html;
Bob Rickard, 'Mars Special: Water,
Water, Everywhere', *Fortean Times*
139, pp.24–5
8. Rabkin, 2005, p.168
9. McDaniel & Paxton, 1998, pp.4–5
10. McDaniel & Paxton, 1998, pp.40–4

11. https://en.wikipedia.org/wiki/ Richard_C._Hoagland
12. http://www.enterprisemission.com/ Corbett.htm
13. Bob Rickard, 'Alternative 3', *Fortean Times* 64, pp.47–9; McGovern & Rickard, 2007, p.21
14. http://www.enterprisemission.com/ catbox.htm
15. http://www.enterprisemission.com/ corbett.htm
16. Crossley, 2011, p.84
17. Ley, 1964, pp.302–3; Crowe, p.525
18. Crowe, 1999, p.534
19. Rabkin, 2005, pp.83–5; Crowe, 1999, pp.485–6, 489
20. Crossley, 2011, p.41
21. Cited in Ley, 1964, p.299
22. Lowell, 1895, pp.202–9
23. Rabkin, 2005, p.95
24. Rabkin, 2005, p.95
25. Crossley, 2011, pp.70–1; Rabkin, 2005, p.93
26. Crossley, 2011, pp.73, 82; Crowe, 1999, p.509
27. Sheehan, 1988, p.187
28. Sheehan, 1988, p.189
29. Lowell, 1895, p.75
30. Crossley, 2011, pp.81, 85
31. Crossley, 2011, p.82
32. Gardner, 1957, p.40
33. Crossley, 2011, pp.76–7
34. Crossley, 2011, p.88
35. Crossley, 2011, p.87
36. Sheehan, 1988, p.193; Crowe 1999, pp.514–15
37. Crossley, 2011, p.111; Ley, 1964, p.296
38. Ley, 1964, pp.294–6
39. Ley, 1964, p.303
40. Crossley, 2011, p.263
41. Crowe, 1999, p.529
42. Crowe, 1999, p.393
43. Crossley, 2011, pp.230–1
44. Rabkin, 2005, pp.119–20

# 7 California Dreaming: Eccentric Orbits around the Desert Sun

1. Crossley, 2011, p.129; Sheehan, 1988, p.170
2. Sheehan, 1988, p.178
3. Crossley, 2011, p.75
4. Lowell, 1894, p.1
5. Lowell, 1911, p.17
6. Lowell, 1911, pp.6–7
7. Crossley, 2011, p.75
8. Lowell, 1911, p.8
9. Lowell, 1911, p.9
10. Lowell, 1911, p.382
11. Sheehan, 1988, p.169
12. Moore, 1972, p.14
13. Moore, 1972, pp.9–15
14. Tucker, 2015, pp.12–13
15. Sagan, 1997, pp.174–7; http:// harpers.org/archive/1954/12/the-jet-propelled-couch/
16. Denis Stacy, 'An Unsolicited Elf' in *Fortean Times* 64, p.56; it is possible that Hale's encounter was a misinterpretation of a figure of speech; see http://articles.adsabs.harvard.edu// full/2000JHA....31...93S/0000106.000. html
17. Ball, 2011, pp.147–51
18. Carter, 2004, pp.8–9
19. Carter, 2004, p.12
20. Carter, 2004, pp.4, 9, 119–53
21. Carter, 2004, pp.123, 126–7
22. Carter, 2004, pp.150–1
23. Carter, 2004, p.130
24. Carter, 2004, p.184
25. Carter, 2004, pp.177–84
26. Carter, 2004, p.192
27. Carter, 2004, pp.84–8

28. Carter, 2004, p.132
29. Carter, 2004, p.135
30. Colin Bennett, 'Rocket in His Pocket', *Fortean Times* 132, p.38
31. McGovern & Rickard, 2207, pp.489–90
32. http://www.bibliotecapleyades.net/ciencia/ciencia_flyingobjects55.htm; Goodrick-Clarke, 2002, pp.155, 167
33. Keel, 1991, p.160
34. Clarke & Roberts, 2007, pp.41–3
35. Colin Bennett 'Stranger Than Fiction', *Fortean Times* 126, pp.28–31
36. Evans & Stacy, 1997, p.67
37. Bennett, 2008, p.23
38. Cited in Evans, 1987, p.132
39. Evans, 1987, p.132, Redfern, 2010, pp.26, 34; Clarke & Roberts, 2007, pp.42–3; Bennett, 2008, pp.23–8
40. Redfern, 2010, p.30
41. Bennett, 2008, p.188
42. Evans, 1987, p.134
43. Redfern, 2010, pp.26–7; Evans, 1987, p.133; Bennett, 2008, pp.33–6
44. Evans & Stacy, 1997, p.56
45. Redfern, 2010, p.65
46. Compiled from Greg Bishop, 'Calling Occupants', *Fortean Times* 118, pp.28–30; Redfern 2010, p.69–83

# 8 Desert Warfare: The Contactees Conquer the World

1. Rabkin, 2005, pp.157–9
2. *The Times*, 4 June 2016
3. *The Times*, 6 May 2016
4. *The Times*, 4 June 2016 & 28 September 2016; *The Sunday Times*, 2 October 2016
5. *The Times*, 4 June 2016
6. *The Sunday Times*, 2 October 2016
7. Keel, 1991, pp.207–8
8. Keel, 1976, p.203
9. Evans & Stacy, 1997, pp.57, 74–5
10. Keel, 1991, pp.50–7, 63–4, 84–5, 112–13, 151, 271; Clark, 2010, pp.162–5
11. Redfern, 2010, pp.119–20; Keel, 1976, pp.209–10
12. Greg Bishop, 'Calling Occupants', *Fortean Times* 118, p.31
13. Devereux & Brookesmith, 1997, p.31
14. Evans, 1987, p.148
15. https://en.wikipedia.org/wiki/Buck_Nelson
16. Nelson, 1956, p.28
17. Nelson, 1956, pp.11–12
18. Nelson, 1956, pp.12–13, 29
19. Nelson, 1956, pp.14–16
20. Nelson, 1956, p.18
21. Nelson, 1956, p.19
22. Nelson, 1956, pp.20–1
23. Nelson, 1956, pp.21–3
24. Nelson, 1956, p.24
25. Nelson, 1956, p.28
26. Nelson, 1956, p.17
27. Nick Redfern, 'A Town Like Aztec', *Fortean Times* 181, p.35
28. 'The Pixels of Roswell' by Colin Bennett in Moore, 1999, pp.76–7
29. Bennett in Moore, 1999, p.80
30. Cited in Redfern, 2010, pp.222–3
31. Bennett, 2008, p.76
32. Bennett, 2008, p.78
33. Bennett in Moore, 1999, pp.100–1
34. Redfern, 2010, pp.30, 43
35. Clark, 2010, p.160
36. Evans & Stacy, 1997, p.59; https://en.wikipedia.org/wiki/Clyde_Tombaugh
37. Evans, 1987, p.140
38. Moore, 1972, p.12
39. Clarke & Roberts, 2007, p.73
40. Clarke & Roberts, 2007, pp.79–81
41. Moore, 1972, pp.111–12, 113

42. Moore, 1972, p.117
43. Moore, 1972, pp.119–20
44. Moore, 1972, pp.115–16
45. Evans, 1987, p.139
46. Redfern, 2010, p.86
47. Evans, 1987, pp.139–40; Clarke & Roberts, 2007, pp.73–4
48. Evans, 1987, pp.140–1
49. Redfern, 2010, pp.97–8
50. Redfern, 2010, pp.98–100

## 9 Red Planets: Communism Throughout the Cosmos?

1. Keel, 1991, pp.216–17
2. Bill Chalker, 'The 1954 UFO Invasion of Australia' in Evans & Stacy, 1997, pp.78–84
3. See Barker, 1997
4. Colin Bennett, 'Stranger Than Fiction', *Fortean Times* 126, pp.28–31
5. All of the above compiled from extracts of Stranges' book online at http://www.bibliotecapleyades. net/bb/stranges.htm; http:// prepareforchange.net/2014/02/01/ valiant-thor-thestranger-at-the-pentagon/
6. http://thepromiserevealed.com/ dr-frank-stranges/
7. Crowe, 1999, pp.50–3
8. Crossley, 2011, p.51
9. Bogdanov, 1984, pp.245–6
10. Crossley, 2011, p.38
11. Cited in Crossley, 2011, p.83
12. Crossley, 2011, p.90
13. Wells, 2005, pp.1–2
14. https://en.wikipedia.org/wiki/ Henry_Olerich
15. Olerich, 1893, pp.22–3
16. Crossley, 2011, p.97
17. Olerich, 1893, pp.56, 59
18. Crossley, 2011, pp.47–9
19. Crossley, 2011, pp.90–1
20. Crossley, 2011, pp.91–4; Rabkin, 2005, pp.121–4
21. Crossley, 2011, p.93
22. Bogdanov, 1984, p.60
23. Bogdanov, 1984, pp.51–2
24. Bogdanov, 1984, pp.110, 112
25. Bogdanov, 1984, p.115
26. Bogdanov, 1984, p.116
27. Wells, 2005, p.9

## 10 Engineering the Future: From Soviet Science Fiction to Science Fact?

1. Rosenthal, 1997, pp.194, 237
2. Rosenthal, 1997, p.259
3. Rosenthal, 1997, pp.249–50, 271–2
4. 'Introduction' in Howell, 2015
5. Smith, 2014, pp.6–7
6. Tucker, 2016, pp.255–60
7. Nikolai Fedorov, 'Karazin: Meteorologist or Meteorurge?' in Howell, 2015
8. Tucker, 2016, pp.260–2
9. Cited in Andrews, 2009, p.65
10. Cited in Noble, 1999, p.119
11. Cited in Andrews, 2009, p.69
12. Andrews, 2009, pp.68–78
13. Andrews, 2009, pp.91–6
14. Bogdanov, 1984, p.165
15. Richard Stites, 'Introduction' in Bogdanov, 1984, pp.15–16
16. Tucker, 2015, pp.271–2
17. Rosenthal, 1997, pp.258–9
18. J. Posadas, 'Flying Saucers …'; online at http://quatrieme-internationale-posadiste.org/pdf_texte/EN/JP-Flying-Saucers-JP-final. pdf
19. Compiled from Matt Salusbury, 'Trots in Space', *Fortean Times* 176, pp.40–5; https://waisworld.org/

go.jsp?id=o2a&o=85038; http://
rationalwiki.org/wiki/Posadism

20. J. Posadas, 'Childbearing in Space
    ...'; online at http://quatrieme-
    internationale-posadiste.org/
    inedit_pdf/EN/JP-Childbearing-in-
    space-JP-final.pdf

21. https://waisworld.org/
    go.jsp?id=o2a&o=85038

22. Keel, 1991, p.217

23. All relevant quotes taken from J.
    Posadas, 'Flying Saucers ...'; online
    at http://quatrieme-internationale-
    posadiste.org/pdf_texte/EN/
    JP-Flying-Saucers-JP-final.pdf

24. Matt Salusbury, 'Trots in
    Space', *Fortean Times* 176,
    pp.40–5; https://waisworld.org/
    go.jsp?id=o2a&o=85038

25. http://exonews.org/true-exopolitics-
    space-people-warn-dangers-social-
    divide/

26. I have accessed the full text of
    Beers' book online at http://www.
    galactic.no/rune/iarga.html; there
    are no original page-numbers
    provided, the above material is all
    taken from Ch 1

27. All quotes and information above
    taken from Ch 2

28. All quotes and information above
    taken from Ch 3

29. Ch 5

30. http://exonews.org/true-exopolitics-
    space-people-warn-dangers-social-
    divide/

## 11. Atomic Aliens: The Interplanetary Campaign for Nuclear Disarmament

1. Appleyard, 2006, pp.21–3
2. Nadis, 2005, p.215
3. Cited at http://exonews.org/true-
   exopolitics-space-people-warn-
   dangers-social-divide/
4. Cited in Bennett, 2008, p.98
5. Cited in Redfern, 2010, p.42
6. Redfern, 2010, pp.176–7
7. Redfern, 2010, pp.34–5
8. Redfern, 2010, p.76
9. Redfern, 2010, pp.77–9
10. Redfern, 2010, p.81
11. Clarke & Roberts, 2007, p.73
12. Moore, 1972, pp.109–10
13. Devereux & Brookesmith, 1997,
    pp.78–9
14. Cited in Clarke & Roberts, 2007, p.82
15. Clarke & Roberts, 2007, pp.81–5
16. Evans, 1987, pp.143–4
17. Evans, 1987, pp.143–5; Redfern,
    2010, pp.18–22
18. Evans, 1987, pp.144; Redfern,
    2010, pp.21–2
19. Carter, 2004, pp.72–3
20. https://en.wikipedia.org/wiki/Clyde_
    Tombaugh
21. Carter, 2004, p.75
22. Evans & Stacy, 1997, p.72
23. Cited in Evans & Stacy, 1997, p.73
24. Appleyard, 2006, p.24
25. Appleyard, 2006, p.29
26. Clarke & Roberts, 2007, pp.23–4
27. Appleyard, 2006, p.30
28. Redfern, 2010, p.137
29. Appleyard, 2006, p.32
30. Robert Bartholomew *et al*, 'The
    Swedish Ghost-Rocket Delusion of
    1946' in Moore, 1999, pp.64–74;
    Evans & Bartholomew, 2009,
    pp.208–11
31. Jung, 1978, pp.22–3
32. Jung, 1978, pp.108–9
33. Jung, 1978, p.111
34. Jung, 1978, p.21
35. Jung, 1978, pp.19–20, 117–18
36. Jung, 1978, p.118

37. Jung, 1978, p.116
38. Redfern, 2010, p.59
39. Evans, 1987, p.137; Jung, 1978, p.117
40. Cited in Evans, 1987, pp.137–8
41. Jung, 1978, p.117
42. Jung, 1978, pp.113, 115

## 12. From Sputniks to Beatniks: Getting Out of This World by Whatever Means

1. Devereux & Brookesmith, 1997, p.79; Moore, 1972, p.112
2. Dickson, 2007, p.251
3. https://bigjelly.net/ten-amazing-facts-about-william-burroughs-e97aa07d2c57
4. http://www.abbesville.com/landing/index.asp
5. Cited at http://prepareforchange.net/2014/02/01/valiant-thor-the-stranger-at-the-pentagon
6. Devereux & Brookesmith, 1997, p.30; Jung, 1978, p.115
7. Redfern, 2010, pp.59–61
8. Redfern, 2010, pp.65–7
9. Cited in Redfern, 2010, pp.184–5
10. Jung, 1978, p.116
11. Huxley, 2004, pp.xii–xvi
12. Huxley, 2004, p.16
13. Redfern, 2010, pp.185–6
14. Devereux & Brookesmith, 1997, p.180
15. Cited in Redfern, 2010, p.170
16. McKenna, 1992, pp.258–9
17. Cited in Devereux, 1997, p.181
18. Huxley, 2004, title-page
19. Clarke & Roberts, 2007, p.181
20. Clarke & Roberts, 2007, p.184
21. Keel, 1976, pp.204–5
22. Eros, 2011, pp.40–1
23. https://en.wikipedia.org/wiki/A_Taste_for_Honey
24. Clarke & Roberts, 2007, pp.22–3
25. http://www.geraldheard.com/bio2.htm
26. Heard, 1950, pp.117, 130
27. Clarke & Roberts, 2007, p.23
28. Heard, 1950, p.97
29. Aimé Michel, 'Do Flying Saucers Originate from Mars?', *Flying Saucer Review* Vol 6, No.2, pp.13–15
30. Heard, 1950, p.117
31. Evans & Stacy, 1997, p.75; Heard, 1950, p.104
32. Heard, 1950, p.120
33. Heard, 1950, pp.121–2
34. Heard, 1950, p.150
35. Heard, 1950, p.150
36. 'Archive Gem', *Fortean Times* 102, p.19; Heard, 1950, p.126
37. Heard, 1950, pp.127–9
38. Heard, 1950, pp.124–5, 126–7
39. Heard, 1950, p.98
40. Heard, 1950, p.123
41. Heard, 1950, pp.131–3
42. Heard, 1950, pp.134–44
43. Heard, 1950, p.139
44. Heard, 1950, pp.103–10
45. Heard, 1950, pp.113–15
46. Heard, 1950, p.115
47. Heard, 1950, p.116
48. Heard, 1950, p.118
49. Heard, 1950, p.106
50. Heard, 1950, p.107
51. Cited in Wilson, 2004, pp.134, 135
52. Wilson, 2004, pp.106–7
53. http://www.geraldheard.com/bio2.htm; http://www.geraldheard.com/writings.htm; https://en.wikipedia.org/wiki/Gerald_Heard
54. Smith, 2014, p.5
55. Smith, 2014, pp.314–15
56. Smith, 2014, pp.119–20
57. Smith, 2014, p.202
58. Smith, 2014, pp.306–8

59. Eros, 2011, pp.20–5, 38
60. Eros, 2011, p.39
61. Eros, 2011, pp.26–7
62. Cited at http://www.geraldheard.com/writings.htm
63. Eros, 2011, pp.26–7

# 13. Venus Has a Penis: The Coital Cosmology of Charles Fourier

1. Marx & Engels, 1985, pp.114–18; Breton, 1997, p.69
2. Polizzotti, 2009, pp.477-8
3. Hugh Doherty, 'Introduction' in Fourier, 1851, pp.i–v
4. Breton, 1997, pp.76–7, 80
5. Hugh Doherty, 'Introduction' in Fourier, 1851, p.x
6. Charles Fourier, 'On the Role of the Passions', available online at http://www.marxists.org/reference/archive/fourier/works/cho1.htm; http://combinedorder.blogspot.co.uk/search/label/gastronomy%20and%20gastrosophy
7. Fourier's essay is online at http://combinedorder.blogspot.co.uk/search/label/gastronomy%20and%20gastrosophy
8. Beecher, 1992, p.340; Riasanovsky, 1969, pp.91–2; http://roebuckclasses.com/201/reform/utopiansocialists%20fourier%20lecture%2019.htm; https://connerhabib.wordpress.com/2015/08/11/thesexradicals-part-4-charles-fouriers-impossible-pleasure/
9. Hugh Doherty, 'Introduction' in Fourier, 1851, pp.iii–v
10. Beecher, 1992, p.334; Hugh Doherty, 'Introduction' in Fourier, 1851, p.iii
11. Fourier, 1996, p.26; http://combinedorder.blogspot.co.uk/search/label/Treatise%20on%20Domestic-Agricultural%20Association
12. Hugh Doherty, 'Introduction' in Fourier, 1851, p.ii; http://pgoh13.com/paris_4theapple.html; http://roebuckclasses.com/201/reform/utopiansocialists%20fourier%20lecture%2019.htm
13. Fourier, 1996, pp.25, 27, 36, 42
14. Crowe, 1999, pp.251–2
15. Fourier, 1996, p.45
16. Fourier, 1996, p.45
17. Fourier, 1996, pp.30–1
18. Beecher, 1992, pp.338–9
19. Fourier, 1996, pp.45–6; Beecher, 1992, pp.337–9, 351, Riasanovsky, 1969, p.88
20. Fourier, 1996, p.54
21. Riasanovsky, 1969, pp.93–4; Godwin, 1844, pp.92, 96, 98; Hugh Doherty, 'Introduction' in Fourier, 1851, pp.xiv–xviii
22. Fourier, 1996, pp.47–50
23. Riasanovsky, 1969, p.91
24. Fourier, 1996, p.50
25. Fourier, 1996, p.51
26. Fourier, 1996, pp.51–3
27. Fourier, 1996, p.37; Beecher, 1992, p.335
28. Fourier, 1996, pp.54–6
29. Beecher, 1992, p.351
30. Beecher, 1992, p.333
31. Fourier, 1996, p.29
32. Beecher, 1992, p.352
33. Fourier, 'Detail of a Creation of the Hypo-Major Keyboard', reproduced in Breton, 1997, pp.77–9
34. Hugh Doherty, 'Introduction', in Fourier, 1851, pp.xx–xxi; Beecher, 1992, p.350 has other examples of this kind of thinking
35. Beecher, 1992, p.333
36. Cited in Beecher, 1992, pp.340–1

37. Fourier's essay is online at http://combinedorder.blogspot.co.uk/search/label/gastronomy%20and%20gastrosophy
38. Fourier, 1996, pp.37–8
39. Beecher, 1992, pp.342–3
40. Riasanovsky, 1969, p.96
41. Beecher, 1992, p.338
42. Beecher, 1992, p.342
43. Cited in Beecher, 1992, p.345
44. Fourier, 'Elephant and Dog', reproduced in Breton, 1997, pp.81–3; Riasanovsky, 1969, pp.96–7
45. Beecher, 1992, p.344
46. All quotes which follow, and those above, are taken from a translation of this text online at http://combinedorder.blogspot.co.uk/2012/05/cosmogony-i.html; no moons of Uranus are called Hebe and Cleopatra today.
47. Hugh Doherty, 'Introduction' in Fourier, 1851, p.xiii
48. Fourier, 1851, pp.120–3
49. Fourier, 1851, pp.124–8
50. Fourier, 1851, pp.130–1, 134–6
51. Fourier, 1851, pp.150–2
52. Fourier, 1851, p.154
53. Fourier, 1851, pp.153–5
54. Fourier, 1851, pp.156–7

## 14. Turn On, Tune In, Drop Out: Talking with Aliens, Talking with Ourselves

1. Cited in Brooks, 2009, p.99.
2. http://www.telegraph.co.uk/news/worldnews/australiaandthepacific/australia/11582733/Strange-outer-space-signal-that-baffled-Australian-scientists-turns-out-to-be-microwave-oven.html
3. Crowe, 1999, pp.204–8; Ley, 1964, p.502; Florence Raulin-Cerceau, 'The Pioneers of Interplanetary Communication: From Gauss to Tesla' in *Acta Astronautica* issue 67; online at www.academia.edu/12935035/The_pioneers_of_interplanetary_communication_From_Gauss_to_Tesla
4. Crowe, 1999, p.207–8
5. Crowe, 1999, pp.203–4
6. Sconce, 2000, p.96
7. Florence Raulin-Cerceau, *op. cit*; Camille Flammarion, 'Shall We Talk with the Men in the Moon?' in *Review of Reviews* 5, p.90
8. Crowe, 1999, p.394; Florence Raulin-Cerceau, *op. cit.*
9. Florence Raulin-Cerceau, *op. cit.*
10. Crowe, 1999, p.396
11. Tucker, 2015, pp.38–41; http://www.york.ac.uk/depts/maths/histstat/bib/galtonkey.htm
12. Crossley, 2011, p.111; Crowe, 1999, p.395–6
13. Francis Galton, 'Intelligible Signals Between Neighbouring Stars' in *The Fortnightly Review* 60, pp.657–64
14. Crowe, 1999, p.399–400; Florence Raulin-Cerceau, *op. cit.*
15. Brookesmith, 1995, p.39
16. Crosse, 1999, p.397; Florence Raulin-Cerceau, *op. cit.*
17. Crowe, 1999, pp.394–5; Ley, 1964, pp.294–5; Florence Raulin-Cerceau, *op. cit.*
18. See Tucker, 2016, pp.148–205 for my own take upon Tesla
19. Tucker, 2016, pp.201–3
20. Childress, 2009, pp.289–91
21. *Collier's Weekly*, 9 February 1901, pp.4–5
22. David Hambling, 'Science or Noise?', *Fortean Times* 346, p.16
23. *Current Opinion*, 1 March 1919

24. http://www.bibliotecapleyades.net/
   tesla/lostjournals/lostjournals04.htm
25. Crowe, 1999, pp.398–9; Crossley,
   2011, p.171
26. *New York Sun Magazine Section*, 25
   January 1920
27. Crossley, 2011, p.173
28. Cited in Florence Raulin-Cerceau,
   *op. cit*
29. https://en.wikipedia.org/wiki/Prix-
   Guzman
30. Crossley, 2011, pp.11–12
31. Crossley, 2011, pp.138–9
32. Sconce, 2000, pp.92, 101
33. Sconce, 2000, p.92
34. Sconce, 2000, p.100, 102, 106
35. Cited in Crossley, 2011, p.172
36. Crowe, 1999, p.399
37. Muirhead, Reeves-Stevens & Reeves-
   Stevens, 2004, pp.122–3; Crossley,
   2011, pp.172–3; https://en.wikipedia.
   org/wiki/David_Peck_Todd
38. http://www.lettersofnote.
   com/2009/11/prepare-for-contact.html
39. Keel, 1999, p.165
40. Brookesmith, 1995, pp.38–41;
   McGovern & Rickard, 2007, pp.610–
   11; Brooks, 2009, pp.96–7, 103
41. *Times*, 5 July 2016, p.33
42. Brooks, 2009, pp.103–106
43. McGovern & Rickard, 2007,
   pp.610–11
44. Sagan, 1996, p.178; Sleevenotes to
   The Byrds' *Younger Than Yesterday*,
   written by Johnny Rogan; 1996 CD
   re-release version
45. Sagan, 1996, pp.178–9
46. Brooks, 2009, pp.97–103
47. *Times*, 18 April 2016, p.21
48. Florence Raulin-Cerceau, *op. cit.*;
   Crowe, 1999, p.400

## 15. London Calling: Dr Robinson's Interplanetary Telegraph of Love

1. http://artuk.org/discover/
   artworks/dr-h-mansfield-robinson-
   vestry-clerk-of-shoreditch-
   18911899-and-town-clerk-of-
   shoreditch-19001911-134620
2. Fodor, 1959, pp.266–7; Clark,
   2010, pp.153–4
3. https://www.paimages.co.uk/image-
   details/2.7788307
4. Cited in Fodor, 1959, p.265
5. http://mars.nasa.gov/allaboutmars/
   nightsky/opposition/; http://www.
   uapress.arizona.edu/onlinebks/
   MARS/APPENDS.HTM
6. Fodor, 1959, pp.259–62; Clark,
   2010, p.154
7. *St Petersburg Times*, (Florida), 29
   Oct 1926
8. *Milwaukee Sentinel*, 11 Mar 1930
9. *Lewiston Evening Journal*, 28 Oct
   1926; http://londonist.com/2014/04/
   londons-edwardian-seti-programme-
   and-the-girl-from-mars
10. *The Evening Independent* (Florida)
   29 October 1926
11. *Montreal Gazette*, 20 October 1928;
   *San Jose News*, 22 October 1928;
   *Pittsburgh Press*, 24 October 1928;
   *Milwaukee Sentinel*, 25 October
   1928
12. *Eugene Register-Guard*, 22
   October 1928 & 24 October 1928;
   *Glasgow Herald*, 22 October 1928;
   *Pittsburgh Press*, 24 October 1928
13. *Eugene Register-Guard*, 22 October
   1928
14. Bleiler & Bleiler, 1998, pp.264–6;
   http://www.bis-space.com/what-we-
   do/the-british-interplanetary-society/

history/a-m-low; https://en.wikipedia. org/wiki/Archibald_Low

15. *The Times*, 24 October 1928; http:// www.ukweatherworld.co.uk/forum/ index.php?/topic/68543-a-tornado- tours-the-sights-of-london/

16. *Eugene Register-Guard*, 24 October 1928; *Glasgow Herald*, 25 October 1928; *Pittsburgh Press*, 22 & 28 October 1928; Clark, 2010, p.155

17. *Pittsburgh Press*, 24 October 1928; *Ottawa Citizen*, 25 October 1928; *Glasgow Herald*, 25 October 1928

18. http://blog.modernmechanix.com/ mars-refuses-to-answer-radio- messages/

19. *The Southeast Missourian*, 13 November 1928; *The Pittsburgh Press*, 18 December 1928; *The Evening Independent* (Florida), 20 December 1928; *The Pittsburgh Press*, 20 December 1928

20. 'The Goblin' Vol IX No 6, Feb 1929; online at https://archive. org/stream/goblinv9n6toro/ goblinv9n6toro_djvu.txt

21. *Milwaukee Sentinel*, 21 March 1930

22. *Berkeley Daily Gazette*, 30 May 1930; *The Sunday Morning Star*, 1 June 1930

23. *Daily Iowan*, 6 April 1933

24. *DeKalb Daily Chronicle*, 30 Dec 1929

25. *Eugene Register-Guard*, 24 October 1928; *Ottawa Citizen*, 25 October 1928

26. *San Jose News*, 22 October 1928

27. *St Petersburg Times* (Florida), 29 October 1926

28. Fodor, 1959, p.262

29. Fodor, 1959, pp.263–4

30. Smith, 2014, p.173

31. Fodor, 1959, pp.260, 264

32. *Pittsburgh Press*, 24 October 1928

33. Fodor, 1959, pp.267–8; Clark, 2010, pp.155–6; *Pittsburgh Press*, 22 October 1928

34. Cited at http://www. harrypricewebsite.co.uk/Biography/ search-for-truth-chapter10.htm

35. Clark, 2010, p.156

36. Fodor, 1959, pp.267–8

## 16. You Look Familiar: The Alien in the Mirror

1. Nicolson, 1960, pp.41–5; Koestler, 2014, pp.200, 203, 353–7

2. Nicolson, 1960, pp.45–7; Koestler, 2014, pp.384–8

3. Koestler, 2014, p.387

4. Nicolson, 1960, p.57

5. Jerome de la Lande, 'Introduction' in Fontenelle, 1804, p.iv; Crowe, 2011, pp.18–19

6. Fontenelle, 1804, p.xiv

7. Fontenelle, 1804, p.44

8. Fontenelle, 1804, pp.64–5

9. Fontenelle, 1804, p.85

10. Fontenelle, 1804, pp.107–8

11. Fontenelle, 1804, p.102

12. Fontenelle, 1804, pp.xiv–xvi

13. Fontenelle, 1804, pp.50–1

14. Fontenelle, 1804, pp.77–8

15. Fontenelle, 1804, p.39

16. Crowe, 1999, p.30

17. Crowe, 1999, pp.47–53

18. Voltaire, 1829, Ch I

19. Voltaire, 1829, Ch I

20. Voltaire, 1829, Ch II

21. Voltaire, 1829, Ch IV

22. Fontenelle, 1804, p.76

23. Bowen, 1977, p.10

24. Bowen, 1977, p.57

25. Bowen, 1977, p.51

26. Bowen, 1977, p.55

27. Keel, 1976, p.211

28. Bowen, 1977, p.156

29. Bowen, 1977, pp.88–9
30. Bowen, 1977, pp.77–80
31. Bowen, 1977, p.103
32. Bowen, 1977, p.104
33. Bowen, 1977, p.158
34. Bowen, 1977, p.120
35. Bowen, 1977, pp.19–20
36. Bowen, 1977, p.32; Brookesmith, 1995, pp.42–3
37. Randles, 1983, pp.65–7
38. Brookesmith, 1995, pp.34–7
39. McGovern & Rickard, 2007, pp.15–16, 284; Appleyard, 2006, p.66
40. Appleyard, 2006, pp.68–9
41. John Rimmer, 'Evaluating the Abductee Experience' in Evans & Spencer, 1988, pp.167–8
42. Devereux & Brookesmith, 1997, p.184
43. Devereux & Brookesmith, 1997, pp.100–2; Brookesmith, 1995, pp.60–4
44. Crossley, 2011, pp.179–81
45. Sagan, 1996, p.133
46. Huygens, 1698, p.3
47. Huygens, 1698, pp.17–18
48. Huygens, 1698, pp.21–4
49. Huygens, 1698, pp.31–3
50. Huygens, 1698, pp.36–7
51. Huygens, 1698, pp.41–3
52. Huygens, 1698, pp.41–8
53. Huygens, 1698, pp.61-3
54. Huygens, 1698, pp.61–6, 78–80
55. Huygens, 1698, pp.70–3
56. Huygens, 1698, 73–4
57. Huygens, 1698, pp.74-7
58. Huygens, 1698, p.77

## 17. Wandering Stars: Astral Travels

1. Crossley, 2011, pp.133–4
2. Fodor, 1959, p.263
3. Dingwall, 1950, p.29
4. Dingwall, 1950, p.19
5. Dingwall, 1950, p.32
6. Dingwall, 1950, pp.16–17
7. Dingwall, 1950, p.34
8. Swedenborg, 1839, p.10–13
9. Swedenborg, 1839, pp.13–31
10. Swedenborg, 1839, pp.32–40
11. Swedenborg, 1839, pp.38–9
12. Swedenborg, 1839, pp.41–56
13. Swedenborg, 1839, p.62
14. Swedenborg, 1839, p.68
15. Swedenborg, 1839, pp.73–4
16. Swedenborg, 1839, p.83
17. Swedenborg, 1839, p.86
18. Swedenborg, 1839, p.103
19. Swedenborg, 1839, p.92
20. Swedenborg, 1839, p.112
21. Swedenborg, 1839, p.108
22. Swedenborg, 1839, p.90
23. T. Peter Park, 'The Poughkeepsie Seer', online at http://www.anomalist.com/features/seer.html
24. Davis, 1871, p.121
25. Davis, 1871, pp.131–6
26. Davis, 1871, p.154
27. Davis, 1871, p.169
28. Davis, 1871, pp. 189, 196, 207
29. Davis, 1871, p.294
30. Davis, 1871, p.585
31. Davis, 1871, pp.736, 748
32. Davis, 1871, p.586
33. Davis, 1871, p.739–41
34. Davis, 1871, p.777
35. Gareth J. Medway, 'Mediums, Mystics and Martians' in *Magonia* magazine issue 99; online at http://magoniamagazine.blogspot.co.uk/2014/02/martians.html
36. Harris, 1855, pp. vii, xii
37. Harris, 1855, pp.200
38. Harris, 1855, p.27
39. Harris, 1855, p.27
40. Harris, 1855, pp.84–5
41. Harris, 1855, pp.81–2

42. Harris, 1855, p.83
43. Harris, 1855, pp.96–8
44. Harris, 1855, p.179
45. Harris, 1855, pp.188–9
46. Harris, 1855, p.189
47. Cited in Crowe, 1999, p.238
48. Leadbeater, 1922, pp.275–6
49. Leadbeater, 1922, pp.277–8
50. Leadbeater, 1922, p.278
51. Leadbeater, 1922, pp.278–9
52. Leadbeater, 1922, pp.279–80
53. Leadbeater, 1922, p.280
54. Leadbeater, 1922, p.281
55. Leadbeater, 1922, pp.281–3
56. Leadbeater, 1922, p.275
57. Leadbeater, 1922, p.133
58. Leadbeater, 1922, pp.184, 211–23
59. Leadbeater, 1922, p.207
60. Leadbeater, 1922, p.230
61. Leadbeater, 1922, p.235
62. Leadbeater, 1922, 236
63. Leadbeater, 1922, pp.236–7
64. Leadbeater, 1922, pp.237–8
65. Leadbeater, 1922, p.239
66. Denton & Denton, 1863, pp.33–6
67. Denton & Denton, 1863, pp.38–9, 40
68. Denton & Denton, 1863, pp.114–18
69. Denton & Denton, 1863, pp.68–9
70. Denton & Denton, 1863, pp.78–9
71. Denton & Denton, 1863, p.80
72. Denton & Denton, 1863, pp.122–4
73. Denton & Denton, 1863, pp.278–9
74. Denton, 1874, pp.147–8
75. Denton, 1874, p.158
76. Denton, 1874, p.339
77. Denton, 1874, pp.159–60
78. Denton, 1874, pp.161–2
79. Denton, 1874, pp.163–4
80. Denton, 1874, pp.281–2, 291
81. Denton, 1874, pp.268–9, 271, 275, 276
82. Denton, 1874, pp.269–70
83. Denton, 1874, pp.169–70
84. Denton, 1874, pp.171–2
85. Denton, 1874, p.178
86. Denton, 1874, pp.172, 174
87. Denton, 1874, pp.173– 5
88. Denton, 1874, pp.175–6, 178
89. Denton, 1874, p.183
90. Denton, 1874, pp.186–7
91. Denton, 1874, p.189
92. Denton, 1874, pp.191–2
93. Denton, 1874, p.195
94. Denton, 1874, p.196
95. Denton, 1874, pp.199–200
96. Denton, 1874, pp.211, 213
97. Denton, 1874, pp.217–9
98. Denton, 1874, p.219
99. Denton, 1874, pp.220–2
100. Denton, 1874, p.223
101. Denton, 1874, pp.227–9
102. Denton, 1874, pp.237–8, 239–40
103. Denton, 1874, pp.241–2
104. Denton, 1874, pp.249–51
105. Denton, 1874, pp.254–6
106. Denton, 1874, p.263
107. Denton, 1874, pp.263–6

## 18. Jung Dreamers: From Freud to Flammarion

1. Jung, 1995, pp.150–2
2. Jung, 1993, pp.24–5
3. Jung, 1993, pp.36–8
4. Jung, 1993, pp.20–1
5. Jung, 1993, pp.80–3; Jung, 1995, p.128
6. Evans, 1987, pp.119–20; Hilary Evans, 'Talking with Martians', *Fortean Times* issue 76, pp.22–8
7. Complied from Hyslop, 1908, pp.222–67
8. Flournoy, 1900, p.1
9. Flournoy, 1900, p.21
10. Flournoy, 1900, p.25
11. Flournoy, 1900, pp.26–7
12. Flournoy, 1900, pp.18–19
13. Flournoy, 1900, p.2

14. Flournoy, 1900, p.142
15. Flournoy, 1900, p.149
16. Flournoy, 1900, pp.145–7
17. Flournoy, 1900, pp.146, 148
18. Flournoy, 1900, p.154
19. Flournoy, 1900, p.153
20. Flournoy, 1900, pp.159, 195–260
21. Flournoy, 1900, p.162
22. Flournoy, 1900, p.170
23. Flournoy, 1900, p.172
24. Flournoy, 1900, p.176
25. Flournoy, 1900, pp.190–4
26. Flournoy, 1900, p.162
27. Flournoy, 1900, pp.177–8
28. Flournoy, 1900, pp.183–5
29. Flournoy, 1900, p.262
30. Flournoy, 1900, p.186
31. Flournoy, 1900, p.269
32. Cited in Flournoy, 1900, p.140
33. Rabkin, pp.86–7; Crowe, 1999, pp.378–9, 385–6
34. Crowe, 1999, p.486
35. Crowe, 1999, p.379
36. Crowe, 1999, p.382
37. Crowe, 1999, p.385
38. Flammarion, 1890, pp.18–34
39. Flammarion, 1890, pp.35–43
40. Flammarion, 1890, pp.116–21
41. Flammarion, 1890, pp.152–7, 204
42. Flammarion, 1890, pp.161–97
43. Flammarion, 1890, p.313
44. Flammarion, 1890, p.162
45. Flammarion, 1890, pp.198–9
46. Flammarion, 1890, pp.227–56
47. Flammarion, 1890, pp.302–13
48. Crowe, 1999, pp.408–11
49. Crowe, 1999, p.381
50. Gareth J. Medway, 'Mediums, Mystics and Martians' in *Magonia* magazine issue 99; online at http://magoniamagazine.blogspot.co.uk/2014/02/martians.html
51. Crowe, 1999, p.381
52. Camille Flammarion, 'Shall We Talk with the Men in the Moon?' in *Review of Reviews* 5, p.90
53. Rabkin, 2005, p.89
54. Crossley, 2011, p.169
55. Crossley, 2011, p.132

## Conclusion: Owning Your Own Space

1. Moore, 1972, p.122
2. *Sunday Times News Review*, 17 January 2016, p.4
3. *The Times*, 21 January 2016, pp.1, 6
4. *Sunday Times News Review*, 27 July 2010, p.4
5. Young, 2012, pp.224–6
6. *Times*, 9 September 2016, p.7
7. *Fortean Times* 103, p.7
8. *Daily Mail*, 12 July 2016
9. *Fortean Times* 104, p.104; *Daily Mail*, 12 July 2016
10. Lewis, 2010, pp.222–3

# Bibliography

NOTE: *The dates of publication given below are for the editions I personally made use of whilst writing this book, and do not necessarily correspond with each title's first printing.*

Andrews, James T., *Red Cosmos: K. E. Tsiolkovskii, Grandfather of Soviet Rocketry* (Texas: Texas A & M University Press, 2009)

Appleyard, Bryan, *Aliens: Why They Are Here* (London: Scribner, 2006)

Bainbridge, William Sims, *The Meaning and Value of Spaceflight: Public Perceptions* (Virginia: Springer, 2015)

Ball, Philip, *Unnatural: The Heretical Idea of Making People* (London: Bodley Head, 2011)

Barker, Gray, *They Knew Too Much About Flying Saucers* (Georgia: IllumiNet Press, 1997)

Beecher, Jonathan, *Charles Fourier: The Visionary and His World* (California: University of California Press, 1992)

Bennett, Colin, *Looking for Orthon* (New York: Cosimo, 2008)

Bleiler, Everett Franklin & Bleiler, Richard, *Science Fiction: The Gernsback Years* (Ohio: Kent State University Press, 1998)

Bogdanov, Alexander, *Red Star* (Indiana: Indiana University Press, 1984)

Bowen, Charles (Ed.), *The Humanoids* (London: Futura, 1977)

Breton, André, *Anthology of Black Humour* (London: Telegram, 1997)

Brueton, Diana, *Many Moons: The Myth and Magic, Fact and Fantasy of Our Nearest Heavenly Body* (New York: Prentice Hall, 1991)

Brookesmith, Peter (Ed.), *Marvels & Mysteries: Aliens* (London: Parallel, 1995)

Brooks, Michael, *13 Things That Don't Make Sense* (London: Profile, 2009)

Buck, Christopher, *God & Apple-Pie: Religious Myths and Visions of America* (New York: Educators' International Press, 2015)

Carter, John, *Sex and Rockets: The Occult World of Jack Parsons* (Washington State: Feral House, 2004)

Childress, David Hatcher, *The Fantastic Inventions of Nikola Tesla* (Illinois: Adventures Unlimited, 2009)

Clark, Jerome, *Hidden Realms, Lost Civilisations and Beings from Other Worlds* (Detroit: Visible Ink Press, 2010)

Clarke, David & Roberts, Andy, *Flying Saucerers; A Social History of Ufology* (Loughborough: Heart of Albion, 2007)

Clegg, Claude Andrew III, *An Original Man: The Life and Times of Elijah Muhammad* (North Carolina: University of North Carolina Press, 2014)

Crossley, Robert, *Imagining Mars: A Literary History* (Connecticut: Wesleyan University Press, 2011)

Crowe, Michael J., *The Extraterrestrial Life Debate, 1750–1900* (New York: Dover Books, 1999)

Davis, Andrew Jackson, *The Principles of Nature* (Boston: William White, 1871)

Davis, Andrew Jackson, *A Stellar Key to the Summer-Land* (Los Angeles: Austin Publishing, 1920)

Denton, William & Denton, Elizabeth, *The Soul of Things: Volumes I & II* (Boston: Walker, Wise & Co, 1863)

Denton, William, *The Soul of Things: Volume III* (Boston: Self-Published, 1874)

Devereux, Paul & Brookesmith, Peter, *UFOs & Ufology: The First 50 Years* (London: Blandford, 1997)

Dickson, Paul, *Sputnik: The Shock of the Century* (New York: Berkley, 2001)

Dingwall, E. J., *Very Peculiar People* (London: Rider, 1950)

Eros, Paul, *One of the Most Penetrating Minds in England: Gerald Heard and the British Intelligentsia of the Interwar Period* (Cambridge: D.Phil Thesis, 2011)

Essers, I., *Max Valier – A Pioneer of Space-Flight* (Virginia: NASA, 1976)

Evans, Hilary, *Gods, Spirits, Cosmic Guardians: Encounters with Non-Human Beings* (Wellingborough: The Aquarian Press, 1987)

Evans, Hilary & Bartholomew, Robert, *Outbreak! The Encyclopedia of Extraordinary Social Behaviour* (Texas: Anomalist Books, 2009)

Evans, Hilary & Spencer, John, (Eds.) *Phenomenon: From Flying Saucers to UFOs – Forty Years of Fact and Research* (London: Macdonald, 1988)

Evans, Hilary & Stacy, Dennis, (Eds.) *UFO 1947–1997: Fifty Years of Flying Saucers* (London: John Brown, 1997)

Flammarion, Camille, *Urania* (Boston: Estes & Lauriat, 1890)

Flournoy, Théodore, *From India to the Planet Mars: A Study of a Case of Somnambulism* (New York: Harper & Bros, 1900)

Fodor, Nandor, *The Haunted Mind: A Psychoanalyst Looks at the Supernatural* (New York: Garrett Publications, 1959)

Fontenelle, Bernard de, *Conversations on the Plurality of Worlds* (London: J. Cundee/T. Hurst, 1804)

Fourier, Charles, *The Passions of the Human Soul and Their Influence on Society and Civilization* (London: Hippolyte Bailliere, 1851)

Fourier, Charles, *The Theory of the Four Movements* (Cambridge: Cambridge University Press, 1996)

Gardell, Mattias, *In the Name of Elijah Muhammad: Louis Farrakhan and the Nation of Islam* (Durham: Duke University Press, 1996)

Gardner, Martin, *Fads & Fallacies in the Name of Science* (New York: Dover Books, 1957)

Garwood, Christine, *Flat Earth: The History of an Infamous Idea* (London: Macmillan, 2007)

Godwin, Parke, *A Popular View of the Doctrines of Charles Fourier* (New York: J. S. Redfield, 1844)

Goodman, Matthew, *The Sun and the Moon* (New York: Basic Books, 2008)

Goodrick-Clarke, Nicholas, *The Occult Roots of Nazism* (London: IB Tauris, 2009)

Goodrick-Clarke, Nicholas, *Black Sun: Aryan Cults, Esoteric Nazism and the Politics of Identity* (New York: New York University Press, 2002)

Graves, Robert, *The White Goddess* (London: Faber & Faber, 1999)

Harley, Rev Timothy, *Moon-Lore* (London: Swan Sonnenschein, 1885)

Harris, Thomas Lake, *An Epic of the Starry Heaven* (New York: Partridge & Brittan, 1855)

Heard, Gerald, *The Riddle of the Flying Saucers: Is Another World Watching?* (London: Carroll & Nicholson, 1950)

Henriet, Gabrielle, *Heaven and Earth* (Arundell: Mitchell & Co, 1958)

Herrick, James A., *Scientific Mythologies: How Science and Science Fiction Forge New Religious Beliefs* (Illinois: InterVarsity Press, 2008)

Howell, Yvonne (Ed.), *Red Star Tales* (Montpelier: Russian Life Books, 2015)

Huygens, Christiaan, *The Celestial Worlds Discovered* (London: Timothy Childe, 1698)

Huxley, Aldous, *The Doors of Perception & Heaven and Hell* (London: Vintage Classics, 2004)

Hyslop, James H., *Psychical Research and the Resurrection* (Boston: Small, Maynard & Co, 1908)

Jung, C. G., *Flying Saucers: A Modern Myth of Things Seen in the Skies* (New York: MJF Books, 1978)

Jung, C. G., *Psychology and the Occult* (London: Ark, 1993)

Jung, C. G., *Memories, Dreams, Reflections* (London: Harper Perennial, 1995)

Keel, John, *UFOs: Operation Trojan Horse* (London: Sphere, 1976)

Keel, John, *The Mothman Prophecies* (Georgia: IllumiNet Press, 1991)

Keel, John, *Our Haunted Planet* (Minnesota: Galde Press, 1999)

Klarer, Elizabeth, *Beyond the Light-Barrier: The Autobiography of Elizabeth Klarer* (Cape Town: New Vision, 2008)

Koestler, Arthur, *The Sleepwalkers* (London: Penguin, 2014)

Kossy, Donna, *Kooks: A Guide to the Outer Limits of Human Belief* (Portland: Feral House, 1994)

Kossy, Donna, *Strange Creations: Aberrant Ideas of Human Origins from Ancient Astronauts to Aquatic Apes* (Los Angeles: Feral House, 2001)

Leadbeater, C. W., *The Inner Life: Volume II* (Chicago: The Theosophical Press, 1922)

Leonard, George H., *Somebody Else is On the Moon* (New York: Pocket Books, 1977)

Lewis, C. S., *The Discarded Image: An Introduction to Medieval and Renaissance Literature* (Cambridge: Cambridge University Press, 2010)

Ley, Willy, *Watchers of the Skies: An Informal History of Astronomy from Babylon to the Space Age* (London: Sidgwick & Jackson, 1964)

Lowell, Percival, *Occult Japan* (Boston: Houghton & Mifflin, 1894)

Lowell, Percival, *Mars* (Boston: Houghton & Mifflin, 1895)

Lowell, Percival, *Mars as the Abode of Life* (New York: Macmillan, 1908)

Lowell, Percival, *Mars and Its Canals* (New York: Macmillan, 1911)

Marx, Karl & Engels, Friedrich, *The Communist Manifesto* (London: Penguin, 1985)

McDaniel, Stanley V. & Paxson, Monica Rix (Eds.), *The Case for the Face* (Illinois: Adventures Unlimited, 1998)

McGovern, Una & Rickard, Bob, *Chambers Dictionary of the Unexplained* (London: Chambers, 2007)

McKenna, Terence, *Food of the Gods* (London: Rider, 1992)

Michell, John, *Eccentric Lives and Peculiar Notions* (Illinois: Adventures Unlimited, 1999)

Moore, Patrick, *Can You Speak Venusian?* (London: David & Charles, 1972)

Moore, Steve (Ed.), *Fortean Studies Volume Six* (London: John Brown, 1999)

Muirhead, Brian, Reeves-Stevens, Judith & Reeves-Stevens, Garfield, *Going to Mars* (New York: Pocket Books, 2004)

Nadis, Fred, *Wonder Shows: Performing Science, Magic and Religion in America* (New Jersey: Rutgers University Press, 2005)

Naylor, Ernest, *Moonstruck: How Lunar Cycles Affect Life* (Oxford: Oxford University Press, 2015)

Nelson, Buck, *My Trip to Mars, the Moon and Venus* (Washington State: Health Research Books, 2006)

Nicolson, Marjorie Hope, *Voyages to the Moon* (New York: Macmillan, 1960)

Noble, David F., *The Religion of Technology* (New York: Penguin, 1999)

Olerich, Henry, *A Cityless and Countryless World: An Outline of Practical Co-Operative Individualism* (Iowa: Gilmore & Olerich, 1893)

Oliver, Kendrick, *To Touch the Face of God: The Sacred, the Profane, and the American Space-Programme, 1957–1975* (Baltimore: Johns Hopkins University Press, 2013)

Pauwels, Louis & Bergier, Jacques, *The Morning of the Magicians* (London: Souvenir Press, 2001)

Plutarch, *The Face Which Appears on the Orb of the Moon* (Winchester: Warren & Son, 1911)

Polizzotti, Mark, *Revolution of the Mind: The Life of André Breton* (Boston: Black Widow Press, 2009)

Puharich, Andrija, *Uri* (London: Futura, 1974)

Pynchon, Thomas, *Gravity's Rainbow* (London: Vintage, 2000)

Rabkin, Eric S., *Mars: A Tour of the Human Imagination* (London: Praeger, 2005)

Randles, Jenny, *UFO Reality: A Critical Look at the Physical Evidence* (London: Robert Hale, 1983)

Redfern, Nick, *Contactees: A History of Alien-Human Interaction* (New Jersey: New Page, 2010)

Riasanovsky, Nicholas V., *The Teaching of Charles Fourier* (Los Angeles: University of California Press, 1969)

Rosenthal, Bernice Glatzer (Ed.), *The Occult in Soviet and Russian Culture* (New York: Cornell University Press, 1997)

Roud, Steve, *The Penguin Guide to the Superstitions of Britain and Ireland* (London: Penguin, 2006)

Russell, Bertrand, *History of Western Philosophy* (London: Routledge, 2006)

Sagan, Carl, *The Demon-Haunted World: Science as a Candle in the Dark* (New York: Ballantine Books, 1996)

Sconce, Jeffrey, *Haunted Media: Electronic Presence from Telegraphy to Television* (London: Duke University Press, 2000)

Sheehan, William, *Planets & Perception: Telescopic Views and Interpretations, 1609–1909* (Tucson: University of Arizona Press, 1988)

Smith, Michael G., *Rockets & Revolution: A Cultural History of Early Spaceflight* (London: University of Nebraska Press, 2014)

Swedenborg, Emanuel, *Concerning the Earths in Our Solar System, Which Are Called Planets* (Boston: Otis Clapp, 1839)

Thurston, Father Herbert, *The Physical Phenomena of Mysticism* (Guildford: White Crow, 2013)

Tucker, S. D., *Great British Eccentrics* (Stroud: Amberley, 2015)

Tucker, S. D., *Forgotten Science* (Stroud: Amberley, 2016)

Voltaire, *The Works of Voltaire Volume XXXIII* (Paris: A. Firmin Didot, 1829)

Wells, H. G., *The War of the Worlds* (London: Penguin, 2005)

Wilson, Bee, *The Hive: The Story of the Honeybee and Us* (London: John Murray, 2004)

Wilson, Colin, *The Occult* (London: Watkins, 2006)

Woolley, Benjamin, *The Queen's Conjuror* (London: Flamingo, 2002)

Young, George M., *The Russian Cosmists* (New York: Oxford University Press, 2012)

# Index

# About the Author

S. D. Tucker is an author and journalist. He has written three books for Amberley and writes a regular column in *Fortean Times* magazine. He lives in Widnes, Cheshire.